Lecture Notes in Artificial Intelligence   10527

Subseries of Lecture Notes in Computer Science

More information about this series at http://www.springer.com/series/1244

Octavio A. Agustín-Aquino · Emilio Lluis-Puebla
Mariana Montiel (Eds.)

Mathematics
and Computation
in Music

6th International Conference, MCM 2017
Mexico City, Mexico, June 26–29, 2017
Proceedings

 Springer

Editors
Octavio A. Agustín-Aquino ⓘ
UNCA and UTM
Oaxaca
Mexico

Mariana Montiel ⓘ
Georgia State University
Atlanta, GA
USA

Emilio Lluis-Puebla
UNAM
Mexico City
Mexico

ISSN 0302-9743   ISSN 1611-3349 (electronic)
Lecture Notes in Artificial Intelligence
ISBN 978-3-319-71826-2   ISBN 978-3-319-71827-9 (eBook)
https://doi.org/10.1007/978-3-319-71827-9

Library of Congress Control Number: 2017959634

LNCS Sublibrary: SL7 – Artificial Intelligence

Printed on acid-free paper

This Springer imprint is published by Springer Nature
The registered company is Springer International Publishing AG
The registered company address is: Gewerbestrasse 11, 6330 Cham, Switzerland

Preface

The 6th Biennial International Conference for Mathematics and Computation in Music (MCM 2017) took place during June 26–29, 2017, at the Faculty of Sciences of the Universidad Nacional Autónoma de México, in Mexico City, Mexico. Additional venues for recitals were kindly provided by the Escuela Superior de Música and the Museo Nacional de Historia "Castillo de Chapultepec".

As the flagship conference of the Society for Mathematics and Computation in Music (SMCM), MCM 2017 provided a dedicated platform for the communication and exchange of ideas among researchers in mathematics, informatics, music theory, composition, musicology, and related disciplines. It brought together researchers from around the world who combine mathematics or computation with music theory, music analysis, composition, and performance.

The program is available at http://www.mcm2017.org and featured three plenary lectures: the first by Guerino Mazzola (who introduced the *musical mathematical game* as a complement to the *mathematical music theory*), the second by Harald Fripertinger (who spoke on the combinatorics of tone-rows and their role in music), and a final one by Julio Estrada (who described some of the mathematical tools and inspirations that underlie his compositions).

There was a panel titled "Contemporary Music Composition in Relation to Mathematics and Computing: Current Perspectives and Approaches", with the participation of four Mexican composers: Juan Sebastián Lach (Conservatorio de las Rosas), Roberto Morales-Manzanares (Universidad de Guanajuato), Gabriel Pareyón (CENIDIM), and Edmar Soria, which was a valuable firsthand testimony of the fruitful interplay of the mathematical and computational approaches in the creation of music.

During three daily one-hour sessions, Octavio Alberto Agustín-Aquino (a member of the Organizing and Scientific Program Committee) delivered a "nano-course" on Guerino Mazzola's mathematical music theory, whose intent was to serve both as an homage to the current SMCM president and to introduce to an audience as wide as possible the techniques, results, and philosophical postures contained in Mazzola's work, with an emphasis on counterpoint.

The program included three evening recitals. The first was a guitar recital by Octavio Alberto Agustín-Aquino, who visited musical landscapes from eight countries (four from Europe and four from America) in a travelling salesman route, while keeping the proportion of the durations in approximate correspondence to that of the landmasses of the continents. The second one was a free jazz recital by Heinz Geisser (drums) and the president of the SMCM, Guerino Mazzola (piano), in the auditorium of the Escuela Superior de Música, which constituted an electrifying dialogue of gestures and mutual spaces of performance. The third one was performed by Harald Fripertinger (flute) and the head of the Organizing and Scientific Program Committee, Emilio Lluis-Puebla (piano), featuring music from Telemann, Beethoven, Schubert,

Gant, Marsh, and Thomas, in the wonderful environment of the *Alcázar* of the Museo Nacional de Historia "Castillo de Chapultepec", as a pure enjoyment of music, mathematics, and life.

The chapters in these proceedings correspond to the papers and two selected posters presented at the conference, following a careful peer-review process, which was optionally double-blind. We received 40 submissions from 62 authors across 10 countries. Each submission was assigned to one or two reviewers. A paper was accepted if and only if the recommendation of the reviewers was positive and a majority of the editors judged it a meritorious contribution; sometimes it required a second round of revisions. A total of 28 papers were accepted following review.

Last, but not least, we thank the following institutions for providing their infrastructure and human resources for the organization and promotion of MCM 2017:

- Facultad de Ciencias de la Universidad Nacional Autónoma de México
- Society for Mathematics and Computation in Music
- Escuela Superior de Música
- Georgia State University
- Museo Nacional de Historia
- Sociedad Matemática Mexicana
- Universidad de la Cañada
- Universidad Tecnológica de la Mixteca

July 2017

Octavio A. Agustín-Aquino
Emilio Lluis-Puebla
Mariana Montiel

Organization

General Organizing Committee

Guerino Mazzola	School of Music, University of Minnesota, USA
Emilio Lluis-Puebla	Faculty of Sciences, UNAM, Mexico
Octavio Alberto Agustín-Aquino	Universidad de la Cañada and Universidad Tecnológica de la Mixteca, Oaxaca, Mexico
Mariana Montiel	Georgia State University, USA
Gabriel Pareyón	National Center for Music Research, Documentation and Information, CENIDIM-INBA, Mexico
Roberto Morales-Manzanares	Laboratorio de Informática Musical, Universidad de Guanajuato, Mexico
Emil Awad	CECDA, Universidad Veracruzana, Mexico
Juan Sebastián Lach	Conservatorio de las Rosas, Michoacán, Mexico

Scientific Program Committee

The Scientific Program Committee was responsible for the scientific content of MCM 2017. It prepared the final list of oral and poster presentations and invited speakers and selected contributed papers based on peer review amongst those submitted for consideration. It consisted of:

Emilio Lluis-Puebla	Faculty of Sciences, UNAM, Mexico
Mariana Montiel	Georgia State University, Georgia, USA
Octavio Alberto Agustín-Aquino	Universidad de la Cañada and Universidad Tecnológica de la Mixteca, Mexico

Local Organizing Committee

The Local Organizing Committee was responsible for functional organization of MCM 2017, including the selection of the most suitable locations; preparation of the internet site and conference software; arrangement of the musical and social program; production and publication of the proceedings volume; organization of book exhibitions; and coordinating the contact between invited speakers, discussants, contributing authors, participants, publishers, and exhibitors. The LOC consisted of

Emilio Lluis-Puebla	Faculty of Sciences, UNAM, Mexico
Octavio Alberto Agustín-Aquino	Universidad de la Cañada and Universidad Tecnológica de la Mixteca, Mexico

with the valuable assistance of the following graduate and undergraduate students, all from the Faculty of Sciences, UNAM:

Alison Barbosa Guzmán
Ruby Almazán Calzada
Yemile Chávez Martínez
Kuauhtemok González Cortés
Bruno Martínez Warnholtz
Julisa Rodríguez Torres

Reviewers

O. A. Agustín-Aquino
Emmanuel Amiot
Gilles Baroin
Chantal Buteau
Clifton Callender
Clément Cannone
Johanna Devaney
Andrée Ehresmann
Alice Eldridge
Francisco Gómez
Yupeng Gu
Julian Hook
Timothy Hsu
Franck Jedrzejewski
M. Kaliakatsos-Papakostas

Maria Mannone
Alan Marsden
Andrew Milne
Mariana Montiel
Thomas Noll
Pablo Padilla
Robert Peck
Richard Plotkin
Alexandre Popoff
David Rappaport
David Temperley
Florian Thalmann
Jason Yust
Marek Žabka

Collaborating Institutions

Facultad de Ciencias de la Universidad Nacional Autónoma de México
Society for Mathematics and Computation in Music
Escuela Superior de Música
Georgia State University
Museo Nacional de Historia
Sociedad Matemática Mexicana
Universidad de la Cañada
Universidad Tecnológica de la Mixteca

Contents

X   Contents

Algebraic Models

Primal-Circular Substitutions

Marcus D. Booth[1,2(✉)]

[1] San Diego State University, San Diego, USA
[2] Tulane University, New Orleans, USA
mbooth1@tulane.edu

Abstract. There are two ongoing tensions in the creation of new musical systems that traditional innovation procedures in the domain of harmony have acknowledged yet left under-explored in the systematic sense from the composer's perspective. Learned terminological constraints and reactionary creative practices often limit the realization of underlying creative logic that connects past influences to present innovations, and creative procedures on one structural unit of a particular musical element to those on another. The following paper will proceed by example from theory toward a compositional end, employing algebraic techniques to create a system of chord substitution which serves as an exemplary solution of the aforementioned issues. Primal-circular substitution has a basis in western tertian harmony, shows compatibility with neo-Riemannian local transformation sequences, and re-envisions substitutions as the realization of globally applicable systems.

Keywords: Chord substitution · Re-harmonization system
Composition practices

1 Introduction

Two major themes have long pervaded music composition across cultural, chronological, and creative boundaries. One is the tendency of creative progress to be of a reactionary rather than systematic nature, and involve a dichotomous choice between slightly editing or wholly rejecting a past creative system. Initially, a composer learns the traditions of the nearby and recent, as a social being learns a language in preparation for the task of being effective socially. Then he or she reacts to these in the process of incorporating them into the worldview that contextualizes the compositional products. A decision is made about how effective or not these traditional approaches are, and action is taken over time to increase the efficacy of the compositional tools and transmit the composer's message.

The second theme at play in the compositional process is consideration of the interplay between local and global musical structures. Focusing on either, the composer must decide which combinations of one or more musical elements (e.g. melody, harmony, timing, timbre, etc...) most strongly serve the overall purpose of the work (or some other large formal construct), and which large scale musical forms and techniques will constrain the treatment of these elements. The orientation described in the first theme gives rise to particular tendencies in this decision making process and becomes

© Springer International Publishing AG 2017
O. A. Agustín-Aquino et al. (Eds.): MCM 2017, LNAI 10527, pp. 3–12, 2017.
https://doi.org/10.1007/978-3-319-71827-9_1

the starting point for implementation. The composer's palette is governed by the evolution of a language and its implied "conditions of possibility" [3]. Over the course of composing a work, making a hierarchy of decisions of the aforementioned nature, the composer intuitively creates a system of patterns of which he or she is only partially conscious in the moment. When one level of patterned abstraction is being acted upon creatively, another is unfolding logically without agency.

In what follows, a method of chord substitution will be introduced in attempt to provide an example of thinking about harmony and altering traditional chord progressions that overcomes the epistemological limitations previously described. The approach, called primal-circular substitution, makes use of cyclical quotient groups in Z_{12} to explain mutually dependent interval circularities and the harmonies they relate for use in re-harmonization procedures.

2   Aesthetic and Philosophical Concerns

While aesthetic judgments in the absolute sense may be left to the individual and the situational, in the case of employing mathematics to compositional ends, a few words are needed in explanation and defense of the approach. Without this, we are left with the trivial case that any finite collection of objects, musical or not, can be placed in an organizational and axiomatic framework that appears mathematically sound because it was initially designed to be so. While not sufficient to justify a new system, it is certainly necessary to view this extreme not as a potential derailment of the theory, but as an indication that the astounding variety of organizational structures that mathematics explains make mathematical thinking central to human intuition [1]. In the case of primal-circular substitution, this defense then has two parts.

The first is that the theory is not a freely generative approach to harmony, but one with existing musical anchors. It is a theory of substitution which uses prime number patterns to catalogue similarities between chords and allow substitution of part or all of a known chord progression. Regarding commonly accepted features of tonality, the example of the process soon to be introduced is centered on circularity and western tertian notions of consonance, maintains chord shapes and distances [4], and introduces a new scope of underlying structural similarity that re-envisions sonic relationships known to work aesthetically for audiences of the past and present. This maintains the composer's opportunity to manipulate emotions rather than just logically connected patterns.

The second point worthy of mention is that mathematics provides a language for relating formal components of different types of inquiry about the world, and is not itself responsible for cases in which the user does not choose to do so. Again, this is not an argument restrictively in favor of a particular teleological position on the use of mathematics in music. It is safe to say though, that representationalism has been a key mode of creative endeavor in music from inspirations acknowledged to form and content decided upon by composers, so it provides one solid base for defending the widespread relevance of the quantitative approach. With mathematics being largely a study of structures and processes independent of content that may conform to its particular models, its use in connecting music to other human endeavors and thought patterns knows few boundaries. Beyond its representational potential by way of mapping relations, it also facilitates specific instances of the reductionist thought that has been pervasive across different types of scientific inquiry at least sense the Age of Enlightenment [1]. From Chemillier's "ethnomathematics" to Xenakis' inspiration drawn from architecture and Cage's interests in chance and choice, many exemplary connections between thought structure and medium-independent creative output can be found which invite mathematical scrutiny as described.

Now that the long surviving relevance of mathematics to time and efficacy tested aesthetic practices has been summarized, let us move back to the plan to introduce the primal-circular substitution framework. In the next section, we will begin by defining the basic tools for creating one version of the system from the starting point of accepted tonal principles: interval cycles (in all such systems) and major/minor chord structures (this version). Following that, we will advance our conceptualization to the codependency of intervallic cycles, populating the chord substitution charts in the process so they are ready for compositional use. In doing this, mathematical generalizations will become apparent and proofs of those will be offered. Finally we will show the application of the system in an especially restrictive formal context: the mapping of one Hamiltonian progression (typically generated by P, L, and R operations) to another, connecting the local (neo-Riemannian, chord to chord) to the global (primal-circular substitution systems, chord/progression to superstructure) with mathematical continuity.

3 Primal-Circular Substitutions

We will begin the example construction of primal-circular systems by noting some behaviors of single intervals, as chord types as defined in the western European tradition contain predefined subsets of the intervallic possibilities. The reader is assumed to be familiar with determining basic cycling of intervals, so this part need not be directly illustrated (numerical results are given exhaustively in Table 1). Interval cycling is isomorphic to dividing a whole number of equal temperament octaves into equal parts, or multiplying an interval out until it finishes one complete cycle through distinct pitch classes. So to go from the culturally dependent notion of interval names to something that can be extended free of terminological needs, we proceed to a mathematical explanation. In this case, to allow both relative and absolute interpretation of the system, we will move from consideration of the interval to consideration of the pitch class numbers that arise. The table below shows the behavior of each pitch class n when cycled multiplicatively by the interval M and adjusted modulo 12 for labeling.

The generating formula for the chart below is the familiar:

$$Pitch\,Category = Mn \bmod 12$$

In the body of the table above, one can observe the mod cycles for a given multiplier horizontally. Tertian chords then, correspond to groupings of three elements.

Table 1. Catalogue of mod cycles: Note numbers n are presented horizontally and multipliers M are catalogued vertically, with the value of the above equation in each interior box

n →	1	2	3	4	5	6	7	8	9	10	11	12
1	1	2	3	4	5	6	7	8	9	10	11	12
2	2	4	6	8	10	0	2	4	6	8	10	0
3	3	6	9	0	3	6	9	0	3	6	9	0
4	4	8	0	4	8	0	4	8	0	4	8	0
5	5	10	3	8	1	6	11	4	9	2	7	0
6	6	0	6	0	6	0	6	0	6	0	6	0
7	7	2	9	4	11	6	1	8	3	10	5	0
8	8	4	0	8	4	0	8	4	0	8	4	0
9	9	6	3	0	9	6	3	0	9	6	3	0
10	10	8	6	4	2	0	10	8	6	4	2	0
11	11	10	9	8	7	6	5	4	3	2	1	0

For a root pitch p, the following sets represent major and minor chords:

$$Major\ chord = (p, p+4, p+7)$$

$$Minor\ chord = (p, p+3, p+7)$$

Multiplication example:

$$4 \times C\,major = 4 \times (1, 5, 8) \bmod 12 = (4, 8, 8)$$

$$8 \times C\,major = 8 \times (1, 5, 8) \bmod 12 = (4, 4, 8)$$

So C Major is type 448/488 then.

Categorizations can then be visualized in the chart above, but for ease of use in compositional practices or experimentation with the theory, they will be organized in a table below with by category labels and chord names before laying out the mathematical reasoning explicitly and discussing the codependency of the three intervals present in a tertian chord. Before we do this, note the following two necessary abstractions:

1. Because this is a theory of chord substitution, which is an operation done on chord progressions rather than on specific voice leadings, permutation of voices is abstracted away by always ordering pitch class numbers from least to greatest after each operation in *modulo 12*. And...

2. Because the substitution approach is based on comparisons of prime factors of M with prime factors of 12, considering that the quotient groups Z_{12}/MZ_{12} and $Z_{12}/(12-M)Z_{12}$ are isomorphic, we categorize chords according to results of multiplication by M and $12-M$ together, since the residue classes are re-orderings, which we said above we abstract away.

The names of the chord types in the charts below correspond to the positions of their individual tones in the mod cycles for M and 12-M. Four steps are now given, explaining how to use the chord charts:

1. Write a traditional western tonal chord progression, e.g. I-vi-IV-V-I in C major:

$$C \; Am \; F \; G \; C$$

2. Choose the 3×8 chart or the 4×6 chart (or other examples you generate via the mathematics) below. (for this example we will use the 3×8 chart)

3. Find the category (column) of each chord by locating it in the chart:

$$C \; Am \; F \; G \; C \; = 448/488, 448/488, 044/088, 004/008, 448/488.$$

4. Write a new progression using chords from each corresponding category (Tables 2 and 3).

$$\text{Example}: \; Cm \; Eb \; Ab \; Fm \; C.$$

Table 2. 3 Substitution Categories of 8 Chords Each: M = 4, 12-M = 8.

Type 448/488	Type 004/008	Type 044/088
C Major	C#/Db Major	C#/Db Minor
C Minor	D Minor	D Major
D#/Eb Major	E Major	E Minor
D#/Eb Minor	F Minor	F Major
F#/Gb Major	G Major	G Minor
F#/Gb Minor	G#/Ab Minor	G#/Ab Major
A Major	A#/Bb Major	A#/Bb Minor
A Minor	B Minor	B Major

Table 3. 4 substitution categories of 6 chords each: M = 3, 12-M = 9.

C Major	C#/Db Major	C#/Db Minor	C Minor
D#/Eb Minor	D Minor	D Major	D#/Eb Major
E Major	F Major	F Minor	E Minor
G Minor	F#/Gb Minor	F#/Gb Major	G Major
G#/Ab Major	A Major	A Minor	G#/Ab Minor
B Minor	A#/Bb Minor	Bb Major	B Major

Any values of M on the interval [1, 12] can be used to generate systems of harmony around points of symmetry dividing the octave as this procedure does. The particular choices for the examples above were made because of their connection to common rhythmic/metric divisions of 3 and 4 as well as observations by the author of their relationship to neo-Riemannian transformations. This in part serves the intent expressed earlier to show that this system aligns with certain traditional features of western music and common creative treatments of other musical elements. Neo-Riemannian transformations were originally devised as an explanation for movement from individual chord to chord, the local [2]. Primal-circular substitution theory connects them to the global. This will be demonstrated following some further mathematical explanations of how the chord charts above are determined.

Briefly outlined earlier, chords are a set of fixed intervallic relationships that we then map onto two of the M systems in the Table 1). The M system takes each circularity, i.e. that of the root, third, and fifth of the chord, and cycles them through the remainder classes. Depending upon the value of M and the position of each chord tone relative to the points of symmetry defined by cycling M itself starting from pitch class 0 and ending at the first pitch p such that the product of M and p is congruent to 0 mod 12, two of the three chord tones with distinct behaviors will define the category. One of these tones is always the root, simply because that is how the positions are defined and the chord is named. This leaves either the 3rd or the 5th of the chord irrelevant to the categorization depending on the mapping (with p as root) of $p + 3$ (for the third of minor chords), or $p + 4$ (for major chords) and $p + 7$ (the fifth) to the corresponding positions in the cycle of Mp mod 12. A proof of this is given below.

First, note that:

\# of substitution categories = length of mod cycle = 12/GCF(12, M)

This follows from the concept of circularity. Because the relative positions p, $p + (3 \text{ or } 4)$, and $p + 7$ are fixed, as soon as one of the chord tones lands on an Mp such that Mp mod 12 = 0, the cycle is complete. Referring to the Table 1), we see the following regarding the particular substitution sets we will continue to discuss (Tables 4 and 5):

Table 4. Number of categories for various M values.

M Value	# of Categories (12/GCF(12, M)
3	4
4	3
5	3
6	4

Table 5. Number of chords per category for various M values.

M Value	# of Categories	# of Chords in Category
3	4	6
4	3	8
8	3	6
9	4	8

Now we are prepared to prove that either the third or the fifth of the chord is irrelevant to its categorization. Cases for multiple M values will be presented as the point applies to each.

Consider again the general major and minor chord structures:

Major: $[p, p+4, p+7]$ *Minor*: $[p, p+3, p+7]$

Given the property:

$$(x+y) \bmod M = x \bmod M + y \bmod M$$

Proofs that the 3rd or the 5th of any chord is irrelevant for categorization purposes:

$$([p, p+4, p+7] \,||\, [p, p+3, p+7]) \bmod (\textit{Size of Category}: [2, 3, 4, 6, \textit{or } 12])$$

Now in order to map them onto the mod cycles for different M values, consider the following:

Mod 2 Example (corresponding to M = 6)

So for a major or minor chord in root position, you have 3 tones with only 2 distinct results modulo 2

$$(p+3) \bmod 2 = (p+7) \bmod 2 \,(p+(3\,||\,7)) \bmod 2 = ((p \bmod 2) + 1) \bmod 2$$

$$3 \bmod 2 = 7 \bmod 2$$

$$4 \bmod 2 = 0 \bmod 2 \implies$$

Depending on the root, either 3rd = 5th or Root = 3rd.

Mod 3 Example (corresponding to M = 4 || M = 8)

$$(p+3) \bmod 3 = p \bmod 3$$
$$(p+7) \bmod 3 = p \bmod 3 + 1$$
$$(p+4) \bmod 3 = p \bmod 3 + 1$$
$$(p+4) \bmod 3 = (p+7) \bmod 3$$

Again for a major or minor chord in root position, you have 3 tones with only two distinct results modulo 3.

$$3 \bmod 3 = 0$$

$$4 \bmod 3 = 7 \bmod 3 \implies$$

Depending on the root, either 3rd = 5th or root = 3rd

Mod 4 Example (corresponding to M = 3 ∥ M = 9)

$$(p+3)\,mod\,4 = p\,mod\,4 + 3$$
$$(p+7)\,mod\,4 = p\,mod\,4 + 3$$
$$(p+4)\,mod\,4 = p\,mod\,4$$
$$(p+3)\,mod\,4 = (p+7)\,mod\,4$$

So for a major or minor chord in root position, you have 3 tones with only 2 distinct results modulo 4.

$$4\,mod\,4 = 0\,mod\,4$$

$$3\,mod\,4 = 7\,mod\,4 \Longrightarrow$$

Depending on the root, either 3rd = 5th or root = 3rd.

Now that the system has been explained mathematically and the construction has been shown, we conclude by showing musical examples on the next page, each of which satisfies the intentions described at the outset. Here, primal-circular substitution is used to alter local harmonies while satisfying the global choice to preserve a particularly restrictive form. For this purpose, the progression chosen for re-harmonization is Moreno Andreatta's Hamiltonian progression entitled "Aprile", from which new Hamiltonian progressions are generated.

Note: "Aprile" was composed using P, L, and R transformations. Primal-circular substitution systems can be understood locally in this way as detailed in **Appendix B**.

3 × 8 (M = 4 or 8) random selection re-harmonization of "Aprile"

Random Chord Selections = 1 8 3 7 6 5 2 4 3 7 1 8 5 6 1 4 5 2 4 3 6 8 2 7 *1

Progression (category, chord #) = (2,1) (3,8) (1,3) (1,7) (3,6) (2,5) (2,2) (3,4) (2,3) (3,7) (1,1) (1,8) (1,5) (1,6) (3,1) (2,4) (3,5) (1,2) (1,4) (3,3) (2,6) (2,8) (3,2) (2,7) *(2,1)

Progression (chord names) = Db major, B major, Eb major, A major, Ab major, G major, D minor, F major, E major, Bb minor, C major, A minor, Gb major, F# minor, C# minor, F minor, G minor, C minor, Eb minor, E minor, G# minor, B minor, D major, Bb major, *Db major

4 × 6 (M = 3 or 9) random selection

Random Chord Selections = 6 3 6 1 5 3 4 5 5 6 1 3 2 3 1 2 6 4 5 2 1 2 4 4 6

Progression (category, chord) = (3,6), (1,3), (4,6), (1,1), (4,5), (4,3), (1,4), (3,5), (2,5), (2,6), (3,1), (2,3), (2,2), (3,3), (2,1), (3,2), (1,6), (4,4), (1,5), (4,2), (4,1), (1,2), (3,4), (2,4), *(3,6)

Progression (chord names) = Bb major, E major, B major, C major, G# minor, E minor, G minor, A minor, A major, Bb minor, C# minor, F major, D minor, F minor, Db major, D major, B minor, G major, Ab major, Eb major, C minor, D# minor, Gb major, F# minor, *Bb major

With an extreme example of primal-circular substitution (for M = 3 and M = 4) in western harmonic context shown above, further experimentation on shorter and more conventional progressions is now left to the readers and composers. This approach to harmonic substitution can now be performed on aesthetically grounded materials according to the user, to novel creative ends serving his or her particular purpose at any harmonic structural level.

Appendix A: Primal-Circular Substitution Charts (Reproductions)

See Tables 6 and 7

Table 6. M = 4, 12-M = 8.

Type 448/488	Type 004/008	Type 044/088
C Major	C#/Db Major	C#/Db Minor
C Minor	D Minor	D Major
D#/Eb Major	E Major	E Minor
D#/Eb Minor	F Minor	F Major
F#/Gb Major	G Major	G Minor
F#/Gb Minor	G#/Ab Minor	G#/Ab Major
A Major	A#/Bb Major	A#/Bb Minor
A Minor	B Minor	B Major

Table 7. M = 3, 12-M = 9.

C Major	C#/Db Major	C#/Db Minor	C Minor
D#/Eb Minor	D Minor	D Major	D#/Eb Major
E Major	F Major	F Minor	E Minor
G Minor	F#/Gb Minor	F#/Gb Major	G Major
G#/Ab Major	A Major	A Minor	G#/Ab Minor
B Minor	A#/Bb Minor	Bb Major	B Major

Appendix B: Neo-Riemannian Analogues of Primal-Circular Substitutions

Assumption: Use each transformation (P, L, R) either 0 or 1 times.

Note: The term **"preserves"** is used below to indicate that a particular transformation on a chord in a particular category yields another chord in that category, whereas other transformations would yield chords outside the category.

Summary of Neo-Riemannian Results

- The category **448/488** preserves: **P || R in any order and place, no L.**
- The category **004/008** preserves: **P & R consecutively in any order and place, L alone.**
- The category **044/088** preserves **P || R first && P || R last.**
- The category **033/099** preserves **P || L first && P || L last.**
- The category **366/669** preserves **L & P consecutively in any order/place, R alone.**
- The category **336/699** preserves **P || L first && P || L last.**
- The category **003/009** preserves **L & P consecutively in any order/place, R alone.**

Examples for clarification (use charts to verify):

Ex. 1: Transformation of a chord in 448/488 by PR or RP yields another chord in 44/.488.

Ex. 2: Transformation of a chord in 004/008 by L alone yields another chord in 004/008. So does transformation of a chord in 004/008 by PR or RP.

Ex. 3: Transformation of a chord in 044/088 by PR, RP, PLR, or RLP yields another chord in 044/088.

References

1. Chemillier, M.: Les Mathématiques Naturelles, pp. 18–22. Odile Jacob, Paris (2007)
2. Cohn, R.: An introduction to Neo-Riemannian theory: a survey and historical perspective. J. Music Theor. **42**(2), 167–180 (1998)
3. Smith, D., Somers-Hall, H.: The Cambridge Companion to Deleuze. Cambridge University Press, Cambridge (2012)
4. Tymoczko, D.: A Geometry of Music: Harmony and Counterpoint in the Extended Common Practice, pp. 1–45. Oxford University Press, Oxford (2014)

On the Group of Transformations of Classical Types of Seventh Chords

Sonia Cannas[1,2]($\boxtimes$), Samuele Antonini[1]($\boxtimes$), and Ludovico Pernazza[1]($\boxtimes$)

[1] Dipartimento di Matematica "Felice Casorati", Università di Pavia, Pavia, Italy
sonia.cannas01@universitadipavia.it,
{samuele.antonini,ludovico.pernazza}@unipv.it
[2] IRMA, Université de Strasbourg, Strasbourg, France

Abstract. This paper presents a generalization of the well-known neo-Riemannian group PLR to the classical five types of seventh chord (dominant, minor, half-diminished, major, diminished) considered as tetrachords with a marked root and proving that it is isomorphic to the abstract group $S_5 \ltimes \mathbb{Z}_{12}^4$. This group includes as subgroups the PLR group and several other groups already appeared in the literature.

Keywords: Transformational theory · neo-Riemannian group
Semi-direct product · Seventh chord

1 Introduction

Since the pioneering works by David Lewin [8,9] and Guerino Mazzola [10,11], the main idea of *transformational theory* is to model musical transformations using algebraic structures. The most famous example is probably the neo-Riemannian group PLR, that acts on the set of all 24 minor and major triads of twelve-tone equal temperament and is abstractly isomorphic to the dihedral group of order 24. It is generated by the P, L and R operations that transform major triads in minor triads (and vice versa) shifting a single note by a semitone or a whole tone. They were introduced by 19th-century music theorist Hugo Riemann [12] for pure intervals. Lewin rediscovered the PLR operations, defined them considering the equal temperament, and gave birth to a branch of the transformational theory called *neo-Riemannian theory*.

Neo-Riemannian transformations can be modelled with several geometric structures, of which the most important is the Tonnetz, first introduced by Euler [4] and later studied by the several musicologists of the 19th century, such as Wilhelm Moritz Drobisch, Carl Ernst Naumann, Arthur von Oettingen and the same Hugo Riemann. From a mathematical point of view the Tonnetz is an infinite two-dimensional simplicial complex which tiles the plane with triangles where 0-simplices represent pitch classes, and 2-simplices identify major and minor triads: the relative position of 2-simplices makes it also a natural tool in the theory of parsimonious voice leading.

© Springer International Publishing AG 2017
O. A. Agustín-Aquino et al. (Eds.): MCM 2017, LNAI 10527, pp. 13–25, 2017.
https://doi.org/10.1007/978-3-319-71827-9_2

In addition to triads, seventh chords are often used in the music literature. A natural question arises: can we define a group similar to the neo-Riemannian group PLR acting on the set of seventh chords (of the twelve-tone equal temperament)? More precisely: can we define a group of transformations between seventh chords to describe parsimonious voice leading, so that the generators fix three notes and move a single note by a semitone or a whole tone? Problems on relationships between seventh chords were studied by Childs [2], Gollin [6], by Fiore and Satyendra [5], by Arnett and Barth [1] and by Kerkez [7] for some of the types of seventh chords. In this paper we will extend their studies considering all five "classical" types of seventh chords: dominant, minor, half-diminished, major, diminished.

In Sect. 2 we provide some preliminaries about the neo-Riemannian group. Section 3 presents briefly the known results about the generalization of the PLR-group to seventh chords, and a classification of all transformations between seventh chords shifting a single note by a semitone or a whole tone. In the fourth and final section we will define the $PLRQ$ group, generalizing the PLR group, and we identify its abstract algebraic structure.

2   The neo-Riemannian Group PLR

The neo-Riemannian group PLR is generated by the following P, L and R operations.

- P ("Parallel"): if the triad is major, P moves the third down a semitone, while if the triad is minor P moves the third up a semitone.
- L ("Leading-Tone"): if the triad is major L moves the root down a semitone, while if the triad is minor L moves the fifth up a semitone.
- R ("Relative"): if the triad is major, R moves the fifth up a whole tone, while if the triad is minor R moves the root down a whole tone.

There exist many ways to represent algebraically or geometrically such transformations. We will denote pitch classes by elements of the cyclic group of 12 elements $\mathbb{Z}/12\mathbb{Z}$ (or, more briefly, $\mathbb{Z}_{12}$) and n-chords by n-ples of pitch classes in brackets, ordered in the ascending direction (as induced by the linear order of pitches) and starting from some pitch class of reference.

In this notation, Crans, Fiore and Satyendra [3] use twelve equally-spaced points on a circle to represent pitch classes and relate the above operations to an inversion operation I_{k+h} as follows. Let S be the set of all 24 minor and major triads $\{[x_1, x_2, x_3] \mid x_1, x_2, x_3 \in \mathbb{Z}_{12}, x_2 = x_1 + 3 \text{ or } x_2 = x_1 + 4, x_3 = x_1 + 7\}$; then

$$P([x_1, x_2, x_3]) = I_{x_1+x_3}([x_1, x_2, x_3]) \tag{1}$$

$$R([x_1, x_2, x_3]) = I_{x_1+x_2}([x_1, x_2, x_3]) \tag{2}$$

$$L([x_1, x_2, x_3]) = I_{x_2+x_3}([x_1, x_2, x_3]) \tag{3}$$

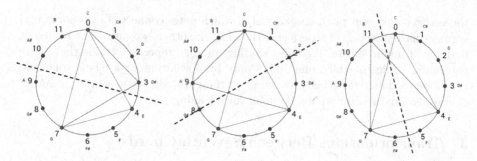

Fig. 1. $P(C) = c$ **Fig. 2.** $R(C) = a$ **Fig. 3.** $L(C) = e$

where I_{k+h} is the reflection of the circle across the axis of the line passing through k and h. As depicted in Figs. 1, 2 and 3, when applied to the triad of C major P gives c (C-minor), R gives a (A-minor) and L gives e (E-minor).

Another way to define the P, L and R operations is proposed by Arnett and Barth [1]:

$$P: M \leftrightarrow m \qquad\qquad P: [x, x + 4, x + 7] \leftrightarrow [x, x + 3, x + 7] \quad (4)$$
$$R: M \leftrightarrow m - 3 \qquad\qquad R: [x, x + 4, x + 7] \leftrightarrow [x, x + 4, x + 9] \quad (5)$$
$$L: M \leftrightarrow m + 4 \qquad\qquad L: [x, x + 4, x + 7] \leftrightarrow [x - 1, x + 4, x + 7] \quad (6)$$

where M represents a major triad, m a minor one, -3 and 4 are the numbers of semitones to be added to each component of the parallel triad (where here, as in the whole paper, the sum is made   mod 12, i.e. in the group $\mathbb{Z}_{12}$).

It is an easy calculation to verify that P is obtained as $RLRLRLR$, therefore the PLR group is in fact generated by R and L. The isomorphism to the dihedral group of order 24 becomes apparent noting that the element $RPLP$ is a translation up a semitone and therefore has order 12. We can visualize these operations in the neo-Riemannian Tonnetz (see Fig. 4), a simplicial complex where

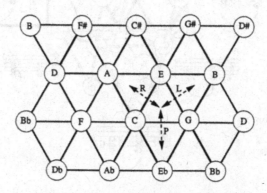

Fig. 4. Reflections preserving a triangle's edge in the Tonnetz represent the P, L and R operations

the vertices represent pitch classes and in which notes connected by a horizontal segment have intervals of a perfect fifth, while the other two directions represent major and minor thirds. Triangles sharing an edge represent triads that share two notes, while the third one differs only by a semitone or a whole tone. We can observe that reflections preserving a triangle's edge represent P, L and R operations (and realize a parsimonious voice leading).

3   Transformations Between Seventh Chords

Childs investigated transformational parsimonious voice leading between dominant and half-diminished sevenths in [2]. In particular he studied transformations that fix two notes and move the other two notes by a semitone or a whole tone.

Gollin also studied the relationships between the same types of sevenths chords [6]. He introduced a possible three-dimensional expansion of the Tonnetz in which horizontal planes contain copies of the traditional Tonnetz, while segments in a chosen direction outside the plane represent intervals of minor seventh. While the Tonnetz tiles the plane with triangles, its three-dimensional expansion tiles the three-dimensional Euclidean space with tetrahedra, representing dominant and half-diminished seventh chords, and triangular prisms (not representing chords). There are six transformations between tetrahedra sharing a common edge: they are represented spatially as a "flip" of the two tetrahedra about their common edge. Each "edge-flip" maintains at least the two notes represented by the two vertices of the common edge, and in one case the two tetrahedra share three notes (see Fig. 5).

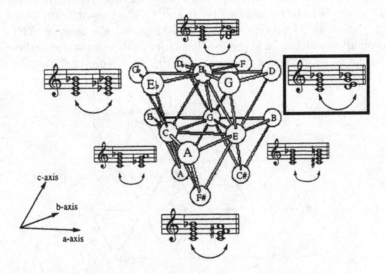

Fig. 5. The six edge-flips between tetrahedra. In the upper right the only flip in which the tetrahedra represent seventh chords sharing three common notes

Arnett and Barth [1] start from the three-dimensional expansion of the Tonnetz introduced by Gollin and observe that Gollin's study does not include the minor seventh chords, very common in the music literature. Therefore they propose to consider a set of 36 chords consisting of all dominant, half-diminished and minor seventh chords and to find the transformations between them that maintain three common notes. They define the following five operations:

$$P1: D \leftrightarrow m \qquad P1: [x, x+4, x+7, x+10] \leftrightarrow [x, x+3, x+7, x+10]$$
$$P2: m \leftrightarrow hd \qquad P1: [x, x+3, x+7, x+10] \leftrightarrow [x, x+3, x+6, x+10]$$
$$R1: D \leftrightarrow m-3 \qquad P1: [x, x+4, x+7, x+10] \leftrightarrow [x, x+4, x+7, x+9]$$
$$R2: m \leftrightarrow hd-3 \qquad R2: [x, x+3, x+7, x+10] \leftrightarrow [x, x+3, x+7, x+9]$$
$$L: D \leftrightarrow hd+4 \qquad P1: [x, x+4, x+7, x+10] \leftrightarrow [x+2, x+4, x+7, x+10]$$

The first four transformations move a single note by a semitone, whereas L shifts a note by a whole tone. In fact, L is the algebraic formalization of the edge-flip between tetrahedra representing seventh chords with three common notes described in Gollin's three-dimensional Tonnetz.

Although this study includes more types of seventh chords than Childs' and Gollin's ones, other important types of seventh chords are not considered and the algebraic structure of these transformations is not analyzed.

Kerkez gives an idea to extend the PLR group to seventh chords in [7].

Let H be the set of major and minor seventh chords, that is,

$$H = \{[x_1, x_2, x_3, x_4] | x_1, x_2, x_3, x_4 \in \mathbb{Z}_{12}, x_2 = x_1 + 4, x_3 = x_1 + 7, x_4 = x_1 + 11\} \cup$$
$$\{[x_1, x_2, x_3, x_4] | x_1, x_2, x_3, x_4 \in \mathbb{Z}_{12}, x_2 = x_1 + 3, x_3 = x_1 + 7, x_4 = x_1 + 10\}$$

Kerkez defines the following two maps $P, S: H \to H$:

$$P[a, b, c, d] = [(type[a, b, c, d]) \cdot 2 + d, a, b, c]$$
$$S[a, b, c, d] = [b, c, d, (-1) \cdot (type[a, b, c, d]) \cdot 2 + a]$$

where

$$type(t) = \begin{cases} 1 & \text{if } t \text{ is a minor seventh} \\ -1 & \text{if } t \text{ is a major seventh} \end{cases}$$

P maps each major seventh to its relative minor seventh moving the seventh down a whole tone. Vice versa, it maps each minor seventh to its relative major seventh moving the root up a whole tone.

S maps each major seventh to the minor seventh having root 4 semitones up, moving its root up a whole tone. Vice versa, it maps each minor seventh to the major seventh having root 4 semitones down, moving its seventh down a whole tone.

Kerkez proves that transformations P and S act on H generating a group again isomorphic to the dihedral group D_{12} of order 24.

In his work, Kerkez considers only major and minor seventh chord. But, as he noted in his conclusions, these transformations are just two of the possible operations between seventh chords.

3.1 Transformations Between Seventh Chords

We want to find all transformations between seventh chords describing parsimonious voice leading, i.e. those that fix three notes and move only one note by a semitone or a whole tone. We consider the following types of seventh chords: dominant (D), minor (m), half-diminished (hd), major (M) and diminished (d), and let $\tilde{H}$ be the set of all seventh chords of these 5 types. We first analyze transformations moving just one note by one semitone: if it exists, let us call Q_{i+} the map that sends each type of seventh chord to another type moving the i-th member up a semitone, where $i = R, T, F, S$ depending on whether the member is considered to be the root (R), the third (T), fifth (F) or seventh (S), respectively. Likewise, let Q_{i-} be the map that moves the i-th member down a semitone. We have the following:

$$
\begin{array}{lllll}
Q_{R+}(D) = d & Q_{R+}(m) = D & Q_{R+}(hd) = m & Q_{R+}(M) = hd & Q_{R+}(d) = hd \\
\cancel{Q_{R-}(D)} & \cancel{Q_{R-}(m)} & Q_{R-}(hd) = M & \cancel{Q_{R-}(M)} & Q_{R-}(d) = D \\
\cancel{Q_{T+}(D)} & Q_{T+}(m) = D & \cancel{Q_{T+}(hd)} & \cancel{Q_{T+}(M)} & Q_{T+}(d) = hd \\
Q_{T-}(D) = m & \cancel{Q_{T-}(m)} & \cancel{Q_{T-}(hd)} & \cancel{Q_{T-}(M)} & Q_{T-}(d) = D \\
\cancel{Q_{F+}(D)} & \cancel{Q_{F+}(m)} & Q_{F+}(hd) = m & \cancel{Q_{F+}(M)} & Q_{F+}(d) = hd \\
\cancel{Q_{F-}(D)} & Q_{F-}(m) = hd & \cancel{Q_{F-}(hd)} & \cancel{Q_{F-}(M)} & Q_{F-}(d) = D \\
Q_{S+}(D) = M & \cancel{Q_{S+}(m)} & \cancel{Q_{S+}(hd)} & \cancel{Q_{S+}(M)} & Q_{S+}(d) = hd \\
Q_{S-}(D) = m & Q_{S-}(m) = hd & Q_{S-}(hd) = d & Q_{S-}(M) = D & Q_{S-}(d) = D
\end{array}
$$

The maps that do not produce any of the classical types of seventh chords have been overstruck. We observe that some transformations are inverse to each other:

$$
\begin{array}{llll}
Q_{R+}(M) = hd & Q_{R-}(hd) = M & \Rightarrow & Q_R : M \leftrightarrow hd \\
Q_{R+}(m) = D & Q_{S-}(D) = m & \Rightarrow & Q_R, Q_S : D \leftrightarrow m \\
Q_{R+}(hd) = m & Q_{S-}(m) = hd & \Rightarrow & Q_R, Q_S : hd \leftrightarrow m \\
Q_{S+}(D) = M & Q_{S-}(M) = D & \Rightarrow & Q_S : D \leftrightarrow M \\
Q_{T+}(m) = D & Q_{T-}(D) = m & \Rightarrow & Q_T : m \leftrightarrow D \\
Q_{F+}(hd) = m & Q_{F-}(m) = hd & \Rightarrow & Q_F : hd \leftrightarrow m
\end{array}
$$

It remains to consider the following operations:

$$
\begin{array}{lllll}
Q_{R+}(D) = d & Q_{R-}(d) = D & Q_{T-}(d) = D & Q_{F-}(d) = D & Q_{S-}(d) = D \\
Q_{S-}(hd) = d & Q_{S+}(d) = hd & Q_{R+}(d) = hd & Q_{T+}(d) = hd & Q_{F+}(d) = sd
\end{array}
$$

$Q_{R+}(D) = d$ is the inverse of $Q_{R-}(d) = D, Q_{T-}(d) = D, Q_{F-}(d) = D$ and $Q_{S-}(d) = D$. This is due to the particular symmetry of the interval structure of diminished sevenths, in which the members of the chord play an identical role: for example the diminished seventh $C\sharp^{o7} = [C\sharp, E, G, B\flat]$ acoustically coincides to the diminished seventh $E\sharp^{o7} = [E, G, B\flat, D\flat]$ because they are enharmonically equivalent. Unlike the other four types, the diminished sevenths would be only three (and not twelve), e.g. $C, C\sharp, D$, because the other nine chords are three by three enharmonic to them. This explains why we have four transformations that have the same inverse. To obtain a set of well-defined musical transformations, we will consider the diminished seventh as 12 distinct chords, using the marked root to distinguish them. Hence we have 4 transformations between diminished and half-diminished seventh chords and 4 transformations between diminished and dominant seventh chords

$$
\begin{array}{lll}
Q_{S-}(hd) = d & Q_{R-}(d) = hd & \Rightarrow & Q_R, Q_S : hd \leftrightarrow d \\
Q_{S-}(hd) = d & Q_{T-}(d) = hd & \Rightarrow & Q_T, Q_S : hd \leftrightarrow d \\
Q_{S-}(hd) = d & Q_{F-}(d) = hd & \Rightarrow & Q_F, Q_S : hd \leftrightarrow d \\
Q_{S-}(hd) = d & Q_{S-}(d) = hd & \Rightarrow & Q_S : hd \leftrightarrow d \\
Q_{R+}(D) = d & Q_{R-}(d) = D & \Rightarrow & Q_R : D \leftrightarrow d \\
Q_{R+}(D) = d & Q_{T-}(d) = D & \Rightarrow & Q_R, Q_T : D \leftrightarrow d \\
Q_{R+}(D) = d & Q_{F-}(d) = D & \Rightarrow & Q_R, Q_F : D \leftrightarrow d \\
Q_{R+}(D) = d & Q_{S-}(d) = D & \Rightarrow & Q_R, Q_S : D \leftrightarrow d
\end{array}
$$

Now we consider the transformations that move a single note by a whole tone. Analogously to what was done above, if they exist let us call Q_{i++} the map which sends each type of seventh chord in another type moving the i-th member up a whole tone, and Q_{i-} the map which moves the i-th member down a whole tone. We obtain another classical type of seventh chords only moving the root up a whole tone and the seventh down a whole tone:

$$
\begin{array}{llll}
Q_{R++}(D)=hd & Q_{R++}(m)=M & \cancel{Q_{R++}(hd)} & Q_{R++}(M) = m & \cancel{Q_{R++}(d)} \\
\cancel{Q_{S--}(D)} & Q_{S--}(m)=M & Q_{S--}(hd) = D & Q_{S--}(M)=m & \cancel{Q_{S--}(d)}
\end{array}
$$

Again, we find some transformations that are the inverse one another:

$$
\begin{array}{lll}
Q_{R++}(D) = hd & Q_{S--}(hd) = D & \Rightarrow & Q_R, Q_S : D \leftrightarrow hd \\
Q_{R++}(m) = M & Q_{S--}(M) = m & \Rightarrow & Q_R, Q_S : m \leftrightarrow M \\
Q_{R++}(M) = m & Q_{S--}(m) = M & \Rightarrow & Q_R, Q_S : M \leftrightarrow m
\end{array}
$$

Overall we have 17 transformations corresponding to a parsimonious voice leading among our 5 types of seventh chords.

We want to define these transformations in a similar way to the neo-Riemannian operations. We will use the Arnett and Barth's notation, but we want to formalize it more precisely.

Definition 1. *We define a cyclic marked chord $[\underline{x_1}, x_2, \ldots, x_n]$ as a chord constituted by the n musical notes $x_1, x_2, \ldots, x_n$, so that acoustically $[\underline{x_1}, x_2, \ldots, x_n] = [x_2, \ldots, x_n, \underline{x_1}] = \cdots = [x_n, \underline{x_1}, \ldots, x_2]$, where $x_i \in \mathbb{Z}_{12}$ and the note corresponding to the root of the chord is underlined.*

As above, in cyclic marked chords all notes will be expressed in terms of a single note by adding or subtracting the appropriate number of semitones.

We start defining a parallel operation P for seventh chords. Let $P_{ij} \colon \tilde{H} \to \tilde{H}$ be the maps which send a i-th type of seventh chord to a j-th type of seventh chord, $1 \le i, j \le 5$ and $i \ne j$, and vice versa, and that fix the other types. 4 of the 17 transformations are parallel operations:

$$Q_T \colon D \leftrightarrow m \quad \Leftrightarrow \quad P_{12} \colon [\underline{x}, x+4, x+7, x+10] \leftrightarrow [\underline{x}, x+3, x+7, x+10]$$
$$Q_S \colon D \leftrightarrow M \quad \Leftrightarrow \quad P_{14} \colon [\underline{x}, x+4, x+7, x+10] \leftrightarrow [\underline{x}, x+4, x+7, x+11]$$
$$Q_F \colon m \leftrightarrow hd \quad \Leftrightarrow \quad P_{23} \colon [\underline{x}, x+3, x+7, x+10] \leftrightarrow [\underline{x}, x+3, x+6, x+10]$$
$$Q_S \colon hd \leftrightarrow d \quad \Leftrightarrow \quad P_{35} \colon [\underline{x}, x+3, x+6, x+10] \leftrightarrow [\underline{x}, x+3, x+6, x+9]$$

Remark 1. P_{12} and P_{23} coincide with $P1$ and $P2$ defined by Arnett and Barth.

Now we consider a relative operation R. We observe that if the triad is major $R = P \circ T_{-3} = T_{-3} \circ P$, if it is minor $R = P \circ T_3 = T_3 \circ P$. Then let $R_{ij} \colon \tilde{H} \to \tilde{H}$ be the maps which send a i-th type of seventh chord to a j-th type of seventh chord transposed three semitones down, a j-th type of seventh to a i-th type of seventh transposed three semitones up, and fix the other types. Then:

$$R_{ij} = T_{\pm 3} \circ P_{ij} = P_{ij} \circ T_{\pm 3} \qquad \forall i, j \in \{1, 2, 3, 4, 5\} \tag{7}$$

Now, 5 of the 17 transformations are relative operations:

$$Q_R, Q_S \colon D \leftrightarrow m - 3 \Leftrightarrow R_{12} \colon [\underline{x}, x+4, x+7, x+10] \leftrightarrow [x, x+4, x+7, \underline{x+9}]$$
$$Q_R, Q_S \colon m \leftrightarrow hd - 3 \Leftrightarrow R_{23} \colon [\underline{x}, x+3, x+7, x+10] \leftrightarrow [x, x+3, x+7, \underline{x+9}]$$
$$Q_R, Q_S \colon M \leftrightarrow m - 3 \Leftrightarrow R_{42} \colon [\underline{x}, x+4, x+7, x+11] \leftrightarrow [x, x+4, x+7, \underline{x+9}]$$
$$Q_R, Q_S \colon hd \leftrightarrow d - 3 \Leftrightarrow R_{35} \colon [\underline{x}, x+3, x+6, x+10] \leftrightarrow [x, x+3, x+6, \underline{x+9}]$$
$$Q_F, Q_S \colon d \leftrightarrow hd - 3 \Leftrightarrow R_{53} \colon [\underline{x}, x+3, x+6, x+9] \leftrightarrow [x, x+3, x+7, \underline{x+9}]$$

Remark 2. R_{12} and R_{23} coincide with $R1$ and $R2$ defined by Arnett and Barth. Moreover R_{42} coincide with the map P defined by Kerkez.

For the operation L we observe that if the triad is major $L = P \circ T_4 = T_4 \circ P$, if it is minor $L = P \circ T_{-4} = T_{-4} \circ P$. Then let $L_{ij} \colon H \to H$ be the maps which send a i-th type of seventh chord to a j-th type of seventh chord transposed four semitones up, a j-th type of seventh to a i-th type of seventh transposed four semitones down, and fix the other types. Then:

$$L_{ij} = T_{\pm 4} \circ P_{ij} = P_{ij} \circ T_{\pm 4} \qquad \forall i, j \in \{1, 2, 3, 4, 5\} \tag{8}$$

This time, 3 of the 17 transformations are L_{ij} operation:

$$Q_{R++} \colon D \leftrightarrow hd + 4 \quad \Leftrightarrow \quad L_{13} \colon [\underline{x}, x+4, x+7, x+10] \leftrightarrow [x+2, \underline{x+4}, x+7, x+10]$$
$$Q_R \colon D \leftrightarrow d + 4 \quad \Leftrightarrow \quad L_{15} \colon [\underline{x}, x+4, x+7, x+10] \leftrightarrow [x+1, \underline{x+4}, x+7, x+10]$$
$$Q_{R++} \colon M \leftrightarrow m + 4 \quad \Leftrightarrow \quad L_{42} \colon [\underline{x}, x+4, x+7, x+11] \leftrightarrow [x+2, \underline{x+4}, x+7, x+11]$$

Remark 3. L_{13} coincides with L defined by Arnett and Barth end the "edge-flip" described by Gollin in his three-dimensional Tonnetz.
L_{42} coincides with S defined by Kerkez.

We have identified 12 of the 17 transformations between seventh chords as operations similar to P, L and R. We now see that the other transformations correspond to new operations obtained by the composition of a parallel transformation and a transposition (with a number of semitones different from 3 and 4).
We denote by:

- Q_{ij} the maps which send a i-th type of seventh chord to a j-th type of seventh chord transposed one semitone up, a j-th type of seventh to a i-th type of seventh transposed one semitone down, and fix the other types;
- RR_{ij} the maps which send a i-th type of seventh chord to a j-th type of seventh chord transposed six semitones, and fix the other types;
- QQ_{ij} the maps which send a i-th type of seventh chord to a j-th type of seventh chord transposed two semitones up, a j-th type of seventh to a i-th type of seventh transposed two semitones down, and fix the other types;
- N_{ij} the maps which send a i-th type of seventh chord to a j-th type of seventh chord transposed five semitones up, a j-th type of seventh to a i-th type of seventh transposed five semitones down, and fix the other types.

With these transformations we can define the missing operations in the following way:

$$Q_R, Q_S : M \leftrightarrow hd + 1 \quad \Leftrightarrow \quad Q_{43} : [x, x+4, x+7, x+11] \leftrightarrow [x+1, x+4, x+7, x+11]$$
$$Q_R : D \leftrightarrow d + 1 \quad \Leftrightarrow \quad Q_{15} : [\underline{x}, x+4, x+7, x+10] \leftrightarrow [\underline{x+1}, x+4, x+7, x+10]$$
$$Q_T, Q_S : hd \leftrightarrow d - 6 \quad \Leftrightarrow \quad RR_{35} : [\underline{x}, x+3, x+6, x+10] \leftrightarrow [x, x+3, \underline{x+6}, x+9]$$
$$Q_R, Q_T : d \leftrightarrow D + 2 \quad \Leftrightarrow \quad QQ_{51} : [\underline{x}, x+3, x+6, x+9] \leftrightarrow [x, \underline{x+2}, x+6, x+9]$$
$$Q_R, Q_F : d \leftrightarrow D + 5 \quad \Leftrightarrow \quad N_{51} : [\underline{x}, x+3, x+6, x+9] \leftrightarrow [x, x+3, \underline{x+5}, x+9]$$

Remark 4. Crans, Fiore and Satyendra define P, L and R as inversions I_n; since inversions are isometries, they leave unchanged lengths and angles, and minor and major triads geometrically are represented by triangles which the edge lengths correspond to $3, 4$ and 5 semitones. This idea could in principle also be used to define transformations between seventh chords, but it can not be applied to all types since the lengths of the edges and the angles of the quadrilaterals that compose them are not equal. We have only 2 quadrilaterals that are isometric: the one representing the dominant sevenths and the one representing half-diminished sevenths. There exists a unique transformation between this types of seventh chords, L_{13}.

To visualize the 17 transformations just defined we can represent them in a graph whose vertices represent the types of seventh chord, and the edges represent the transformations between them. Therefore we have 5 vertices and 17 edges Fig. 6.

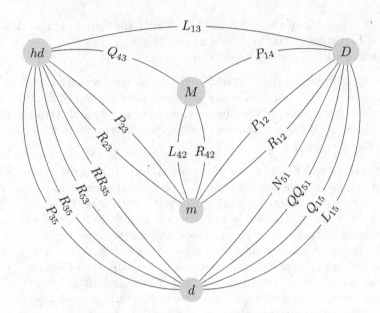

Fig. 6. The graph representing the 17 transformations between seventh chords.

4   The *PLRQ* Group

Let *PLRQ* be the group generated by the 17 transformations among seventh chords. Each transformation $t \in PLRQ$ exchanges two types of sevenths and fixes the others, thus we can associate to it a permutation of S_5 (more precisely, a transposition). This information is not sufficient to identify the transformation: to identify it, we add a vector $v \in \mathbb{Z}_{12}^5$, in which the i-th component, $i \in \{1, \ldots, 5\}$, is the number of semitones of which the root of the chord of type i has to be shifted to become the root of the chord of type j. It is easy to see that in this way no ambiguity is possible.

We write the 17 transformations between seventh chords as pairs of elements $(\sigma, v) \in S_5 \times \mathbb{Z}_{12}^5$ explicitly:

$P_{12}\colon [\underline{x}, x+4, x+7, x+10] \leftrightarrow [\underline{x}, x+3, x+7, x+10]$ $(\sigma, v) = ((12), (0, 0, 0, 0, 0))$

$P_{14}\colon [\underline{x}, x+4, x+7, x+10] \leftrightarrow [\underline{x}, x+4, x+7, x+11]$ $(\sigma, v) = ((14), (0, 0, 0, 0, 0))$

$P_{23}\colon [\underline{x}, x+3, x+7, x+10] \leftrightarrow [\underline{x}, x+3, x+6, x+10]$ $(\sigma, v) = ((23), (0, 0, 0, 0, 0))$

$P_{35}\colon [\underline{x}, x+3, x+6, x+10] \leftrightarrow [\underline{x}, x+3, x+6, x+9]$ $(\sigma, v) = ((35), (0, 0, 0, 0, 0))$

$R_{12}\colon [\underline{x}, x+4, x+7, x+10] \leftrightarrow [x, x+4, x+7, \underline{x+9}]$ $(\sigma, v) = ((12), (-3, 3, 0, 0, 0))$

$R_{23}\colon [\underline{x}, x+3, x+7, x+10] \leftrightarrow [x, x+3, x+7, \underline{x+9}]$ $(\sigma, v) = ((23), (0, -3, 3, 0, 0))$

$R_{42}\colon [\underline{x}, x+4, x+7, x+11] \leftrightarrow [x, x+4, x+7, \underline{x+9}]$ $(\sigma, v) = ((42), (0, 3, 0, -3, 0))$

$R_{35}\colon [\underline{x}, x+3, x+6, x+10] \leftrightarrow [x, x+3, x+6, \underline{x+9}]$ $(\sigma, v) = ((35), (0, 0, -3, 0, 3))$

$R_{53}\colon [\underline{x}, x+3, x+6, x+9] \leftrightarrow [x, x+3, x+7, \underline{x+9}]$ $(\sigma, v) = ((53), (0, 0, 3, 0, -3))$

$L_{13}: [\underline{x}, x+4, x+7, x+10] \leftrightarrow [x+2, \underline{x+4}, x+7, x+10] \quad (\sigma, v) = ((13), (4, 0, -4, 0, 0))$

$L_{15}: [\underline{x}, x+4, x+7, x+10] \leftrightarrow [x+1, \underline{x+4}, x+7, x+10] \quad (\sigma, v) = ((15), (4, 0, 0, 0, -4))$

$L_{42}: [\underline{x}, x+4, x+7, x+11] \leftrightarrow [x+2, \underline{x+4}, x+7, x+11] \quad (\sigma, v) = ((42), (0, -4, 0, 4, 0))$

$Q_{43}: [\underline{x}, x+4, x+7, x+11] \leftrightarrow [\underline{x+1}, x+4, x+7, x+11] \quad (\sigma, v) = ((43), (0, 0, -1, 1, 0))$

$Q_{15}: [\underline{x}, x+4, x+7, x+10] \leftrightarrow [\underline{x+1}, x+4, x+7, x+10] \quad (\sigma, v) = ((15), (1, 0, 0, 0, -1))$

$RR_{35}: [\underline{x}, x+3, x+6, x+10] \leftrightarrow [x, x+3, \underline{x+6}, x+9] \quad (\sigma, v) = ((35), (0, 0, -6, 0, 6))$

$QQ_{51}: [\underline{x}, x+3, x+6, x+9] \leftrightarrow [x, \underline{x+2}, x+6, x+9] \quad (\sigma, v) = ((51), (-2, 0, 0, 0, 2))$

$N_{51}: [\underline{x}, x+3, x+6, x+9] \leftrightarrow [x, x+3, \underline{x+5}, x+9] \quad (\sigma, v) = ((51), (-5, 0, 0, 0, 5))$

More precisely, we can represent each transformation $t \in PQRL$ as an element of

$$S_5 \times Z \qquad \text{where } Z = \{v \in \mathbb{Z}_{12}^5 | \sum_{i=1}^{5} v_i = 0\},$$

since this is clearly true for all the 17 generators. The mapping thus defined becomes a group homomorphism if we define on this set the following operation:

$$(\sigma_k, v_k) \circ \cdots \circ (\sigma_1, v_1)$$
$$= (\sigma_k \cdots \sigma_1, v_1 + \sigma_1^{-1}(v_2) + (\sigma_2 \sigma_1)^{-1}(v_3) + \cdots + (\sigma_{k-1} \cdots \sigma_1)^{-1}(v_k)) \qquad (9)$$
$$= (\sigma_k \cdots \sigma_1, v_1 + \sigma_1^{-1}(v_2) + \sigma_1^{-1} \sigma_2^{-1}(v_3) + \cdots + \sigma_1^{-1} \cdots \sigma_{k-1}^{-1}(v_k))$$

We want to prove that $PLRQ$ is isomorphic to $S_5 \ltimes Z$. We remind the definition of semidirect product of two subgroups.
Let G be a group. If G contains two subgroups H and K such that

(i) $G = HK$;
(ii) $K \trianglelefteq G$;
(iii) $H \cap K = 1$;

G is the *semidirect product* of H and K. Conversely, given two groups H and K and a group homomorphism $\phi: H \to Aut(K)$, we can construct a new group $H \ltimes K$ defining in the cartesian product $H \times K$ the following operation:

$$(h_1, k_1)(h_2, k_2) = (h_1 h_2, \phi_{h_2}(k_1) \cdot k_2)$$

Theorem 1. *The group $PLRQ$ is isomorphic to $S_5 \ltimes \mathbb{Z}_{12}^4$.*

Proof. First of all we prove that $PLRQ$ is isomorphic to $S_5 \ltimes Z$.
   We observe that the subgroup formed by the elements (Id, v) is normal. In fact, for all $(\sigma, v) \in S_5 \times Z, (Id, v') \in \{Id\} \times Z$, we have

$$(\sigma, v)(Id, v')(\sigma, v)^{-1} = (\sigma\sigma^{-1}, -v + \sigma(v') + \sigma(v)) = (Id, v'') \in \{Id\} \times Z$$

On the other hand, since S_5 is generated by transpositions, it is easy to see that, calling O the origin in $\mathbb{Z}_{12}^5$, $S_5 \times \{O\} < PLRQ$, since we already have in it $(P_{12}, O), (P_{14}, O), (P_{23}, O), (P_{35}, O)$.

To prove our thesis we are only left to see that there is a subgroup isomorphic to Z in $PLRQ$ having trivial intersection with $S_5 \times \{O\}$. But this is exactly the subgroup of the elements of type (Id, v). In fact we compute the permutations and vectors associated to $R_{42}L_{42}$, $P_{14}L_{42}P_{14}R_{12}$, $P_{12}L_{13}P_{12}R_{23}$:

$$R_{42}L_{42} = (\sigma', v')$$
$$\sigma' = \sigma_2\sigma_1 = (42)(42) = Id$$
$$v' = v_1 + \sigma_1^{-1}(v_2)$$
$$= (0, -4, 0, 4, 0) + (0, -3, 0, 3, 0)$$
$$= (0, -7, 0, 7, 0)$$
$$P_{14}L_{42}P_{14}R_{12} = (\sigma'', v'')$$
$$\sigma'' = \sigma_4\sigma_3\sigma_2\sigma_1 = (14)(42)(14)(12) = Id$$
$$v'' = v_1 + \sigma_1^{-1}(v_2) + \sigma_1^{-1}\sigma_2^{-1}(v_3) + \sigma_1^{-1}\sigma_2^{-1}\sigma_3^{-1}(v_4)$$
$$= (-3, 3, 0, 0, 0) + (0, 0, 0, 0, 0) + (-4, 4, 0, 0, 0) + (0, 0, 0, 0, 0)$$
$$= (7, -7, 0, 0, 0)$$
$$P_{12}L_{13}P_{12}R_{23} = (\sigma''', v''')$$
$$\sigma''' = \sigma_4\sigma_3\sigma_2\sigma_1 = (12)(13)(12)(23) = Id$$
$$v''' = v_1 + \sigma_1^{-1}(v_2) + \sigma_1^{-1}\sigma_2^{-1}(v_3) + \sigma_1^{-1}\sigma_2^{-1}\sigma_3^{-1}(v_4)$$
$$= (0, -3, 3, 0, 0) + (0, 0, 0, 0, 0) + (0, -4, 4, 0, 0) + (0, 0, 0, 0, 0)$$
$$= (0, 7, -7, 0, 0)$$

With the following elements just computed

$$R_{42}L_{42} = (Id, (0, -7, 0, 7, 0)) \tag{10}$$
$$P_{14}L_{42}P_{14}R_{12} = (Id, (7, -7, 0, 0, 0))$$
$$P_{12}L_{13}P_{12}R_{23} = (Id, (0, 7, -7, 0, 0))$$

we can generate each element $(Id, (v_1, v_2, v_3, v_4, 0))$, with $(v_1, v_2, v_3, v_4, 0) \in \mathbb{Z}_{12}^5$ such that $\sum_1^4 v_i = 0$. To see this, taken $a, b, c \in \mathbb{Z}$, we have to solve

$$a(0, -7, 0, 7, 0) + b(7, -7, 0, 0, 0) + c(0, 7, -7, 0, 0) \equiv (v_1, v_2, v_3, v_4, 0) \pmod{12}$$
$$(-7b, -7a + 7b - 7c, 7c, 7a) \equiv (v_1, v_2, v_3, v_4, 0) \pmod{12}$$

$$\begin{cases} -7b \equiv v_1 \\ -7a + 7b - 7c \equiv v_2 \\ 7c \equiv v_3 \\ 7a \equiv v_4 \end{cases} \Rightarrow \begin{cases} -7b \equiv v_1 \\ 7b - 7c \equiv v_2 + 7a \\ 7c \equiv -v_3 \\ 7a \equiv v_4 \end{cases} \Rightarrow \begin{cases} 7b \equiv -v_1 \\ -v_1 - v_3 \equiv v_2 + v_4 \\ 7c \equiv -v_3 \\ 7a \equiv v_4 \end{cases}$$

which is solvable because 7 is coprime with 12.

To obtain all elements $(Id, (v_1, v_2, v_3, v_4, v_5))$, with $(v_1, v_2, v_3, v_4, v_5) \in \mathbb{Z}_{12}^5$ such that $\sum_1^5 v_i = 0$, it is sufficient add to the 3 generators listed in 10 the generator $P_{1235}R_{23}P_{12}L_{15}L_{13} = (Id, (7, 0, 0, 0, -7))$.

But it is evident that $Z \simeq \mathbb{Z}_{12}^4$, hence $PLRQ \simeq S_5 \ltimes \mathbb{Z}_{12}^4$. □

References

1. Arnett, J., Barth, E.: Generalizations of the Tonnetz: Tonality Revisited. http://personal.denison.edu/~lalla/MCURCSM2011/10.pdf
2. Childs, A.: Moving beyond Neo-Riemannian triads: exploring a transformational model for Seventh Chords. J. Music Theory 42(2), 181–193 (1998)
3. Crans, A., Fiore, T.M., Satyendra, R.: Musical actions of Dihedral Groups. Am. Math. Monthly 116(6), 479–495 (2009)
4. Euler, L.: Tentamen novae theoriae musicae, Opera Omnia I(III), Teubner, Stuttgart (1739)
5. Fiore, T.M., Satyendra, R.: Generalized contextual groups. Music Theory Online, 11(3) (2005)
6. Gollin, E.: Aspects of Three-Dimensional Tonnetze. J. Music Theory 42(2), 195–206 (1998)
7. Kerkez, B.: Extension of Neo-Riemannian PLR-group to Seventh Chords. Mathematics, Music, Art, Architecture, Culture (2012)
8. Lewin, D.: Transformational techniques in atonal and other music theories. Perspect. New Music 21, 312–371 (1982)
9. Lewin, D.: Generalized Musical Intervals and Transformations. Yale University Press, New Haven (1987)
10. Mazzola, G.: Gruppen und Kategorien in der Musik: Entwurf einer mathematischen Musiktheorie. Heldermannr, Lemgo (1985)
11. Mazzola, G.: Geometrie der Tone. Birkhuser, Basel (1990)
12. Riemann, H.: Handbuch der Harmonielehre. Breitkopf und Hrtel, Leipzig (1887)
13. Rotman, J.J.: Advanced modern algebra. Am. Math. Soc. (2010)

Pairwise Well-Formed Modes and Transformations

David Clampitt[1(✉)] and Thomas Noll[2]

[1] School of Music, Ohio State University, Columbus, USA
clampitt.4@osu.edu
[2] Escola Superior de Música de Catalunya, Departament de Teoria,
Composició i Direcció, Barcelona, Catalonia, Spain
thomas.mamuth@gmail.com

Abstract. One of the most significant attitudinal shifts in the history of music occurred in the Renaissance, when an emerging triadic consciousness moved musicians towards a new scalar formation that placed major thirds on a par with perfect fifths. In this paper we revisit the confrontation between the two idealized scalar and modal conceptions, that of the ancient and medieval world and that of the early modern world, associated especially with Zarlino. We do this at an abstract level, in the language of algebraic combinatorics on words. In scale theory the juxtaposition is between well-formed and pairwise well-formed scales and modes, expressed in terms of Christoffel words or standard words and their conjugates, and the special Sturmian morphisms that generate them. Pairwise well-formed scales are encoded by words over a three-letter alphabet, and in our generalization we introduce special positive automorphisms of $F3$, the free group over three letters.

Keywords: Pairwise well-formed scales and modes
Well-formed scales and modes · Well-formed words · Christoffel words
Standard words · Central words · Algebraic combinatorics on words
Special Sturmian morphisms

1 Introduction: Authentic and Triadic Modes

Figure 1 shows a C-major scale with two different interpretations of its step interval pattern. In the annotation $aaba|aab$ (above the staff) the two letters a and b designate the major and minor steps, respectively. The vertical stroke $|$ designates the authentic divider of the mode into a species of the fifth $aaba$ and a species of a fourth aab. This pattern is called the *Authentic division of the Ionian Mode*. In the annotation $ac|ba||cab$ (below the staff) the three letters a, c and b designate the greater and lesser major and the minor steps, respectively. Together they divide the major mode triadically into a species of the major third ac, a species of the minor third ba and a species of the fourth cab. This pattern shall be called the *Triadic Division of the Ionian Major Mode*.

© Springer International Publishing AG 2017
O. A. Agustín-Aquino et al. (Eds.): MCM 2017, LNAI 10527, pp. 26–37, 2017.
https://doi.org/10.1007/978-3-319-71827-9_3

Fig. 1. Authenic division of the Ionian and triadic division of the Ionian major mode.

The contrasting juxtaposition evokes several open questions of historical and systematic nature about the particular relevance of these modes for different types of music and music analysis. Within the discourse of mathematical music theory they point to the theory of well-formed scales and modes [2,5,6,10], on the one hand, and to the theory of pairwise well-formed scales [3,4], on the other. In the present article we therefore extend some transformational innovations within the theory of well-formed modes in order to make them fruitful within a theory of pairwise well-formed modes. These investigations eventually contribute to a deeper theoretical understanding of the juxtaposition in Fig. 1.

2   Non-singular Pairwise Well-Formed Modes

In this section we revisit some important results from [4] about the structure of non-singular pairwise well-formed scales and re-interpret them in a word-theoretic context. The 3-letter word *acbacab* describes the species of the octave of the Ionian Major mode, and thereby it is the step interval pattern of a *pairwise well-formed scale*. The motivation behind this concept is the following: The two-letter word *aabaaab*, describing the Ionian species of the octave can be obtained from *acbacab* by an identification of the letter c with the letter a: $\pi_{c \to a}(acbacab) = aabaaab$. In traditional music-theoretical terms this letter projection describes *syntonic identification*, i.e., neglecting the difference between the greater and lesser major steps. There are two more such letter identifications, both of which lead to well-formed modes, whence the term *pairwise well-formed*. One of them is $\pi_{b \to c}(bacabac) = cacacac$. It describes an identification of the minor step with the lesser major step, i.e., neglecting their difference, what we may call the (harmonic) *apotome*. This mode neutralizes also the difference between the major and minor thirds and can be seen as a modal refinement of the generic third-generated scale. The third letter projection $\pi_{a \to b}(bacabac) = bbcbbbc$ identifies the minor step with the greater major step. The neglected interval is the sum of the two previously mentioned ones, and so we can formally speak of an *apo-syntonic identification*. It is arguable, though, whether this third projection bears a direct musical meaning. As an auxiliary construction it proves to be very useful on a theoretical level. This becomes clear in the course of the article.

We will represent non-singular pairwise well-formed (PWWF) scales by words over a three-letter alphabet $A = \{a, b, c\}$, as in [3], and we will henceforth refer to PWWF words and drop the qualifier "non-singular" (the singular case is represented by the word *abacaba* and its word-theoretical conjugates). We denote the set of all PWWF words by $\mathfrak{A} \subset \{a,b,c\}^*$. From [2,4], $v \in \mathfrak{A}$ if and only if each of the three projections $\pi_{x \to y}$ (described above) results in a well-formed word, i.e., the conjugate of a Christoffel word, equivalently, of a standard word (see [1,7] for the relevant word theory background). For every word $v \in \mathfrak{A}$, the length $|v|$ is odd, and the multiplicities of two of the letters are the same: $|v|_b = |v|_c$. It follows that $|v|_a$ is odd. In light of these facts, given a special standard word, we can always construct a PWWF word, by the *bisecting substitution* defined below.

Definition 2.1. *Consider a special standard morphism f acting on the word monoid $\{a,c\}^*$ and consider the word $w = f(ac) = w_1 w_2 \ldots w_n$. We further suppose that $|w| = n$ is odd and that $|w|_c$ is even. Then we define the* bisection *of the word w as $w_{\prec} = v = v_1 v_2 \ldots v_n \in \{a,b,c\}^*$ with*

$$
v_k := \begin{cases} a \;\; if\, w_k = a, \\ b \;\; if\, w_k = c \;\, and \;\, |w_1 \ldots w_k|_c \;\, is\;\, odd \\ c \;\; if\, w_k = c \;\, and \;\, |w_1 \ldots w_k|_c \;\, is\;\, even. \end{cases}
$$

The bisecting substitution $\sigma : \{a,b,c\}^* \to \{a,b,c\}^*$ is then defined as

$$\sigma(a) = v_1 \ldots v_m, \quad \sigma(b) = v_{m+1} \ldots v_{2m}, \quad \sigma(c) = v_{2m+1} \ldots v_n, \; where \; m = |f(a)|.$$

Remark: PWWF scales have distinct inversions, whereas the inversion of a mode of a well-formed scale is a mode of that scale (e.g., Ionian inverted is Phrygian). In word-theoretical language, if w is a standard word, the reversal of w is in the conjugacy class of w. For $w \in \mathfrak{A}$, the reversal of w is in its own conjugacy class, distinct from that of w. For $k \in \mathbb{N}$, there are $\phi(k)/2$ distinct conjugacy classes of standard words of length k; for PWWF words (k odd), there are $\phi(k)$ distinct conjugacy classes [4]. The defining projections are insensitive to reversal, however, up to trivial replacements of the letters, so we may pair w with its reversal, and choose whichever is convenient as representative. There is thus a bijection between conjugacy classes of standard words and classes of PWWF words of odd length k.

For example, $w = bacabac \in \mathfrak{A}$, and its reversal is $w' = cabacab$. $\pi_{c \to a}(w) = baaabaa$ (representing Phrygian), $\pi_{b \to a}(w) = aacaaac$ (representing Ionian), and $\pi_{b \to c}(w) = cacacac$ (representing, e.g., Dorian thirds); while $\pi_{c \to a}(w') = aabaaab$ (Ionian), $\pi_{b \to a}(w') = caaacaa$ (Phrygian), and $\pi_{b \to c}(w) = cacacac$. Therefore, we may depart from either PWWF representative. (A future investigation will be dedicated to the musical interpretation of the distinction between the representatives).

Proposition 2.2. *Consider a PWWF substitution σ and let $f(a) = \pi_{b \to c}(\sigma(a)), f(c) = \pi_{b \to c}(\sigma(bc))$ denote its apotomic projection. Let further $M_f = \begin{pmatrix} |f(a)|_a & |f(c)|_a \\ |f(a)|_c & |f(c)|_c \end{pmatrix}$ denote the incidence matrix of f. Then the incidence matrix of σ is given as*

$$
\tilde{M}_\sigma = \begin{pmatrix} |f(a)|_a & |f(a)|_a & |f(c)|_a - |f(a)|_a \\ \left\lfloor \dfrac{|f(a)|_c}{2} \right\rfloor & \left\lceil \dfrac{|f(a)|_c}{2} \right\rceil & \dfrac{|f(c)|_c - |f(a)|_c}{2} \\ \left\lceil \dfrac{|f(a)|_c}{2} \right\rceil & \left\lfloor \dfrac{|f(a)|_c}{2} \right\rfloor & \dfrac{|f(c)|_c - |f(a)|_c}{2} \end{pmatrix}.
$$

3   Apo-syntonic Conversion

In addition to the letter projections $\pi_{b\to c}, \pi_{a\to b}, \pi_{c\to a}$ on words $v \in A^*$, we apply them to substitutions as follows (using the same symbols):

Definition 3.1. *Consider a substitution (a monoid morphism) $\sigma : A^* \to A^*$. Then we obtain three induced substitutions $\pi_{b\to c}(\sigma) : \{a, c\}^* \to \{a, c\}^*$, $\pi_{a\to b}(\sigma) : \{b, c\}^* \to \{b, c\}^*$ and $\pi_{c\to a}(\sigma) : \{a, b\}^* \to \{a, b\}^*$ by virtue of:*

$$[\pi_{b\to c}(\sigma)](a) := \pi_{b\to c}(\sigma(a)) \quad [\pi_{b\to c}(\sigma)](c) := \pi_{b\to c}(\sigma(bc))$$
$$[\pi_{a\to b}(\sigma)](b) := \pi_{a\to b}(\sigma(ab)) \quad [\pi_{a\to b}(\sigma)](c) := \pi_{a\to b}(\sigma(c))$$
$$[\pi_{c\to a}(\sigma)](a) := \pi_{c\to a}(\sigma(ab)) \quad [\pi_{c\to a}(\sigma)](b) := \pi_{c\to a}(\sigma(c))$$

We say that σ is an authentic PWWF substitution iff all three projections $\pi_{b\to c}(\sigma)$, $\pi_{a\to b}(\sigma)$ and $\pi_{c\to a}(\sigma)$ are Special Sturmian morphisms.

In the rest of this section we will assume that σ is the bisecting substitution associated with a suitable special standard morphism f and hence $f = \pi_{b\to c}(\sigma)$. Further we use the symbols g and $\tilde{g}$ for the projections $g = \pi_{a\to b}(\sigma)$ and $\tilde{g} = \pi_{c\to a}(\sigma)$ The diagram in Fig. 2 shows the interplay of σ with its three projections.

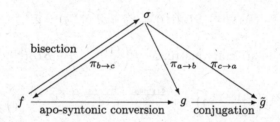

Fig. 2. Interplay of a PWWF substitution σ with its projections f, g and $\tilde{g}$.

Our goal is now to understand the interdependence between f and g.

Proposition 3.2. *Consider an authentic (non-singular) PWWF mode $\sigma \in \mathfrak{A}$ with the projections $f = \pi_{b\to c}(\sigma), g = \pi_{a\to b}(\sigma)$ and $\tilde{g} = \pi_{c\to a}(\sigma)$. Then the common incidence matrix $M_g = M_{\tilde{g}}$ of g and $\tilde{g}$ can be expressed in terms of the coefficients of the incidence matrix $M_f = \begin{pmatrix} |f(a)|_a & |f(c)|_a \\ |f(a)|_c & |f(c)|_c \end{pmatrix}$ as follows:*

$$M_g = M_{\tilde{g}} = \begin{pmatrix} 2|f(a)|_a + |f(a)|_c & |f(c)|_a - |f(a)|_a + \frac{1}{2}(|f(c)|_c - |f(a)|_c) \\ |f(a)|_c & \frac{1}{2}(|f(c)|_c - |f(a)|c_b) \end{pmatrix}.$$

Proof. The incidence matrix M_g can be obtained from adding the first two columns and the first two rows of $\tilde{M}_\sigma$. Thus, after Proposition 2.2 the upper left entry of M_g becomes $|f(a)|_a + |f(a)|_a + \lfloor \frac{|f(a)|_c}{2} \rfloor + \lceil \frac{|f(a)|_c}{2} \rceil = 2|f(a)|_a + f(a)|_c$.

The upper right entry becomes $|f(c)|_a - |f(a)|_a + \dfrac{|f(c)|_c - |f(a)|_c}{2}$, the lower left entry becomes $\lceil \dfrac{|f(a)|_c}{2} \rceil + \lfloor \dfrac{|f(a)|_c}{2} \rfloor = |f(a)|_c$. The lower right entry remains $\dfrac{|f(c)|_c - |f(a)|_c}{2}$.

In order to understand the connection between f and g more directly, rather than via the substitution σ, we have a closer look at the structure of the linear map: $\tilde{\beta} : GL_2(\mathbb{R}) \to GL_2(\mathbb{R})$ with

$$\tilde{\beta}\left(\begin{pmatrix} a_{11} & a_{12} \\ a_{21} & a_{22} \end{pmatrix}\right) := \begin{pmatrix} 2a_{11} + a_{21} & a_{12} - a_{11} + (a_{22} - a_{21})/2 \\ a_{21} & (a_{22} - a_{21})/2 \end{pmatrix}.$$

With $R = \begin{pmatrix} 1 & 1 \\ 0 & 1 \end{pmatrix}$ let Σ^2 and Σ_2 denote the following two subsets of $SL_2(\mathbb{N})$:

$$\Sigma^2 = \left\{ R \cdot \begin{pmatrix} 2b_{11} & b_{12} \\ b_{21} & b_{22} \end{pmatrix} \mid b_{11}, b_{12}, b_{21}, b_{22} \in \mathbb{N}, 2b_{11}b_{22} - b_{12}b_{21} = 1 \right\}$$

$$\Sigma_2 = \left\{ \begin{pmatrix} b_{11} & b_{12} \\ b_{21} & 2b_{22} \end{pmatrix} \cdot R \mid b_{11}, b_{12}, b_{21}, b_{22} \in \mathbb{N}, 2b_{11}b_{22} - b_{12}b_{21} = 1 \right\}$$

Lemma 3.3. $SL_2(\mathbb{N}) \cap \tilde{\beta}^{-1}(SL_2(\mathbb{N})) = \Sigma_2$ and $\tilde{\beta}(\Sigma_2) = \Sigma^2$.

Proof. Consider an arbitrary matrix $X = \begin{pmatrix} c_{11} & c_{12} \\ c_{21} & c_{22} \end{pmatrix} \in SL_2(\mathbb{N})$. Then we have

$$\tilde{\beta}^{-1}\left(\begin{pmatrix} c_{11} & c_{12} \\ c_{21} & c_{22} \end{pmatrix}\right) = \begin{pmatrix} \frac{c_{11}-c_{21}}{2} & \frac{c_{11}-c_{21}}{2} + c_{12} - c_{22} \\ c_{21} & c_{21} + 2c_{22} \end{pmatrix}.$$

The entry $c_{21} + 2c_{22}$ is larger than c_{21} and therefore $\tilde{\beta}^{-1}(X) \in SL_2(\mathbb{N})$ iff $Y = \tilde{\beta}^{-1}(X) \cdot R^{-1} = \begin{pmatrix} \frac{c_{11}-c_{21}}{2} & c_{12} - c_{22} \\ c_{21} & 2c_{22} \end{pmatrix} \in SL_2(\mathbb{N})$. But in order to have a positive entry $c_{12} - c_{22}$ it turns out that $X \in SL_2(\mathbb{N})$ cannot be arbitrary. Also $Z = R^{-1} \cdot X = \begin{pmatrix} c_{11} - c_{21} & c_{12} - c_{22} \\ c_{21} & c_{22} \end{pmatrix}$ must be in $SL_2(\mathbb{N})$. And this implies $c_{11} \geq c_{21}$. Finally, we have to deal with the condition that the upper left entry $\frac{c_{11}-c_{21}}{2}$ of $\tilde{\beta}^{-1}(X)$ needs to be an integer. This implies that $c_{11} - c_{21}$, which is also the upper left entry of Z, is even, i.e., $RZ = X \in \Sigma^2$.

Corollary 3.4. *The set Σ_2 parametrizes the conjugation classes of all authentic PWWF substitutions σ in terms of the incidence matrices M_f of their associated apotomic projections: the special standard morphisms $f = \pi_{b \to c}(\sigma)$. Also the set Σ^2 parametrizes these same conjugation classes by virtue of the incidence matrices M_g of their associated apo-syntonic projections $g = \pi_{a \to b}(\sigma)$.*

This motivates the following definition:

Definition 3.5. *The restriction of $\tilde{\beta}$ to the subset Σ_2 is called the* apo-syntonic conversion:

$$\beta : \Sigma_2 \to \Sigma^2.$$

Let $\delta : SL_2(\mathbb{N}) \to SL_2(\mathbb{N})$ denote the main-diagonal-flip

$$\delta\left(\begin{pmatrix} b_{11} & b_{12} \\ b_{21} & b_{22} \end{pmatrix}\right) := \begin{pmatrix} b_{22} & b_{12} \\ b_{21} & b_{11} \end{pmatrix}.$$

Lemma 3.6. *δ mutually exchanges Σ^2 and Σ_2, i.e. $\delta(\Sigma^2) = \Sigma_2$ and $\delta(\Sigma_2) = \Sigma^2$.*

Proposition 3.7. *We have the following commutative diagram*

$$
\begin{array}{ccc}
\Sigma_2 & \xrightarrow{\quad \beta \quad} & \Sigma^2 \\
\Big\uparrow{\delta} & & \Big\uparrow{\delta} \\
\Sigma^2 & \xleftarrow{\quad \beta \quad} & \Sigma_2
\end{array}
$$

Proposition 3.8. *Under the convention that the apotomic projection f is a standard morphism, the apo-syntonic projection g also turns out to be a special standard morphism.*

Proof. As f is special standard we have a decomposition $w = f(ac) = f(a)f(c) = tcasac$ with a negative (=plagal) standard word tca and a positive (=authentic) standard word sac. We have $g(bc) = u_1 u_2 \ldots u_{|w|} \in \{b, c\}^*$ with

$$
u_k := \begin{cases}
b \text{ if } w_k = a, \\
b \text{ if } w_k = c \text{ and } |w_1 \ldots w_k|_c \text{ is odd} \\
c \text{ if } w_k = c \text{ and } |w_1 \ldots w_k|_c \text{ is even.}
\end{cases}
$$

The final letter of u is c, because $|w|_c$ is even, so by definition of u above, the last letter c is fixed. Let $v = cuc^{-1} \in \{b, c\}^*$ denote the result of conjugating u with c^{-1}. In order to show that g is a special standard morphism, it is sufficient to show that $v = v_1 v_2 \ldots v_{|w|}$ is the bad conjugate of u. Let $m = |g(b)|$ denote the length of $g(b)$. It is sufficient to show that $|v_1 \ldots v_m|_b$ differs from $|u_1 \ldots u_m|_b = |g(b)|_b$. Knowing that $f(a)$ is a prefix of $f(c)$ we write $w = f(a)f(a)f(a^{-1}c) = tcatcarac$ and we may conclude that $w_m = a$, and hence $u_m = b = v_{m+1}$. But in the light of $v_1 = c$ this implies $|v_1 \ldots v_m|_b = |g(b)|_b - 1$, i.e. v is the bad conjugate.

4 PWWF Substitutions and Automorphisms of F_3

In the two-dimensional situation of well-formed modes we may interpret the Sturmian morphisms as positive automorphisms of the free group F_2. In the world of substitutions on words in three letters their analogues constitute different transformational concepts. The PWWF substitutions are well-adopted to the family of (non-singular) pairwise well-formed modes. Still there is a small subfamily of pairwise well-formed modes, where these substitutions are also automorphisms of the free group F_3. This final section is dedicated to their study. To get a concrete idea about the role of F_3-automorphisms, we look into the Authentic and Triadic Divisions of the Phrygian mode (see Fig. 3). In the theory of well-formed modes one describes the species $baaa$ and baa of the fifth and fourth as images of a and b under a word transformation $\tilde{g}(a) = baaa$, $\tilde{g}(b) = baa$. The left side of Fig. 3 shows a decomposition of this compound transformation into three elementary ones. First, the fifth a is filled with a fourth and a major step $a \mapsto ba$, then both fourths are filled with a minor third and a major step $b \mapsto ba$ and finally, both minor thirds are filled with a minor step and a major step $b \mapsto ba$.

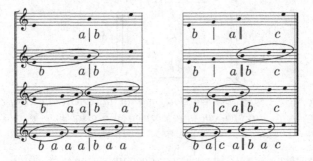

Fig. 3. Construction of the Authentic Phrygian and the Triadic Phrygian Minor modes through substitutions.

The right side of Fig. 3 shows an analogous procedure for the construction of the Phrygian Minor mode $ba|ca||bac$. The fourth is filled with a minor third and a lesser major step: $c \mapsto bc$. Then the major third is filled with a lesser and and greater major step: $a \mapsto ca$ and finally both minor thirds are filled with a minor step followed by a greater major step: $b \mapsto ba$. This final act in the generation of the Phrygian Minor mode does not work analogously for the Ionian Major mode, because there we find two different species of the minor third: ba and ab.

The automorphism group $Aut(F_n)$ of the free group $F_n = \langle x_1, x_2, \ldots, x_n \rangle$ is redundantly generated by the elementary *Nielsen transformations* (see [8,9], p. 162 ff.), namely the letter transpositions $x_i \mapsto x_k, x_k \mapsto x_i$, cyclic letter permutations $x_1 \mapsto x_2 \mapsto \ldots \mapsto x_n \mapsto x_1$, letter inversions $x_i \mapsto x_i^{-1}$ and the substitutions of the types $x_i \mapsto x_i x_k$ or $x_i \mapsto x_k x_i$. The automorphisms of F_3, which potentially coincide with PWWF substitutions, are necessarily positive, so we don't need to consider letter inversions here. This leads to the following definition:

Definition 4.1. *An authentic PWWF substitution is called morphic, if it is a positive automorphism of F_3.*

For two different letters $x, y \in \{a, b, c\}$ let $E_{x,y}, A_{xy}, P_{xy} \in Aut(F_3)$ denote the following positive automorphisms of the free group F_3: Let $z \in \{a, b, c\}$ denote the third letter, respectively.

$$E_{xy}(x) = y, \quad E_{xy}(y) = x, \ E_{xy}(z) = z,$$
$$A_{xy}(x) = xy, \ A_{xy}(y) = y, \ A_{xy}(z) = z,$$
$$P_{xy}(x) = yx, \ P_{xy}(y) = y, \ P_{xy}(z) = z.$$

Hence, a PWWF substitution $\sigma : \{a, b, c\}^* \rightarrow \{a, b, c\}^*$ is morphic, if it can be written as a composition of a finite number of letter permutations E_{xy} and substitutions of the types A_{xy} and/or P_{xy}. Now we inspect the seven conjugates of the Phrygian minor mode to which also the Ionian major mode belongs. The following proposition is thus the portrait of a very special conjugation class of modes. As in the two-letter case it contains bad conjugates.

Proposition 4.2. *Let* $\sigma : \{a, b, c\}^* \rightarrow \{a, b, c\}^*$ *with* $\sigma(b) = ba$, $\sigma(a) = ca$ *and* $\sigma(c) = bac$ *denote the PWWF substitution, which is associated with the Phrygian minor mode. The cycle of the single letter conjugations* $ba|ca||bac \rightarrow ac|ab||acb \rightarrow ca|ba||cba \rightarrow ab|ac||bac \rightarrow ac|ba||cab \rightarrow cb|ac||aba \rightarrow ba|ca||bac$ *contains two bad conjugates and—accordingly—five PWWF modes. Four of these PWWF modes are morphic. The exceptional good, but amorphous instance, is the Ionian major mode. These seven conjugates together with their associated projections under* $\pi_{b \rightarrow c}$, $\pi_{c \rightarrow a}(\sigma)$ *and* $\pi_{a \rightarrow b}(\sigma)$ *are listed below:*

authentic triadic mode	apotomic projection	syntonic projection	apo-syntonic projection	transform. type
$ba|ca||bac$	$ca|cacac$	$baaa|baa$	$bbcb|bbc$	$morphic^{**}$
$ac|ab||acb$	$ac|acacc$	$aaab|aab$	$bcbb|bcb$	$morphic$
$ca|ba||cba$	$ca|cacca$	$aaba|aba$	$cbbb|cbb$	$morphic$
$ab|ac||bac$	$ac|accac$	$abaa|baa$	$bbbc|bbc$	$morphic$
$ba|cb||aca$	$ca|ccaca$	$baab|aaa$	$bbcb|bcb$	bad^{*}
$ac|ba||cab$	$ac|cacac$	$aaba|aab$	$bcbb|cbb$	$good^{*}$
$cb|ac||aba$	$cc|acaca$	$abaa|aba$	$cbbc|bbb$	bad^{**}

Proof. In the left column of the table $ba|ca||bac$ undergoes the full cycle of letter-by-letter conjugations. In parallel the projections $\pi_{b \rightarrow c}(\sigma)$, $\pi_{c \rightarrow a}(\sigma)$, $\pi_{a \rightarrow b}(\sigma)$ run through their corresponding conjugations. By virtue of Proposition 3.8 the conjugations of the apotomic and the apo-syntonic projections are "in sync", i.e. they start with special standard modes (marked as morphic**) and they end with bad modes (marked as bad**). As a consequence there are only two bad conjugates among the triadic modes. The generation of the four morphic modes – up to letter permutations – is given below:

$$A_{ba}P_{ac}P_{cb}(b|a||c) = A_{ba}P_{ac}(b|a||bc) = A_{ba}(b|ca||bc) = ba|ca||bac$$
$$P_{ca}A_{ab}P_{bc}(c|a||b) = P_{ca}A_{ab}(c|a||cb) = P_{ca}(c|ab||cb) = ac|ab||acb$$
$$A_{ba}P_{ac}A_{cb}(a|b||c) = A_{ba}P_{ac}(a|b||cb) = A_{ba}(ca|b||cb) = ca|ba||cba$$
$$P_{ca}A_{ab}P_{cb}(b|a||c) = P_{ca}A_{ab}(b|a||bc) = P_{ca}(ab|c||bc) = ab|ac||bac$$

The obstacle for the "good" PWWF Ionian major mode to be generated by an F_3-automorphism is the co-existence of the two factors ab and ba.

In addition to the bad[*] Locrian mode (with a bad syntonic projection) there is the bad[**] Dorian minor mode (with bad apotomic and apo-syntonic projections), whose structural defects have been discussed in 19th-century treatises, such as Moritz Hauptmann's 1853 *Die Natur der Harmonik und der Metrik*.

We show now that a similar picture arises in connection with a certain family $\mathcal{M}$ of morphic PWWF modes and we conclude the article with the conjecture that this family actually exhausts the morphic PWWF modes entirely. It is useful to start the investigation of this family from the authentic standard modes, whose bisecting transformations then yield the associated PWWF modes. In this particular case, the general form of the standard morphism f, generating the single-divider mode $f(a)|f(c)$ is $f = G^k DG^{2n}$ with $n > 0$ and $k \geq 0$. The corresponding apo-syntonic conversion of f, the standard morphism g, generating the first of the two double divider modes $g(b)|g(c)$ is $g = G^{2k+2}DG^{n-1}$. Thus, we have the two sets

$$\mathcal{F} = \{G^k DG^{2n} \mid n > 0, k \geq 0\} \text{ and } \mathcal{G} = \{G^{2n} DG^k \mid n > 0, k \geq 0\}$$

together with the *apo-syntonic conversion* $\theta : \mathcal{F} \rightarrow \mathcal{G}$, where $\theta(G^k DG^{2n}) := G^{2k+2}DG^{n-1}$. Furthermore, we consider reversal map, i.e. the unique anti-automorphism of $rev : \langle G, D\rangle \rightarrow \langle G, D\rangle$ fixing both G and D. This map rev sends $\mathcal{F}$ to $\mathcal{G}$ and vice versa.

The following proposition specifies the commutative diagram for matrices in Proposition 3.7 to the elements of the sets $\mathcal{F}$ and $\mathcal{G}$:

Proposition 4.3. $\theta \circ rev \circ \theta = rev$.

Proof. Although the relation is a corollary of Proposition 3.7 we give a direct proof here:

$$\theta(rev(\theta(G^k DG^{2n}) = \theta(rev(G^{2k+2}DG^{n-1}))$$
$$= \theta(G^{n-1}DG^{2k+2})$$
$$= G^{2n}DG^{2k+2} = rev(G^k DG^{2n})$$

The third corresponding special Sturmian morphism $\tilde{g}$, generating the syntonic projection $\tilde{g}(a|b)$ of our PWWF mode has the following form:

$$\tilde{g} = \tilde{\theta}(f) = \tilde{\theta}(G^k DG^{2n}) = G^k \tilde{G}^{k+2}DG^{n-1}.$$

Thus, the morpism $\tilde{g}$ is preceded by precisely $k + 2$ conjugate morphisms in the Zarlino ordering, namely by $G^{k+l}\tilde{G}^{k+2-l}DG^{n-1}$ for $l = 1, \ldots, k + 2$. For the corresponding incidence matrices we have:

$$M_f = \begin{pmatrix} k+1 & 2n(k+1)+k \\ 1 & 2n+1 \end{pmatrix} \text{ and } M_g = M_{\tilde{g}} = \begin{pmatrix} 2k+3 & n(2k+3)-1 \\ 1 & n \end{pmatrix}$$

Proposition 4.4. *Consider the authentic PWWF three-letter mode*

$$v_{k,n} := a^k ba | a^k ca || (a^k ba\, a^k ca)^{n-1} a^k ba\, a^k c$$

Its apotomic, apo-syntonic and syntonic projections are $\pi_{b\to c}(v_{k,n}) = f(a|c)$, $\pi_{a\to b}(v_{k,n}) = g(b|c)$ *and* $\pi_{c\to a}(v_{k,n}) = \tilde{g}(a|b)$, *respectively.*

Proof.

$$
\begin{aligned}
f(a|c) &= G^k DG^{2n}(a|c) = G^k D(a|a^{2n}c) = G^k(ca|(ca)^{2n}c) \\
&= a^k ca | (a^k ca)^{2n} a^k c = \pi_{b\to c}(v_{k,n}) \\
g(b|c) &= G^{2k+2} DG^{n-1}(b|c) = G^{2k+2} D(b|b^{n-1}c) = G^{2k+2}(cb|(cb)^{n-1}c) \\
&= b^{2k+2} cb | (b^{2k+2} cb)^{n-1} b^{2k+2} c = \pi_{a\to b}(v_{k,n}). \\
\tilde{g}(a|b) &= G^k \tilde{G}^{k+2} DG^{n-1}(a|b) = G^k \tilde{G}^{k+2} D(a|a^{n-1}b) = G^k \tilde{G}^{k+2}(ba|(ba)^{n-1}b), \\
&= G^k \tilde{G}^{k+2}(ba|(ba)^{n-1}b) = G^k(ba^{k+3}|(ba^{k+3})^{n-1} ba^{k+2}) \\
&= a^k ba^{k+3} | (a^k ba^{k+3})^{n-1} a^k ba^{k+2} = \pi_{c\to a}(v_{k,n}),
\end{aligned}
$$

The subsequent proposition provides an explicit portrait of the full conjugation class of the authentic PWWF mode $v_{k,n}$.

Proposition 4.5. *The table below lists all letter-by-letter conjugations of the authentic PWWF mode $v_{k,n}$ and characterizes them as morphic, good or bad. The segment between the bad** mode (with bad apotomic and apo-syntonic projections) and the bad*-mode (with bad syntonic projection) is exclusively occupied by morphic modes. The opposite segment between the bad*-mode and the bad** mode is exclusively occupied by good modes:*

$a^k ba$	$a^k ca$	$\| (a^k ba\, a^k ca)^{n-1}(a^k ba)(a^k c)$	morphic**
$a^{k-1} ba^2$	$a^{k-1} ca^2$	$\| (a^{k-1} ba^2\, a^{k-1} ca^2)^{n-1}(a^{k-1} ba^2)(a^{k-1} ca)$	morphic
...			
$a^{k-l} ba^{l+1}$	$a^{k-l} ca^{l+1}$	$\| (a^{k-l} ba^{l+1}\, a^{k-l} ca^{l+1})^{n-1}(a^{k-l} ba^{l+1})(a^{k-l} ca^l)$	morphic
...			
ba^{k+1}	ca^{k+1}	$\| (ba^{k+1}\, ca^{k+1})^{n-1}(ba^{k+1})(ca^k)$	morphic
$a^{k+1}c$	$a^{k+1}b$	$\| (a^{k+1}c\, a^{k+1}b)^{n-1}(a^{k+1}c)(a^k b)$	morphic
...			
ca^{k+1}	ba^{k+1}	$\| (ca^{k+1}\, ba^{k+1})^{n-1}(ca^k)(ba^{k+1})$	morphic
$a^{k+1}b$	$a^{k+1}c$	$\| (a^{k+1}b\, a^{k+1}c)^{n-1}(a^k b)(a^{k+1}c)$	morphic
...			
ba^{k+1}	ca^{k+1}	$\| (ba^{k+1}\, ca^{k+1})^{n-2}(ba^{k+1}\, ca^k)(ba^{k+1} ca^{k+1})$	morphic
$a^{k+1}c$	$a^{k+1}b$	$\| (a^{k+1}c\, a^{k+1}b)^{n-2}(a^{k+1}c\, a^k b)(a^{k+1} ca^{k+1}b)$	morphic
...			
ca^{k+1}	ba^{k+1}	$\| (ca^{k+1}\, ba^{k+1})^{n-2}(ca^k\, ba^{k+1})(ca^{k+1} ba^{k+1})$	morphic
$a^{k+1}b$	$a^{k+1}c$	$\| (a^{k+1}b\, a^{k+1}c)^{n-2}(a^k b\, a^{k+1}c)(a^{k+1} ba^{k+1}c)$	morphic
...			
aba^k	aca^k	$\| ba^k aca^k (aba^k aca^k)^{n-1}$	morphic
ba^{k+1}	$ca^k b$	$\| (a^{k+1} ca^{k+1}b)^{n-1} a^{k+1} ca^{k+1}$	bad*
$a^{k+1}c$	$a^k ba$	$\| (a^k ca^{k+1} ba)^{n-1} a^k ca^{k+1}b$	good*
...			
aca^k	ba^{k+1}	$\| (ca^{k+1} ba^{k+1})^{n-1} ca^{k+1} ba^k$	good
$ca^k b$	$a^{k+1}c$	$\| (a^{k+1} ba^{k+1}c)^{n-1} a^{k+1} ba^{k+1}$	bad**

Proof. The following calculation shows that $v_{k,n}$ is morphic. The calculations for the other morphic modes are analogous:

$$
\begin{aligned}
P_{ba}^k P_{ca}^k A_{ba} P_{ac} (P_{cb} P_{ca})^{n-1} P_{cb} E_{ab}(a|b||c) &= P_{ba}^k P_{ca}^k A_{ba} P_{ac} (P_{cb} P_{ca})^{n-1} P_{cb}(b|a||c) \\
&= P_{ba}^k P_{ca}^k A_{ba} P_{ac} (P_{cb} P_{ca})^{n-1}(b|a||bc) \\
&= P_{ba}^k P_{ca}^k A_{ba} P_{ac}(b|a||(ba)^{n-1}bc) \\
&= P_{ba}^k P_{ca}^k A_{ba} P_{ac}(b|a||(ba)^{n-1}bc) \\
&= P_{ba}^k P_{ca}^k A_{ba}(b|ca||(bca)^{n-1}bc) \\
&= P_{ba}^k P_{ca}^k (ba|ca||(ba\,ca)^{n-1}ba\,c) \\
&= P_{ba}^k (ba|a^k ca||(ba\,a^k ca)^{n-1}ba\,a^k c) \\
&= a^k ba|a^k ca||(a^k ba\,a^k ca)^{n-1}a^k ba\,a^k c \\
&= v_{k,n}
\end{aligned}
$$

We try to write the good**-mode $\gamma_{k,n} = a^{k+1}c|a^k ba||(a^k ca^{k+1}ba)^{n-1}a^k ca^{k+1}b$ as an image $f(a|b||c)$ under an automorphism $f = f_m f_{m-1} \cdots f_1 f_0$, where the $f_i(i = 1, ..., m)$ are supposed to be productions of the type A_{xy} or P_{xy} and where f_0 is letter permutation. First we observe that f_m, the last of these morphisms, cannot be a production of b's or c's: First of all, A_{bc}, A_{cb}, P_{bc} and P_{bc} are excluded because there are no letters c and b neighboring each other (even in the case $k = 0$). But furthermore, not all instances of the letter a are followed or preceded by either exclusively b or exclusively c, and so also the productions A_{ac}, A_{ab}, P_{ac} and P_{ac} can be excluded. The only remaining possibilities are productions of the letter a. Among these the append-transformations A_{ba} and A_{ca} both excluded, as the single-divider prefix $a^{k+1}c$ of $\gamma_{k,n}$ ends on c and the double-divider suffix $(a^k ca^{k+1}ba)^{n-1}a^k ca^{k+1}b$ ends on b. For $k > 0$ the prepend-transformations P_{ba} and P_{ca} are suitable in order to produce $\gamma_{k,n}$ from shorter words. To be more precise: P_{ba} and P_{ca} commute with each other and both can be applied k times in any order to the triple $\gamma_{0,n} = ac|ba||(caba)^{n-1}cab$ to produce $\gamma_{k,n}$. Here is one of them:

$$
\begin{aligned}
a^{k+1}c|a^k ba||(a^k ca^{k+1}ba)^{n-1}a^k ca^{k+1}b &= P_{ba}^k(a^{k+1}c|ba||(a^k caba)^{n-1}a^k cab) \\
&= P_{ca}^k(P_{ba}^k(ac|ba||(caba)^{n-1}cab))
\end{aligned}
$$

A closer look at $\gamma_{0,n}$ shows that it is not an image of a shorter word-triple under any of the 8 transformations of the type A_{xy} or P_{xy}.

The similar line of argument works for any of the good modes. For $0 < l \leq k$ the general form of a good mode is

$$
\begin{aligned}
& aa^{k-l}ca^l|a^{k-l}ba^l a||(a^{k-l}ca^l aa^{k-l}ba^l a)^{n-1}a^{k-l}ca^l aa^{k-l}ba^l \\
&= P_{ca}^{k-l}(P_{ba}^{k-l}(A_{ca}^l(A_{ba}^l(\gamma_{0,n}))))
\end{aligned}
$$

So all the good modes are images of the mode $\gamma_{0,n}$. And this is the only possibility to generate them from a shorter triple.

Conjecture 4.6. Consider an authentic PWWF substitution σ in the sense of Definition 3.1. The substitution σ is morphic (i.e. is a positive automorphism of F_3) iff its generated mode $\sigma(a)|\sigma(b)||\sigma(c)$—up to letter permutation—is an instance of a morphic mode in Proposition 4.5.

5   Conclusion

Since the rise of the triad in its role as a governing concept in the music of harmonic tonality, music theorists have been in a quandary as to how to adjust the immemorial diatonic scale so as to be compatible with the triad's new primacy. The interval of the major third challenges the status of the perfect fifth as a scale generator, and in the course of this competition it undermines the validity of other properties that are consequences of the fifth-generatedness of the diatonic scale, such as the well-formedness property. The concept of pairwise well-formedness offers a reconciliation, insofar as it implements the idea of a coexistence of three well-formed scale structures within one parent scale. The letter projections mediate between the competing interpretations of a fifth- and a third-generated scale. The present paper offers a transformational upgrade to that earlier basic insight. Following the pattern of the investigation of well-formed modes through automorphisms of the free group F_2 it clarifies the combinatorial behavior of all non-singular pairwise well-formed modes. For the concrete case of the Ionian major mode $ac|ba||cab$, it turns out that the underlying substitution is not an automorphism of the free group F_3, which is an exception in comparison to the Phrygian minor, Lydian major, Mixolydian major and Aeolian minor modes. This exceptional status corresponds to the musical fact that the species of major third ac and ca as well as those of the minor third ba and ab are different. The mathematical status of the Dorian minor mode as a bad mode has its musical counterpart in the fact that the species $cbac$ doesn't form a proper fifth. This fact in turn is reflected in the special treatment some nineteenth-century theorists accorded the supertonic ii harmony in major, as a sort of cousin to the diminished supertonic triad in minor.

References

1. Berthé, V., de Luca, A., Reutenauer, C.: On an involution of Christoffel words and Sturmian morphisms. Eur. J. Comb. **29**(2), 535–553 (2008)
2. Clampitt, D., Noll, T.: Modes, the height-width duality, and Handschin's tone character. Music Theor. Online **17**(1) (2011)
3. Clampitt, D.: Mathematical and musical properties of pairwise well-formed scales. In: Klouche, T., Noll, T. (eds.) MCM 2007. CCIS, vol. 37, pp. 464–468. Springer, Heidelberg (2009). https://doi.org/10.1007/978-3-642-04579-0_46
4. Clampitt, D.: Pairwise well-formed scales: structural and transformational properties. Ph.D. dissertaion, SUNY at Buffalo (1997)
5. Carey, N., Clampitt, D.: Self-similar pitch structures, their duals, and rhythmic analogues. Perspect. New Music **34**(2), 62–87 (1996)
6. Carey, N., Clampitt, D.: Aspects of well-formed scales. Music Theor. Spectr. **11**(2), 187–206 (1989)
7. Lothaire, M.: Algebraic Combinatorics on Words. Cambridge University Press, Cambridge (2002)
8. Magnus, W., Karrass, A., Solitar, D.: Combinatorial Group Theory. Dover Publications, New York (2004)
9. Nielsen, J.: Die Isomorphismengruppe der freien Gruppen. Math. Ann. **91**, 169–209 (1924)
10. Noll, T.: Ionian theorem. J. Math. Music **3**(3), 137–151 (2009)

Homometry in the Dihedral Groups: Lifting Sets from $\mathbb{Z}_n$ to D_n

Grégoire Genuys[1]([⊠]) and Alexandre Popoff[2]([⊠])

[1] IRCAM/CNRS/UPMC, Paris, France
gregoire.genuys@ircam.fr
[2] IRCAM/119 Rue de Montreuil, Paris, France
al.popoff@free.fr

Abstract. The paper deals with the question of homometry in the dihedral groups D_n of order $2n$. These groups are non-commutative, leading to new and challenging definitions of homometry, as compared to the well-known case of homometry in the commutative group $\mathbb{Z}_n$. We give here a musical interpretation of homometry in D_{12} using the well-known neo-Riemannian groups, some results on a complete enumeration of homometric sets for small values of n, and some properties disclosing the deep links between homometry in $\mathbb{Z}_n$ and homometry in D_n.

Keywords: Homometry · Interval vector · T/I-group · PLR-group
Dihedral groups · Semi-direct product · Discrete Fourier transform

1 Introduction

The concept of homometry first appeared in the 1930s in the field of cristallography. The question was to determine the structure of a crystal from its X-ray diffraction pattern. This type of measurement is directly related to the intensity of the Fourier transform of the crystallographic structure, but the phase information is lost in the process. The problem was therefore to know if a complete reconstruction of the structure of the crystal was possible. This problem later found applications in various fields, such as music theory where the question was to characterize a set of notes (a chord, a melody) from the intervals that compose it. Homometry has been studied through different approaches: group theory [1], Fourier transform [2], distribution theory [3,4], etc. and is an open field of research.

The classical way to model the n-tone equal temperament in musical set theory is to consider the cyclic group $\mathbb{Z}_n = \mathbb{Z}/n\mathbb{Z}$ as the set of pitch classes, for instance $\mathbb{Z}_{12} = \{0 = C, 1 = C^\sharp, ..., 11 = B\}$. Following David Lewin's constructions described and systematized in [5], for any subsets A and B in $\mathbb{Z}_n$, one can consider the interval function $\mathbf{ifunc}(A, B)$ whose components are

$$\mathbf{ifunc}(A, B)(k) = \sharp\{(a, b) \in A \times B \mid b - a = k\}$$

© Springer International Publishing AG 2017
O. A. Agustín-Aquino et al. (Eds.): MCM 2017, LNAI 10527, pp. 38–49, 2017.
https://doi.org/10.1007/978-3-319-71827-9_4

for $k \in \mathbb{Z}_n$, and the interval vector $\mathbf{iv}(A)$ whose components are defined by $\mathbf{iv}(A)(k) = \mathbf{ifunc}(A, A)(k)$. Two sets A and B in $\mathbb{Z}_n$ are *homometric* if they have the same interval vector ($\mathbf{iv}(A) = \mathbf{iv}(B)$), meaning that they contain the same set of intervals. This is traditionally called *Z-relation* and was mainly presented by Forte [6]. In this paper we will only use the word homometry which refers to the same concept. The actions of transposition and inversion clearly do not change the interval vector of a set, hence two homometric sets which do not belong to the same set class modulo transpositions and inversions will be called *non-trivial homometric sets*. A well-known example of a non-trivial homometric pair in $\mathbb{Z}_{12}$ is $(\{C, D^\flat, E, G^\flat\}, \{C, D^\flat, E^\flat, G\})$, for which the interval vector is $[1, 1, 1, 1, 1, 2, 1, 1, 1, 1, 1]$. More detailed explanations can be found in [3,7].

The concept of group action of $(\mathbb{Z}_n, +)$ on itself by translation (where $n \in \mathbb{Z}_n$ acts on $a \in \mathbb{Z}_n$ by $n{+}a$) can also be used to define the interval vector. Thus we call *interval* between a and b, written $\mathbf{int}(a, b)$, the element n such that $n + a = b$, and $\mathbf{iv}(A)(k) = \sharp\{(a, b) \in A^2 \mid \mathbf{int}(a, b) = k\}$ for $k \in \mathbb{Z}_n$. In a more general setting and following Lewin's idea of generalized intervals [5], one can consider homometry in the context of any simply transitive group action. For instance it is well-known that the T/I-group and the neo-Riemannian PLR-group both act simply transitively on the set S of major and minor triads. We recall that the T/I-group is generated by the transpositions $T_p(x) = p + x$ and the inversions $I_p(x) = T_p I_0(x)$ i.e. $I_p(x) = -x + p$, for $p \in \mathbb{Z}_n$. The PLR-group is generated by P, L, and R which correspond respectively to the operations parallel, leading tone exchange and relative (see [8] for more details). The interval between two triads s_1 and s_2 is the unique element of the group sending s_1 to s_2 for the chosen group action. If we use upper-case letters for major triads and lower-case letters for minor triads (C is C-major and c is C-minor) we obtain for instance in the context of the T/I-group: $\mathbf{int}_{T/I}(c, E^\flat) = I_{10}$ and in the context of PLR-group: $\mathbf{int}_{PLR}(c, E^\flat) = R$. For a given Generalized Interval System $(S, \text{IVLS}, \mathbf{int})$, we can thus define the interval vector of a subset A of S as

$$\mathbf{iv}(A)(k) = \sharp\{(a, b) \in A^2 \mid \mathbf{int}(a, b) = k\}$$

for k in IVLS, and two subsets of S will be called homometric if they have the same interval vector. As an example of homometry for both the actions of the T/I-group and the PLR-group, consider the pair of sets $\{c, D^\flat, E^\flat, e, a^\flat\}$ and $\{c, E^\flat, e, F, a^\flat\}$. Figure 1 shows some of the intervals between the elements of these sets, for the action of the T/I-group and the PLR-group respectively. It can clearly be seen that the same intervals $\{T_2, T_4, T_4, T_8, I_0, I_2, I_4, I_6, I_8, I_{10}\}$ are present in both sets, hence they have the same interval vector (all other intervals can be deduced by composition and/or inversion). Similarly, the same intervals $\{R, PL, PL, PL, LRL, LRP, RLP, LPR, PRP, PRLR\}$ are present in both sets leading to an identical conclusion for the interval vector.

It has been showed in [9] that the actions of the T/I-group and of the PLR-group on the set of major and minor triads can be understood respectively as the left and the right actions of the dihedral group D_{12} on this set. Moreover it is well-known that D_{12} is the semi-direct product $(\mathbb{Z}_{12}, +) \rtimes (\mathbb{Z}_2, \times)$. This lead

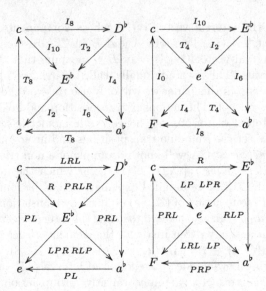

Fig. 1. Intervals in the T/I-group (top) and the PLR-group (bottom) for the two homometric sets $\{c, D^b, E^b, e, a^b\}$ and $\{c, E^b, e, F, a^b\}$.

us to the general topic of this paper, namely the study of homometry in the non-commutative dihedral groups of order $2n$, or equivalently the semi-direct products $(\mathbb{Z}_n, +) \rtimes (\mathbb{Z}_2, \times)$.

This paper is divided into four parts. We first recall the definition of D_n as a semi-direct product and we define homometry for the right and left actions of this group. The link with the well-known musical case $n = 12$ will be explained in the second part. In the third part we give the equations that characterize homometry and some results concerning enumeration. Finally we define in the last part the concept of *lift* which bridges homometry in $\mathbb{Z}_n$ and homometry in D_n, and we give our main results using the discrete Fourier transform.

2   The Dihedral Group D_n as a Semi-direct Product

The dihedral group D_n is the group of symmetries of a regular n-gon, using rotations and reflections. It can be expressed as the semi-direct product $(\mathbb{Z}_n, +) \rtimes (\mathbb{Z}_2, \times)$, where $\mathbb{Z}_2 = \{\pm 1\}$. Its elements are the pairs (k, ϵ) where $k \in \mathbb{Z}_n$ and $\epsilon = \pm 1$, with the identity element being $(0, 1)$, and multiplication between two elements being given by the equation

$$(k, \epsilon)(l, \eta) = (k + \epsilon l, \epsilon \eta). \tag{1}$$

As a non-commutative group D_n acts on itself by right or left multiplication: thus (l, η) acting on (k, ϵ) on the right leads us to Eq. (1), whereas (l, η) acting on (k, ϵ) on the left leads us to

$$(l, \eta)(k, \epsilon) = (l + \eta k, \eta \epsilon). \tag{2}$$

This allows us to define the intervals between any two pairs (k_1, ϵ_1) and (k_2, ϵ_2) of D_n. The left interval is the unique element (l, η) in D_n such that $(l, \eta)(k_1, \epsilon_1) = (k_2, \epsilon_2)$, whereas the right interval is the unique element (l, η) such that $(k_1, \epsilon_1)(l, \eta) = (k_2, \epsilon_2)$. Thus we obtain two functions $^l\mathrm{int}: D_n \times D_n \to D_n$ and $^r\mathrm{int}: D_n \times D_n \to D_n$, called *interval functions* and defined as

$$^l\mathrm{int}((k_1, \epsilon_1), (k_2, \epsilon_2)) = (k_2 - \epsilon_2/\epsilon_1 k_1, \epsilon_2/\epsilon_1), \tag{3}$$

$$^r\mathrm{int}((k_1, \epsilon_1), (k_2, \epsilon_2)) = ((k_2 - k_1)/\epsilon_1, \epsilon_2/\epsilon_1). \tag{4}$$

The left interval vector $^l\mathrm{iv}(A)$ and right interval vector $^r\mathrm{iv}(A)$ of a set A in D_n are then defined as

$$^{l,r}\mathrm{iv}(A)((l, \eta)) = \sharp\{((k_1, \epsilon_1), (k_2, \epsilon_2)) \in A^2 \mid {}^{l,r}\mathrm{int}((k_1, \epsilon_1), (k_2, \epsilon_2)) = (l, \eta)\}$$

for $(l, \eta) \in D_n$. We say that two sets in D_n are *homometric for the left (resp. for the right) action* (or simply *left-/right-homometric*) if they have the same left (resp. right) interval vector. It is easy to see that any left (resp. right) action preserves the right (resp. left) intervals. As a consequence, two left-homometric (resp. right-homometric) sets will be called *non trivially homometric*, if they are not related by right (resp. left) translation. In the rest of the paper 'homometric' will mean 'non-trivially homometric'.

3   Link with the T/I and the PLR-groups in the Case $n = 12$

As mentioned in [9], the actions of the T/I-group and the PLR-group on the set of major and minor triads can be considered as the left and right actions of D_{12} on S, but also as the actions of D_{12} on itself. To understand why, we use a (non canonical) bijection between D_{12} and S. The element $(s, +1)$ of D_{12} will be identified to the major triad whose root is $s \in \mathbb{Z}_{12}$, whereas the element $(s, -1)$ of D_{12} will be considered as the minor triad whose root is $s \in \mathbb{Z}_{12}$. For instance $(0, 1)$ corresponds to C, $(0, -1)$ corresponds to c, $(8, -1)$ corresponds to $a^\flat$, and so on. The set $\{c, D^\flat, E^\flat, e, a^\flat\}$ given in the introduction can be then identified with the set $\{(0, -1), (1, 1), (3, 1), (4, -1), (8, -1)\}$.

Notice that this bijection with the elements of D_{12} is not limited to the set of major and minor triads, but may be applied on any set on which D_{12} acts simply transitively, as mentioned in the introduction. For instance, Lewin used an action of D_{12} on set-class 5-4 in his analysis of Stockhausen's Klavierstück III [10]. This is also true for homometry in D_n in general. Nevertheless, we have decided to focus in this paper on the actions of D_{12} on major and minor triads as it belongs to the most musically relevant examples of group actions.

If we consider the left action of D_{12} on itself we have the following isomorphism between D_{12} (as the acting group) and the T/I-group: $(p, +1)$ corresponds to T_p and $(p, -1)$ corresponds to I_{p-5}. For instance the image of C by T_7 is calculated in D_{12} as the element corresponding to $(7, 1)(0, 1) = (7, 1)$, i.e. the major chord G. Similarly the image of C by I_2 is calculated to be the element corresponding to $(7,-1)(0,1) = (7,-1)$, i.e. the minor chord g.

If we consider the right action of D_{12} on itself we have the following bijection between D_{12} and the PLR-group: P corresponds to $(0,-1)$, L corresponds to $(4,-1)$ and R corresponds to $(9,-1)$. For instance $P(C)$ is calculated as the element corresponding to $(0,1)(0,-1) = (0,-1)$ which is c, $L(d^b)$ corresponds to $(1,-1)(4,-1) = (9,1)$ which is A, and $R(F)$ to $(4,1)(9,-1) = (1,-1)$ which is d^b. We obtain on Fig. 2 a new version of Fig. 1 with the left and the right intervals in D_{12}.

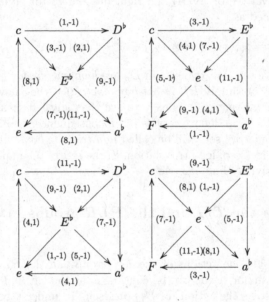

Fig. 2. Left (top) and right (bottom) intervals in D_{12} for the two sets $\{c, D^b, E^b, e, a^b\}$ and $\{c, E^b, e, F, a^b\}$.

Since we are interested in the general case of homometry in D_n, we will work from now on using the general point of view of semi-direct products and their group elements (l, η).

4   Homometry in D_n: Formulas and Enumeration

In order to avoid confusions we will adopt the notation '$\mathcal{A}$' for subsets in D_n, and the notation 'A' for subsets in $\mathbb{Z}_n$. Given the form of the group elements of D_n as pairs (l, η), a set $\mathcal{A} \in D_n$ is the disjoint union of two (possibly empty) subsets $\mathcal{A}_+$ and $\mathcal{A}_-$, with $\mathcal{A}_+ = \{(l, \eta) \in \mathcal{A} \mid \eta = 1\}$, and $\mathcal{A}_- = \{(l, \eta) \in \mathcal{A} \mid \eta = -1\}$. For instance in D_{12} the set

$$\mathcal{A} = \{c, D^b, E^b, e, a^b\} = \{(0,-1), (1,1), (3,1), (4,-1), (8,-1)\}$$

given in the introduction is the union of the major chords $\{D^\flat, E^\flat\} = \{(1,1),(3,1)\}$ and the minor chords $\{c, e, a^\flat\} = \{(0,-1),(4,-1),(8,-1)\}$. Let $\pi : D_n \longrightarrow \mathbb{Z}_n$ be the projection on the first factor, i.e. $\pi((l, \eta)) = l$. We define the sets A_+ and A_- as $A_+ = \pi(\mathcal{A}_+) = \{\pi((l, \eta)) \mid (l, \eta) \in \mathcal{A}_+\}$, and $A_- = \pi(\mathcal{A}_-) = \{\pi((l, \eta)) \mid (l, \eta) \in \mathcal{A}_-\}$. In the example above, we have $A_+ = \{1, 3\} \subset \mathbb{Z}_{12}$ and $A_- = \{0, 4, 8\} \subset \mathbb{Z}_{12}$. Remark that we have $\pi(\mathcal{A}) = \pi(\mathcal{A}_+) \cup \pi(\mathcal{A}_-) = A_+ \cup A_-$. If there is no ambiguity we will call $A = \pi(\mathcal{A})$.

The purpose of the following theorem is to give a characterization of homometry in D_n using **iv** and **ifunc**.

Theorem 1. *Two sets $\mathcal{A}$ and $\mathcal{B}$ in D_n are homometric for the right action if and only if the following two equations hold:*

$$\begin{cases} iv(A_+) + iv(A_-) = iv(B_+) + iv(B_-) \\ ifunc(A_+, A_-) = ifunc(B_+, B_-) \end{cases} \tag{5}$$

Two sets $\mathcal{A}$ and $\mathcal{B}$ in D_n are homometric for the left action if and only if the following two equations hold:

$$\begin{cases} iv(A_+) + iv(A_-) = iv(B_+) + iv(B_-) \\ ifunc(I_0 A_+, A_-) = ifunc(I_0 B_+, B_-) \end{cases} \tag{6}$$

Proof. Let $\mathcal{A}, \mathcal{B}$ be two right homometric sets in D_n. Let us recall that

$$^r\mathbf{int}((k_1, \epsilon_1), (k_2, \epsilon_2)) = ((k_2 - k_1)/\epsilon_1, \epsilon_2/\epsilon_1) \tag{7}$$

for (k_1, ϵ_1) and (k_2, ϵ_2) in $\mathcal{A}$. We then have to consider two cases, corresponding to the two equations of (5).

In the first case, $\epsilon_2/\epsilon_1 = 1$, i.e. $\epsilon_1 = \epsilon_2$. Then,

- either $\epsilon_1 = 1 = \epsilon_2$ in which case we have $^r\mathbf{int}((k_1, \epsilon_1), (k_2, \epsilon_2)) = (k_2 - k_1, 1)$ for k_1 and k_2 in A_+, meaning that we have to calculate $iv(A_+)$ to obtain all the intervals of that type, or
- $\epsilon_1 = -1 = \epsilon_2$ then $^r\mathbf{int}((k_1, \epsilon_1), (k_2, \epsilon_2)) = (k_1 - k_2, 1)$ for k_1, k_2 in A_-, meaning that we have to calculate $iv(A_-)$ to obtain all the intervals of that type.

We then must have $iv(A_+) + iv(A_-) = iv(B_+) + iv(B_-)$.

In the second case, $\epsilon_2/\epsilon_1 = -1$, i.e. $\epsilon_1 = -\epsilon_2$. Then,

- either $\epsilon_1 = 1, \epsilon_2 = -1$, thus we have $^r\mathbf{int}((k_1, \epsilon_1), (k_2, \epsilon_2)) = (k_2 - k_1, -1)$ for $k_1 \in A_+$ and $k_2 \in A_-$, meaning that we have to calculate $ifunc(A_+, A_-)$ to obtain all the intervals of that type, or
- $\epsilon_1 = -1, \epsilon_2 = 1$, then $^r\mathbf{int}((k_1, \epsilon_1), (k_2, \epsilon_2)) = (k_1 - k_2, 1)$ for $k_1 \in A_-$ and $k_2 \in A_+$, meaning that we have to calculate again $ifunc(A_+, A_-)$ to obtain all the intervals of that type.

We then must have $\mathbf{ifunc}(A_+, A_-) = \mathbf{ifunc}(B_+, B_-)$, which leads to the two equations of (5). Reciprocally if two sets verify (5), then they have the same right interval vector. This works similarly with left intervals, the difference being that when $\epsilon_2/\epsilon_1 = -1$, we obtain $^l\mathbf{int}((k_1, \epsilon_1), (k_2, \epsilon_2)) = (k_1 + k_2, -1)$, and we thus calculate $\mathbf{ifunc}(I_0 A_+, A_-)$.   □

We can notice that the first equations in both (5) and (6) are identical, but the second one shows an important difference. For left homometry it is symmetric between A_+ and A_- (and between B_+ and B_-), whereas it is not for right homometry, due to the fact that $\mathbf{ifunc}(I_0 A_+, A_-) = \mathbf{ifunc}(I_0 A_-, A_+)$.

We now describe a simple way to build left homometric sets from right homometric sets and reciprocally, using the inversion operator I. Since we have $(k, 1)^{-1} = (-k, 1)$ and $(k, -1)^{-1} = (k, -1)$ for any k in $\mathbb{Z}_n$, it is easy to calculate $I(\mathcal{A})$ for a set $\mathcal{A}$ in D_n by taking the inverse of A_+ and keeping A_- unchanged. For example, if $\mathcal{A} = \{(0, -1), (1, 1), (3, 1), (4, -1), (8, -1)\} \in D_{12}$, we obtain $I(\mathcal{A}) = \{(0, -1), (11, 1), (9, 1), (4, -1), (8, -1)\}$. This leads to the following result, the proof of which we omit here.

Proposition 2. *Let $\mathcal{A}$ and $\mathcal{B}$ be two sets in D_n. $\mathcal{A}$ and $\mathcal{B}$ are non-trivially right homometric if and only if $I(\mathcal{A})$ and $I(\mathcal{B})$ are non-trivially left homometric.*

Corollary 3. *For all $n \in \mathbb{N}$, the number of right homometric sets in D_n is equal to the number of left homometric sets. Besides, we can deduce all the left homometric sets from the right homometric sets (and reciprocally) using the inversion I.*

This result is useful for the problem of enumerating left- and right-homometric sets in D_n (which, similarly to $\mathbb{Z}_n$, is an open problem [7]) since the calculation needs only to be done for right (or left) homometric sets. However homometries for the right and for the left actions work very differently concerning a specific point, which the following proposition shows.

Proposition 4. *If $\mathcal{A}$ and $\mathcal{B}$ are right homometric in D_n, then their projection $A = \pi(\mathcal{A})$ and $B = \pi(\mathcal{B})$ are homometric in $\mathbb{Z}_n$. Besides, if the homometry is trivial in D_n, the homometry is also trivial between the projections in $\mathbb{Z}_n$.*

Proof. We just prove the first part of the proposition. If $\mathcal{A}$ and $\mathcal{B}$ are right homometric in D_n, we have

$$\mathbf{iv}(A) = \mathbf{iv}(A_+) + \mathbf{iv}(A_-) + \mathbf{ifunc}(A_+, A_-) + \mathbf{ifunc}(A_-, A_+)$$
$$= \mathbf{iv}(B_+) + \mathbf{iv}(B_-) + \mathbf{ifunc}(B_+, B_-) + \mathbf{ifunc}(B_-, B_+)$$
$$= \mathbf{iv}(B)$$ □

In other words, homometry for the right action in D_n "implies" homometry in $\mathbb{Z}_n$. However left homometry does not, as the pair of sets

$$(\{(0, 1), (1, -1), (2, 1), (5, -1), (7, -1)\}, \{(0, 1), (1, -1), (6, 1), (7, -1), (8, 1)\})$$

in D_{20} shows. These sets are left homometric but their projection $\{0, 1, 2, 5, 7\}$ and $\{0, 1, 6, 7, 8\}$ are not homometric in $\mathbb{Z}_{10}$. The Proposition 4 raises the question of whether left- or right-homometric sets in D_n can be found from homometric sets in $\mathbb{Z}_n$. In other words, can we split two homometric sets A and B in $\mathbb{Z}_n$ into subsets (A_+, A_-) and (B_+, B_-) such that the corresponding sets in D_n are homometric? This question will be considered in the following section with the definition of the concept of *lift*.

Before moving to this section we give some computational results concerning the enumeration of homometric sets in D_n. By a brute-force approach, a complete enumeration of such sets was performed, with cardinality equal to 4, 5, or 6 for $n \leq 18$, and with cardinality equal to 7 for $n \leq 15$. The first homometric pair appears for $n = 8$, $p = 4$ (a direct result of Proposition 4 and known results about homometry in $\mathbb{Z}_n$). Homometric t-uples with $t > 2$ also exist, the first triple appearing for $n = 12$ and $p = 5$ (interestingly, the first homometric triple in $\mathbb{Z}_n$ only appears for $n = 16$ and $p = 6$). The first simultaneously right- and left- homometric pair appears for $n = 8$ and $p = 5$. The Table 1 gives a complete list of left and right homometric pairs and triples written in musical form for $n = 12$ with $p = 4$ and $p = 5$. Notice that the first two pairs with $p = 5$ in this are both left- and right-homometric.

Table 1. Left- and right-homometric sets of cardinality p in D_{12}, given in musical form.

$n = 12$	Type	Homometric sets for the action of the T/I-group (left action)	Homometric sets for the action of the PLR-group (right action)
$p = 4$	Pairs	$\{C, d, e^b, G^b\}$ & $\{C, c, g^b, A\}$	$\{C, c, e^b, G^b\}$ & $\{C, c, E^b, g^b\}$
		$\{C, d^b, e, G^b\}$ & $\{C, d^b, g, A\}$	$\{C, d^b, e, G^b\}$ & $\{C, d^b, E^b, g\}$
		$\{C, d, f, G^b\}$ & $\{C, d, a^b, A\}$	$\{C, d, f, G^b\}$ & $\{C, d, E^b, a^b\}$
$p = 5$	Pairs	$\{C, c, d, E, A^b\}$ & $\{C, d, e, E, A^b\}$	$\{C, c, d, E, A^b\}$ & $\{C, d, e, E, A^b\}$
		$\{C, d^b, e^b, E, A^b\}$ & $\{C, e^b, E, f, A^b\}$	$\{C, d^b, e^b, E, A^b\}$ & $\{C, e^b, E, f, A^b\}$
		$\{C, c, d^b, f, G^b\}$ & $\{C, c, g^b, G, B\}$	$\{C, c, d^b, f, G^b\}$ & $\{C, c, D^b, F, g^b\}$
		$\{C, c, e^b, f, G^b\}$ & $\{C, c, E^b, g^b, B\}$	$\{C, c, e, f, G^b\}$ & $\{C, c, D^b, g^b, A^b\}$
		$\{C, c, D^b, g^b, A^b\}$ & $\{C, c, g^b, G, A^b\}$	$\{C, c, E, F, g^b\}$ & $\{C, c, E, g^b, B\}$
		$\{C, d^b, D^b, g, A^b\}$ & $\{C, d^b, g, G, A^b\}$	$\{C, d^b, E, F, g\}$ & $\{C, d^b, E, g, B\}$
		$\{C, D^b, d, a^b, A^b\}$ & $\{C, d, G, a^b, A^b\}$	$\{C, d, E, F, a^b\}$ & $\{C, d, E, a^b, B\}$
		$\{C, D^b, e^b, A^b, a\}$ & $\{C, e^b, G, A^b, a\}$	$\{C, e^b, E, F, a\}$ & $\{C, e^b, E, a, B\}$
	Triples	$\{C, c, d, e^b, G^b\}$ & $\{C, c, D, g^b, B^b\}$ & $\{C, c, g^b, A^b, B^b\}$	$\{C, c, d, e, G^b\}$ & $\{C, c, D, E, g^b\}$ & $\{C, c, D, g^b, B^b\}$
		$\{C, d^b, e^b, f, G^b\}$ & $\{C, d^b, D, g, B^b\}$ & $\{C, d^b, g, A^b, B^b\}$	$\{C, d^b, e^b, f, G^b\}$ & $\{C, d^b, D, E, g\}$ & $\{C, d^b, D, g, B^b\}$

5   The Concept of Lift – Using the Fourier Transform

We begin this section with a definition motivated by Proposition 4. The notation $\mathcal{P}(E)$ corresponds to the power set of the set E.

Definition 5. *A lift is an application $l : \mathcal{P}(\mathbb{Z}_n) \longrightarrow \mathcal{P}(D_n)$ such that $\pi \circ l = id$. We call lift of a set $A \in \mathbb{Z}_n$ for the lift l, the set $l(A)$.*

The question raised in the previous Section can then be formulated as follows: given two homometric sets A and B in $\mathbb{Z}_n$, is there a lift l such that $l(A)$ and $l(B)$ are left- or right-homometric in D_n?

We use the Fourier transform to express these conditions, since it provides a very convenient way to deal with the functions **ifunc** and **iv** for subsets in $\mathbb{Z}_n$, as explained in the work of Amiot [2]. Let us recall that for A and B two subsets in $\mathbb{Z}_n$ we have for $t \in \mathbb{Z}_n$

$$\textbf{ifunc}(A, B)(t) = \mathbb{1}_{-A} \star \mathbb{1}_B(t) = \sum_{k \in \mathbb{Z}_n} \mathbb{1}_A(k)\mathbb{1}_B(t + k) \qquad (8)$$

If we apply the Fourier transform (notated as $\mathcal{F}_A := \mathcal{F}(\mathbb{1}_A) : t \mapsto \sum_{k \in A} e^{-2i\pi kt/n}$) to this convolution product, we obtain the classical result for $t \in \mathbb{Z}_n$

$$\mathcal{F}(\textbf{ifunc}(A, B))(t) = \mathcal{F}_{-A}(t)\mathcal{F}_B(t) \qquad (9)$$

As $\textbf{iv}(A) = \textbf{ifunc}(A, A)$ and $\mathcal{F}_{-A}(t) = \overline{\mathcal{F}_A(t)}$ we deduce from Eq. (9) that $\mathcal{F}(\textbf{iv}(A)) = |\mathcal{F}_A|^2$ and we get the following well-known characterization of homometry, A and B being two subsets of $\mathbb{Z}_n$.

$$A \text{ is homometric with } B \iff |\mathcal{F}_A| = |\mathcal{F}_B| \qquad (10)$$

The use of the Fourier transform gives a new formulation of Theorem 1.

Theorem 6. *Two sets $\mathcal{A}$ and $\mathcal{B}$ in D_n are homometric for the right action if and only if the two following equations hold:*

$$\begin{cases} |\mathcal{F}_{A_+}|^2 + |\mathcal{F}_{A_-}|^2 = |\mathcal{F}_{B_+}|^2 + |\mathcal{F}_{B_-}|^2 \\ \overline{\mathcal{F}_{A_+}}\mathcal{F}_{A_-} = \overline{\mathcal{F}_{B_+}}\mathcal{F}_{B_-} \end{cases} \qquad (11)$$

Two sets $\mathcal{A}$ and $\mathcal{B}$ in D_n are homometric for the left action if and only if the two following equations hold:

$$\begin{cases} |\mathcal{F}_{A_+}|^2 + |\mathcal{F}_{A_-}|^2 = |\mathcal{F}_{B_+}|^2 + |\mathcal{F}_{B_-}|^2 \\ \mathcal{F}_{A_+}\mathcal{F}_{A_-} = \mathcal{F}_{B_+}\mathcal{F}_{B_-} \end{cases} \qquad (12)$$

Recall that we want to decompose each set of a homometric pair in $\mathbb{Z}_n$ into two subsets in order to lift them in D_n. The following proposition gives a specific characterization of homometry in $\mathbb{Z}_n$ for such a decomposition, using the Fourier transform.

Proposition 7. *Let A and B be two sets in $\mathbb{Z}_n$ such that $A = A_1 \cup A_2$ and $B = B_1 \cup B_2$. A and B are homometric if and only if*

$$|\mathcal{F}_{A_1}|^2 + |\mathcal{F}_{A_2}|^2 + 2\mathcal{R}e(\overline{\mathcal{F}_{A_1}}\mathcal{F}_{A_2}) = |\mathcal{F}_{B_1}|^2 + |\mathcal{F}_{B_2}|^2 + 2\mathcal{R}e(\overline{\mathcal{F}_{B_1}}\mathcal{F}_{B_2}) \qquad (13)$$

Proof. We use Eq. (10) and the fact that

$$|\mathcal{F}_A|^2 = |\mathcal{F}_{A_1} + \mathcal{F}_{A_2}|^2 = |\mathcal{F}_{A_1}|^2 + |\mathcal{F}_{A_2}|^2 + 2\mathcal{R}e(\overline{\mathcal{F}_{A_1}}\mathcal{F}_{A_2}) \qquad \square$$

We now provide with the main result of this paper which solves the question of lift in a special case.

Theorem 8. *Let A and B be two homometric sets in $\mathbb{Z}_n$ such that $A = A_1 \cup A_2$ and $B = B_1 \cup B_2$ with $\mathbf{iv}(A_1) = \mathbf{iv}(B_1)$ and $\mathbf{iv}(A_2) = \mathbf{iv}(B_2)$. We can always lift A and B into (non trivial) right homometric sets in D_n.*

Proof. Let A and B be two homometric subsets verifying the conditions of the theorem. We know from Proposition 7 that

$$|\mathcal{F}_{A_1}|^2 + |\mathcal{F}_{A_2}|^2 + 2\mathcal{R}e(\overline{\mathcal{F}_{A_1}}\mathcal{F}_{A_2}) = |\mathcal{F}_{B_1}|^2 + |\mathcal{F}_{B_2}|^2 + 2\mathcal{R}e(\overline{\mathcal{F}_{B_1}}\mathcal{F}_{B_2})$$

As $\mathbf{iv}(A_1) = \mathbf{iv}(B_1)$ and $\mathbf{iv}(A_2) = \mathbf{iv}(B_2)$ we deduce

$$|\mathcal{F}_{A_1}| = |\mathcal{F}_{B_1}| \text{ and } |\mathcal{F}_{A_2}| = |\mathcal{F}_{B_2}| \qquad (14)$$

so we get

$$\mathcal{R}e(\overline{\mathcal{F}_{A_1}}\mathcal{F}_{A_2}) = \mathcal{R}e(\overline{\mathcal{F}_{B_1}}\mathcal{F}_{B_2})$$

We remark also that $|\overline{\mathcal{F}_{A_1}}\mathcal{F}_{A_2}| = |\overline{\mathcal{F}_{B_1}}\mathcal{F}_{B_2}|$ thanks to (14). We obtain finally the two following equations:

$$\begin{cases} \mathcal{R}e(\mathcal{F}_{A_1}\overline{\mathcal{F}_{A_2}}) = \mathcal{R}e(\mathcal{F}_{B_1}\overline{\mathcal{F}_{B_2}}) \\ |\mathcal{F}_{A_1}\overline{\mathcal{F}_{A_2}}| = |\mathcal{F}_{B_1}\overline{\mathcal{F}_{B_2}}| \end{cases} \qquad (15)$$

These equations are of the form $\mathcal{R}e(z) = \mathcal{R}e(z')$ and $|z| = |z'|$, which implies $z = z'$ or $z = \overline{z'}$ i.e.

$$\mathcal{F}_{A_1}\overline{\mathcal{F}_{A_2}} = \mathcal{F}_{B_1}\overline{\mathcal{F}_{B_2}}$$
$$\text{or } \mathcal{F}_{A_1}\overline{\mathcal{F}_{A_2}} = \overline{\mathcal{F}_{B_1}}\mathcal{F}_{B_2}$$

In the first case ($\mathcal{F}_{A_1}\overline{\mathcal{F}_{A_2}} = \mathcal{F}_{B_1}\overline{\mathcal{F}_{B_2}}$) if we choose $A_+ = A_2$, $A_- = A_1$, $B_+ = B_2$ and $B_- = B_1$ we get right homometric sets in the dihedral group since (11) is verified. In the second case ($\mathcal{F}_{A_1}\overline{\mathcal{F}_{A_2}} = \overline{\mathcal{F}_{B_1}}\mathcal{F}_{B_2}$) if we choose $A_+ = A_2$, $A_- = A_1$, $B_+ = B_1$ and $B_- = B_2$ we get also right homometric sets. Thanks to Proposition 4 we know that this homometry is not trivial. $\square$

This result proves not only the existence of right homometric lifts but gives also a practical way to build these lifts, as shown in the example below. We finish with two interesting corollaries of Theorem 8.

Corollary 9. *In $\mathbb{Z}_{4n}$ ($n \geq 2$), we can always lift homometric sets with cardinality equal to 4 into right homometric sets in D_{4n}.*

Proof. Rosenblatt [11] proved that if A and B are homometric in $\mathbb{Z}_{4n}$ with $n \geq 2$ and $\sharp(A) = \sharp(B) = 4$ then there exists $a \in \{1, 2, ..., n-1\}$ such that

$$A = \{0, a, a+n, 2n\} \text{ and } B = \{0, a, n, 2n+a\} \tag{16}$$

If we choose $A_1 = \{0, 2n\}$, $A_2 = \{a, a+n\}$, $B_1 = \{a, a+2n\}$ and $B_2 = \{0, n\}$ we are in the situation of Theorem 8 then we can lift these sets into right homometric sets in D_{4n}. It is easy to verify that we have $\mathcal{F}_{A_1}\overline{\mathcal{F}_{A_2}} = \overline{\mathcal{F}_{B_1}}\mathcal{F}_{B_2}$, thus we know from Theorem 8 that we have to choose $A_+ = A_2$, $A_- = A_1$, $B_+ = B_1$ and $B_- = B_2$ or equivalently $A_+ = A_1$, $A_- = A_2$, $B_+ = B_2$ and $B_- = B_1$. $\square$

Corollary 10. *We can lift all the homometric sets in $\mathbb{Z}_{12}$ into right homometric sets in D_{12}.*

Proof. Goyette classifies the homometric sets in $\mathbb{Z}_n$ in four types [12]. This classification is based on the existence of cyclic subsets contained in the sets we consider. Goyette claims that homometric sets in $\mathbb{Z}_{12}$ are only of type 1 and 2, which satisfy the conditions of Theorem 8. $\square$

In order to give an application of these two corollaries and an explicit construction of lifts, we will consider the example mentioned in the introduction based on the famous "all interval tetrachords" $S_1 = \{0, 1, 4, 6\}$ and $S_2 = \{0, 1, 3, 7\}$ in $\mathbb{Z}_{12}$. These sets are homometric of the form of Eq. (16) with $a = 1$ (here $n = 3$). From the proofs of Corollary 9 and Theorem 8 we know that if we choose $A_+ = \{0, 6\}$, $A_- = \{1, 4\}$, $B_+ = \{0, 3\}$ and $B_- = \{1, 7\}$ we can lift S_1 and S_2 in D_{12} into the two right homometric sets

$$\mathcal{S}_1 = \{(0, 1), (1, -1), (4, -1), (6, 1)\}$$
$$\mathcal{S}_2 = \{(0, 1), (1, -1), (3, 1), (7, -1)\}$$

From a musical point of view the two homometric melodies $S_1 = \{C, D^\flat, E, G^\flat\}$ and $S_2 = \{C, D^\flat, E^\flat, G\}$ lift into the two right-homometric chord sequences $\mathcal{S}_1 = \{C, d^\flat, e, G^\flat\}$ and $\mathcal{S}_2 = \{C, d^\flat, E^\flat, g\}$.

Corollary 10 is interesting when considering musical applications. Every homometric melody in $\mathbb{Z}_{12}$ can be transformed into right-homometric chord progressions with identical roots.

6   Conclusion

The present paper is a first attempt to study homometry in the dihedral groups, which has the specificity to be a non-commutative and an interesting group for musical applications. In the case $n = 12$ we interpreted it as homometry between sets of major and minor triads (which can be for instance chord progressions), the left intervals being elements of the T/I-group, and the right intervals being elements of the *PLR*-group. As already mentioned it is however possible to replace triads by any subsets on which D_n acts simply transitively, which is interesting for tonal (but also atonal) music since we can use seventh chords or more generally chords with k notes ($k > 3$).

We showed that there are some similarities between homometry for the right and for the left actions, and deep links between homometry in D_n and homometry in $\mathbb{Z}_n$: the formulations require the same functions (namely **ifunc** and **iv**), right-homometry in D_n implies homometry in $\mathbb{Z}_n$, and conversely in some cases we can build right homometric sets in D_n from homometric sets in $\mathbb{Z}_n$. However there are still open questions:

- A complete enumeration of homometric sets for small n and p based on a brute-force approach was performed, but the general problem of the possibility of an efficient enumeration is open;
- Can we lift homometric sets in $\mathbb{Z}_n$ into left-homometric sets in D_n, or into both right-and left-homometric sets?
- We can use a similar approach with the time-spans group (cf. [5]), which is the semi-direct product $(\mathbb{R}, +) \rtimes (\mathbb{R}_+^*, .)$, for specific cases but the general question of homometry is still open in this group.

References

1. Lachaussée, G.: Théorie des ensembles homométriques. Master's thesis, Ecole Polytechnique (2010). http://recherche.ircam.fr/equipes/repmus/moreno/MasterLachaussee.pdf
2. Amiot, E.: David Lewin and maximally even sets. J. Math. Music **1**(3), 157–172 (2007)
3. Mandereau, J., Ghisi, D., Amiot, E., Andreatta, M., Agon, C.: Z-relation and homometry in musical distributions. J. Math. Music **5**(2), 83–98 (2011)
4. Mandereau, J., Ghisi, D., Amiot, E., Andreatta, M., Agon, C.: Discrete phase retrieval in musical structures. J. Math. Music **5**(2), 99–116 (2011)
5. Lewin, D.: General Musical Intervals and Transformations. Yale University Press, New Haven (1987)
6. Forte, A.: The Structure of Atonal Music. Yale University Press, New Haven (1977)
7. Jedrzejewski, F., Johnson, T.: The structure of Z-related sets. In: Yust, J., Wild, J., Burgoyne, J.A. (eds.) MCM 2013. LNCS (LNAI), vol. 7937, pp. 128–137. Springer, Heidelberg (2013). https://doi.org/10.1007/978-3-642-39357-0_10
8. Crans, A., Fiore, T., Satyendra, R.: Musical actions of dihedral groups. Am. Math. Mon. **116**(6), 479–495 (2009)
9. Popoff, A.: Building generalized neo-Riemannian groups of musical transformations as extensions. J. Math. Music **7**(1), 55–72 (2013)
10. Lewin, D.: Making and using a Pcset network for Stockhausen's Klavierstück III. In: Musical Form and Transformation: Four Analytic Essays, pp. 16–67. Yale University Press (1993)
11. Rosenblatt, J.: Phase retrieval. Commun. Math. Phys. **95**, 317–343 (1984)
12. Goyette, J.S.: The Z-relation in theory and practice. Ph.D. Thesis, University of Rochester (2012)

A Symmetric Quantum Theory of Modulation in $\mathbb{Z}_{20}$

J. David Gómez-Téllez[1(✉)], Emilio Lluis-Puebla[1], and Mariana Montiel[2]

[1] Universidad Nacional Autónoma de México, México City, México
david.gomez@ciencias.unam.mx, lluisp@unam.mx
[2] Georgia State University, Atlanta, USA
mmontiel@gsu.edu

Abstract. A 20-note scale is revisited (from Balzano and Zweifel) and endowed with a version of Mazzola's theory of modulation based on the symmetry group of the scale. Mazzola's theory has been applied also by Muzzulini in the context of the usual 12-note equally tempered chromatic scale. A modulation for a 7 note exotic scale, based on this model, is presented in Sect. 4 to exemplify the algorithm by which the modulation quanta are computed, that is, the sets of notes that permit the calculation of the pivot progressions that lead from one scale to another. Then, the modulation model based on symmetries is applied to a 11 note diatonic scale, immersed within the 20 note scale, which shows the viability of the symmetry model for this microtonal case. This work is based on the premise that musical expression has an underlying mathematical structure.

Keywords: Modulation by symmetries · Group theory
Microtonal scales · Microtonal music · Modulation quanta

1 Introduction

In 1980, Balzano [1] used group theory to describe the usual 12-note chromatic scale as well as some microtonal scales. In 1996, Zweifel [9] used Balzano's methods to study with greater detail some of those scales, describing their harmonic structure as well. On the other hand, in 1985–1990 Mazzola [3,4] developed his theory of modulation by symmetries. In 1995 Muzzulini [7] applied Mazzola's theory to 7-note subsets (scales) contained within the 12-note chromatic scale; among other scales, he applied the theory to the major and minor scales.

The central result of the present work is a version, in a microtonal context, of Mazzola's theory of modulation by symmetries. To be more specific, we apply the theory to an 11-note diatonic scale, immersed in a 20-note chromatic scale. Such scale was studied by Zweifel [9], based on the group theoretical properties brought to light by Balzano [1]. To the best of our knowledge, Mazzola's theory of modulation by symmetries has not been applied to microtonal scales before.

© Springer International Publishing AG 2017
O. A. Agustín-Aquino et al. (Eds.): MCM 2017, LNAI 10527, pp. 50–62, 2017.
https://doi.org/10.1007/978-3-319-71827-9_5

We present a table with complete information on how to modulate to most keys in this microtonal context.

In Sect. 2 we recall, from Clough and Myerson [2], that the diatonic major scale can be characterised as a subset of 7 notes, contained in a set of 12 notes (the chromatic scale), and fulfills some properties, namely: cardinality equals variety (CV), structure determines multiplicity (SM) and Myhill's property (MP). One can generalise these properties in order to detect the existence of microtonal scales that possess them. Using language from Clough and Myerson [2], a *generalised chromatic scale* is introduced as a division of the octave in c equal parts. By taking a subset of size d from this generalised chromatic scale, one can form a d-note scale, immersed in the c-note scale, which is called a *generalised diatonic scale*. Specifically, we present the construction of an 11-note diatonic scale immersed within a 20-note chromatic scale.

In a similar manner, in Sect. 3, following Balzano and Zweifel's approach, we generalise concepts that, in their origin, only applied to the division of the octave in twelve equal parts. We say that a microtonal scale that shares some characteristic properties of the usual major scale, is a *microtonal diatonic scale*, and we also refer to the corresponding *microtonal chromatic scale*. In the same section we relate the harmonic structure of microtonal scales to group theory.

In Sect. 4 we present Mazzola and Muzzulini's model of modulation by symmetries, for $\mathbb{Z}_{12}$ and in the context of equal temperament (it is necessary to highlight this point, given that Mazzola's model also contemplates other tuning systems). Here we illustrate the algorithm that appears in [7] by means of an example with an *exotic* 7 note scale. This algorithm describes how to compute quanta of modulation which are, in essence, the sets of notes from which the chords that permit a smooth transition from one scale to another, are taken.

Finally, in the concluding section, we present a version of Mazzola's theory of modulation by symmetries for the aforementioned 11-note microtonal diatonic scale.

2   Construction of the Diatonic 11-Note Microtonal Scale

In this section we will very briefly recall Clough and Myerson's approach in [2]. We don't describe their approach thoroughly, since our objective here is only to highlight the fact that the main scale studied in this article, a diatonic 11-note scale immersed in a chromatic 20-note scale, complies with the conditions and properties studied in [2]. For a complete account of results concerning the study of diatonic microtonal scales under this approach, please refer to [2].

Clough and Myerson [2] study certain mathematical properties of the usual 7-note diatonic major scale immersed within the chromatic 12-note scale. These properties are: cardinality equals variety (CV), structure determines multiplicity (SM) and Myhill's property (MP). Clough and Myerson generalise these properties to microtonal diatonic scales immersed in arbitrary-size chromatic scales. This amounts to say that, instead of the usual 12-note chromatic scale, we consider a set of c notes, which we call a *chromatic scale*. From this chromatic scale, we take a d-note subset, which we call a *diatonic scale*. Chromatic notes are noted $C_0, C_1, C_2, \ldots, C_{c-1}$, whereas diatonic notes are noted: $D_1, D_2, \ldots, D_d$. Note that chromatic notes correspond to elements of the cyclic group Z_c.

The shortest way to synthesise Clough and Myerson's results is by way of example. We build a scale with parameters $c = 20$ and $d = 11$. That is, a diatonic 11-note scale, immersed within a chromatic 20-note scale. These two integers are co-primes, so we may apply the following construction theorem, adapted from [2]:

Theorem 1 *(Clough and Myerson). Given two integers c and d such that $(c, d) = 1$, define $a_k = \left[\frac{kc}{d}\right]$ (mod c), where $k = 0, \pm 1, \pm 2, \ldots$. Then integers a_k are the positions (within the c-note chromatic scale) of the d notes which form a diatonic scale.*

Now we apply this theorem to the construction of the aforementioned diatonic 11-note scale. Let k take values from 0 to 11, then we get the following values for $\frac{kc}{d}$:

$$\frac{0}{11}, \frac{20}{11}, \frac{40}{11}, \frac{60}{11}, \frac{80}{11}, \frac{100}{11}, \frac{120}{11}, \frac{140}{11}, \frac{160}{11}, \frac{180}{11}, \frac{200}{11}, \frac{220}{11}. \tag{1}$$

By taking the integer part of these numbers (mod 20), we get the sequence:

$$0, 1, 3, 5, 7, 9, 10, 12, 14, 16, 18, 0. \tag{2}$$

These integer numbers represent the positions of the 11 diatonic notes within the 20-note chromatic scale. In analogy with the C major scale, it looks as if the scale was written from B to B, since there is a one-chromatic-note distance from 0 to 1, so the note 0 would be similar to a leading note. That is, in order to write the scale from C to C (i.e. with the leading note as the last one in the scale), note "1" should be the first note in the scale, thus renamed "0"; and in general all notes' positions should be displaced by 1. The scale is then re-written as follows:

$$0, 2, 4, 6, 8, 9, 11, 13, 15, 17, 19, 0. \tag{3}$$

The scale built by using this theorem has the general properties studied by Clough and Myerson: *Cardinality equals Variety* (CV), *Structure determines Multiplicity* (SM), *Myhill's Property* (MP), and is a *reduced scale*. The meaning and consequences of these properties can be stated in a condensed form:

1. Between two consecutive diatonic notes there is either one chromatic note, or none at all (that is, the diatonic interval of a second has only two possible chromatic sizes); and
2. There is a *generalised circle of fifths*, which in the case of the 11-note diatonic scale is a *circle of sevenths*, since such *generalised fifth* is the chromatic note 11, which is a diatonic seventh from the note 0.

By defining a *tone* as a distance two chromatic notes apart, and a *semi-tone* as a distance one chromatic note apart, the distances between the notes of this diatonic scale form the following structure:

$$\underline{T\,T\,T\,T\,S}\ \ T\ \ \underline{T\,T\,T\,T\,S}, \tag{4}$$

which is very similar to the structure of the usual major scale:

$$\underbrace{T\,T\,S}\ T\ \underbrace{T\,T\,S}, \tag{5}$$

in the sense that both scales consist of a structure that appears repeated at a distance of a whole tone. Such structure is, in the case of the usual major scale, the major tetrachord T T S, and for the 11-note scale, a *hexachord*, which we may interpret as a *generalised major tetrachord*: T T T T S. The structure of the usual major scale, written by tetrachords, is:

$$\underbrace{0,2,4,5,}\ \underbrace{7,9,11,0,} \tag{6}$$

whereas the structure of the 11-note diatonic scale, written by hexachords, is:

$$\underbrace{0,2,4,6,8,9,}\ \underbrace{11,13,15,17,19,0}\ . \tag{7}$$

3   Group Theory and the Harmonic Structure of Microtonal Scales

In this section we present some results by Balzano [1] and Zweifel [9]. Since we are particularly interested in the scale identified with $\mathbb{Z}_{20}$, here we only give a short account of this topic. For a complete presentation, please refer to [1,6,9].

The fundamental fact is that the usual chromatic scale may be represented as the cyclic group $\mathbb{Z}_{12}$. In fact, it may be represented as any one of three isomorphic groups: $\mathbb{Z}_{12}$ generated either by its element 1 or by its inverse 11; $\mathbb{Z}_{12}$ generated either by its element 7 or by its inverse 5 (the sequence of notes generated by the element 1 is the ascending chromatic scale, and the sequence generated by 11 is the descending chromatic scale; whereas the element 7 generates the ascending circle of fifths and the element 5 generates the descending circle of fifths). The remaining representation of the chromatic scale is as the product group $\mathbb{Z}_3 \times \mathbb{Z}_4$.

Following Balzano [1], the group $\mathbb{Z}_{12}$, as generated by its element 1, is called *semi-tone group*, and it reflects melodic relations among notes of the scale. If the group is seen as generated by its element 7, which corresponds to the diatonic interval of a perfect fifth, we call it *group of fifths*.

On the other hand, the group $\mathbb{Z}_3 \times \mathbb{Z}_4$ is called *group of thirds*, since $\mathbb{Z}_3$ represents the augmented chord, composed by two major thirds, and $\mathbb{Z}_4$ represents the diminished chord, composed by two minor thirds. So, we can say that this representation of the scale models its harmonic structure.

Analogous phenomena can be described when these ideas are applied to the 11-note diatonic scale immersed within the 20-note chromatic scale, constructed at the end of the preceding section. Note that the element of $\mathbb{Z}_{20}$ that corresponds to the *generalised fifth*, as described in Sect. 2 (which is the note 11), is the same element that generates the *group of fifths*, as defined in this section. The group $\mathbb{Z}_4 \times \mathbb{Z}_5$, isomorphic to $\mathbb{Z}_{20}$, hints at the harmonic structure of the scale.

Please note, from now on we sometimes use, either the number 12 or 20, as a subscript to indicate that we are making reference to the 12-note chromatic scale, identified with $\mathbb{Z}_{12}$, or to the 20-note chromatic scale, identified with $\mathbb{Z}_{20}$.

The following table identifies by letter each one of all 11 notes belonging to the diatonic scale immersed within the 20-note chromatic scale (Table 1).

Table 1. Diatonic notes by letter in $\mathbb{Z}_{20}$.

Number	0	2	4	6	8	9	11	13	15	17	19	0
Letter	C	D	E	F	G	H	I	J	K	A	B	C

At the end of the previous section, the structure of this scale was described in terms of *hexachords* and also in terms of *whole-tone* and *semi-tone* distances. In the case of the usual diatonic scale, the major mode determines which intervals are called major: the ones measured from the tonic C_{20} to each one of the notes of the diatonic scale.

For the 11-note diatonic scale immersed within $\mathbb{Z}_{20}$, we similarly identify the major intervals through the major mode, measuring distances from the note 0. Table 2 summarizes the intervals: the letter M indicates major intervals; on the other hand, the letter P indicates perfect intervals, namely the unison and the twelfth$_{20}$, which is equivalent to an octave$_{20}$, plus the *generalised fourth* and the *generalised fifth* which, for this scale, are the diatonic sixth$_{20}$ and seventh$_{20}$, respectively. Chromatic size, measured in semi-tones$_{20}$, is also indicated for each interval.

Table 2. Diatonic major intervals in $\mathbb{Z}_{20}$.

Interval	1P	2M	3M	4M	5M	6P	7P	8M	9M	10M	11M	12P
Semi-tones	0	2	4	6	8	9	11	13	15	17	19	20

Notice that, after defining the corresponding *minor intervals*, all of the notes belonging to $\mathbb{Z}_{20}$ will have been used, with exception of the tritone 10_{20}, which may be called *augmented sixth*$_{20}$ or *diminished seventh*$_{20}$, similarly to the *tritone* 6_{12} in $\mathbb{Z}_{12}$, which is called *augmented fourth*$_{12}$ or *diminished fifth*$_{12}$.

Finally, we describe three-note chords in $\mathbb{Z}_{20}$ built by fourths$_{20}$. Table 3 summarises these chords; degree corresponds to the fundamental note of the chord, and an indication is provided on whether the chord is major, minor, or the *diminished* chord whose fundamental is the *leading* note.

Table 3. Chords in the scale of C_{20} Major in $\mathbb{Z}_{20}$.

Seventh	11	13	15	17	19	0	2	4	6	8	9
Fourth	6	8	9	11	13	15	17	19	0	2	4
Fundamental	0	2	4	6	8	9	11	13	15	17	19
Degree	I	II	*iii*	*iv*	*v*	VI	VII	VIII	*ix*	*x*	*xi$_o$*

A proposal for a notation with letters, for all 20 chromatic notes, is as follows (Table 4):

Table 4. Chromatic notes by letter in $\mathbb{Z}_{20}$.

note	0	1	2	3	4	5	6	7	8	9	10	11	12	13	14	15	16	17	18	19
letter	C	C$^{\#}$ D^{b}	D	D$^{\#}$ E^{b}	E	E$^{\#}$ F^{b}	F	F$^{\#}$ G^{b}	G	H	H$^{\#}$ I^{b}	I	I$^{\#}$ J^{b}	J	J$^{\#}$ K^{b}	K	K$^{\#}$ A^{b}	A	A$^{\#}$ B^{b}	B

We conclude by enumerating all 20 keys in $\mathbb{Z}_{20}$, with their respective key-signatures, in Table 6. Just as in $\mathbb{Z}_{12}$, they should be ordered according to the *generalised circle of fifths*, which in this case is a *circle of sevenths$_{20}$*. In Sect. 5 we address the main contribution of this paper, namely how to modulate from one key to another in the microtonal scale $\mathbb{Z}_{20}$. But before that, in the next section we review the theory of modulation in $\mathbb{Z}_{12}$.

4   Symmetry and Modulation in $\mathbb{Z}_{12}$

In [3] and [5], Guerino Mazzola developed a theory that describes modulation between two keys, or two scales that belong to the same translation class, that is, two scales that are equivalent under musical transposition. The modulations computed by Mazzola, for the traditional degrees of relation between keys, agree with classical theory, so the results can be interpreted as a generalisation of established music theory. The following definitions are adapted from Muzzulini [7].

Definition 1. *A seven note scale s is any subset of $\mathbb{Z}_{12}$ which has seven elements.*

In Z_{12}, we can choose an arbitrary note, say D_1, as the *tonic* of scale s, and consequently we enumerate the notes of scale s as: $D_1, D_2, \ldots, D_7$. In order to consider chords and harmony, we sometimes refer to diatonic notes as *degrees* of scale s. For example, note D_1 is called the *first degree*, note D_2 is called the *second degree*, etc.

Definition 2. *For $n = 1, \ldots, 7$, define the chord or triad beginning on the nth degree of scale s as the set $s_n = \left\{ D_n, D_{[(n+1) \bmod 7]+1}, D_{[(n+3) \bmod 7]+1} \right\}$.*

That is, $s_1 = \{D_1, D_3, D_5\}$, $s_2 = \{D_2, D_4, D_6\}$, $\ldots$, $s_7 = \{D_7, D_2, D_4\}$. Also, we want to be able to refer to the set of all chords of scale s:

Definition 3. *The covering $\{s_1, s_2, s_3, s_4, s_5, s_6, s_7\}$ of scale s by its triads is called the triadic interpretation of s, and is denoted by $s^{(3)}$.*

We want to be able to talk about transposing a scale s. Of course, this corresponds to a translation in Z_{12}.

Definition 4. *Two scales r and s belong to the same translation class if there exists a translation defined in Z_{12} that transforms r to s.*

We will also want to refer to cadences, or more precisely, to sets of chords that constitute a cadence, thus called cadential sets.

Definition 5. *A subset μ of triads in $s^{(3)}$ is a cadential set of scale s if there does not exist any other scale r, in the same translation class as s, such that all the elements of μ are also triads of $r^{(3)}$. The cadential set μ is a minimal cadential set if no proper subset of μ is a cadential set.*

The importance of minimal cadential sets is that they allow to distinguish among scales that belong to the same translation class.

Example 1. A major diatonic scale has the following minimal cadential sets: $\{ii, iii\}$, $\{iii, IV\}$, $\{IV, V\}$, $\{ii, V\}$ and $\{vii_o\}$. The set $\{I, V\}$ is not a cadential set in C Major because its chords also belong to the triadic interpretation of G Major. The set $\{I, IV, V\}$ is a cadential set but it is not minimal (Muzzulini [7]).

In this context, *invertible affine transformations* of Z_{12} onto itself are called symmetries, and are written as $T_n I_0$, where n is an element of Z_{12}, T_n is a translation by n and I_0 is the inversion (for example, the inverse of 3 is 9). An example of a translation is:

$$T_3 : \{0, 4, 7\} \to \{0 + 3, 4 + 3, 7 + 3\} = \{3, 7, 10\},$$

and an inversion:

$$I_0 : \{0, 4, 7\} \to \{0, 8, 5\}.$$

Then, an invertible affine transformation is: $T_3 I_0$. Now we define symmetries of a scale s:

Definition 6. *The internal symmetries of a scale s are the symmetries of $\mathbb{Z}_{12}$ that leave s invariant.*

The concept of internal symmetry of a scale can be extended to its triadic interpretation:

Definition 7. *An internal symmetry of the triadic interpretation $s^{(3)}$, is an internal symmetry of s applied to the triads of $s^{(3)}$.*

It is shown in [4] that the only non trivial internal symmetries of a triadic interpretation are inversions.

Now, in order to present Mazzola's theory, we need to have a clear concept of musical modulation. In his 1911 work *Theory of Harmony (Harmonielehre)* [8], Arnold Schönberg defines modulation as a three step process, as highlighted by Mazzola. Schönberg's three steps are:

1. First, some neutral chords, those that are common to both keys, should appear;
2. Then the harmonic *pivot* progressions that introduce the new key, should appear;
3. Finally, the cadence confirms the new key.

In analogy with particle physics, Mazzola interprets modulation as the result of the action of a symmetry based on a *modulation quantum Q*. Explicit construction of the modulation quantum allows for calculation of the pivots. A *modulator* is a transformation $g = T_m f$, where f is an internal symmetry of $s^{(3)}$. Q is defined as a subset of Z_{12}, and μ is defined as a minimal cadence set of the target scale r. The following properties result in the modulation algorithm as it appears in Muzzulini's [7], which is inspired in Mazzola's [4] and [5].

1. There exists modulator g for $(s^{(3)}, r^{(3)})$, which is an internal symmetry of Q, that is, $g(Q) = Q$.
2. All triads of μ are subsets of Q.
3. The only internal symmetry of $r^{(3)}$ that is also an internal symmetry of $\tau = r \cap Q$ is the identity and τ is covered by triads of $r^{(3)}$.
4. Q is minimal with respect to properties (1) and (2).

We will exemplify the modulation algorithm by using an exotic scale of seven notes in $\mathbb{Z}_{12}$.

Example 2. In this example we will use the scale classified as #62 in [7]:

$$\{C, C^{\#}, D, E, F^{\#}, G^{\#}, A^{\#}\} = \{0, 1, 2, 4, 6, 8, 10\},$$

with triadic interpretation:

$$S(C) = \{\{0, 2, 6\}, \{1, 4, 8\}, \{2, 6, 10\}, \{4, 8, 0\}, \{6, 10, 1\}, \{8, 0, 2\}, \{10, 1, 4\}\}.$$

The translations of the scale and their triadic interpretations are easily calculated, which then leads to identifying the minimal cadential sets:

$$\{0, 2, 6\} = \{C, D, F^{\#}\};$$
$$\{1, 4, 8\} = \{C^{\#}, E, G^{\#}\};$$
$$\{6, 10, 1\} = \{F^{\#}, A^{\#}, C^{\#}\};$$
$$\{8, 0, 2\} = \{G^{\#}, C, D\};$$
$$\{10, 1, 4\} = \{A^{\#}, C^{\#}, E\}.$$

All of them consist of just one chord; however, this situation arises from the special characteristics of this scale. It is not that way in the general case.

We present an example of modulation from the *key* of C to the *key* of F. We find that, in this case, the modulator is $g = T_5(T_2 I_0)$, as $m = 5$ and $T_2 I_0$ is an inner symmetry of the scale. Thus we apply it to the first minimal cadential set:

$$T_5(T_2 I_0(\{0, 2, 6\})) = T_5 T_2(\{0, 10, 6\}) = T_5(\{2, 0, 8\}) = \{7, 5, 1\} = \{1, 5, 7\}.$$

Following the algorithm, we make the union with $g(\{1, 5, 7\})$, that is,

$$\{1, 5, 7\} \cup T_5(T_2 I_0(\{1, 5, 7\})) = \{1, 5, 7\} \cup \{6, 2, 0\} = \{0, 1, 2, 5, 6, 7\}.$$

This the modulation quantum Q for the minimal cadential set $\{0, 2, 6\}$. The intersection of Q with scale F is: $\{1, 5, 6, 7\}$. In this case there is only one pivot: $\{1, 5, 7\}$. Once again, the fact that there is only one pivot is a characteristic of this scale, but does not represent the general situation. In the following section the modulation algorithm is applied to a microtonal scale.

5   Symmetry and Modulation in $\mathbb{Z}_{20}$

In this section we present a version of Mazzola's theory of modulation adapted to the microtonal 11-note diatonic scale immersed in $\mathbb{Z}_{20}$; such scale was described in previous sections. We denote this diatonic scale as s and select $D_1 = 0$, as the *tonic* of the scale; the other notes are numbered in increasing order: $D_1, D_2, \ldots,$ D_{11}. As already discussed, in this scale we will use chords constructed by fourths:

Definition 8. *For $n = 1, \ldots, 11$, we define a chord or triad on the nth degree as the set:* $s_n = \{D_n, D_{[(n+2) \bmod 11]+1}, D_{[(n+5) \bmod 11]+1}\}$

Other concepts are defined in a similar way to the previous section. The triadic interpretation $s^{(3)}$ of scale s is the set of all chords constructed with notes from s. After constructing and comparing the cadential sets, it can be seen that there are five minimal cadential sets, namely: $\{iii, v\}$, $\{v, VI\}$, $\{iii, VIII\}$, $\{VI, VIII\}$ and $\{xi_o\}$.

There are evident similarities with the usual major scale: minimal cadential sets are combinations of only four chords: two minor chords, two major chords, and the diminished chord. One minimal cadential set is formed by two minor chords, another one is formed by two major chords, two are formed by one major chord and one minor chord and the other one contains only the diminished chord. We may call *subdominant* and *dominant* the two major chords that appear as elements of the minimal cadential sets. Their importance comes also, as will be seen, from their repeated appearance as pivots of the modulations. In the case of the degree VIII (dominant), its harmonic importance is also related to the fact that this chord contains the leading note $D_{11} = 19$.

The only non trivial symmetry of $s^{(3)}$ is $T_8 I_0$, which acts on triads as shown next (Table 5):

Table 5. Transformation of chords under the symmetry $T_8 I_0$ in $\mathbb{Z}_{20}$.

I	$\leftrightarrow$	x
II	$\leftrightarrow$	ix
iii	$\leftrightarrow$	VIII
iv	$\leftrightarrow$	VII
v	$\leftrightarrow$	VI
xi_o	$\leftrightarrow$	xi_o

There is a clear similarity with the symmetry of the usual major scale: major chords are transformed into minor chords and vice versa, while the diminished chord is left invariant.

In the case of the usual major scale, the internal symmetry of the scale transforms the C major chord into the A minor chord, its relative minor. For the 11-note diatonic scale s, immersed within $\mathbb{Z}_{20}$, the internal symmetry $T_8 I_0$ transforms the major chord whose fundamental is the degree I into the minor chord whose fundamental is the degree x. This is the main reason why we propose that the mode that begins on the tenth degree (the note 17_{20}), should be called the relative minor of s. But, what justifies that the major mode of the scale should begin with the note 0_{20}? The answer lies in the similarities that exist between this mode and the major mode of the usual diatonic scale. To be specific: the structure of the hexachords (generalised tetrachords) and the position of the semitones, which allow for the existence of the diminished chord whose fundamental note is the leading note.

We have computed the modulation quanta for modulations from the scale s to its translations $r = T_p(s)$, for $p = 1, 2, \ldots, 19$. Table 7 shows all quantised modulations. First of all, note that, contrary to the case of $\mathbb{Z}_{12}$, in $\mathbb{Z}_{20}$ not every translation of s admits a quantised modulation: there are no quantised modulations for values of p equal to 8 or 12.

For each value of p and for each quantised modulation from s to $r = T_p(s)$, Table 7 shows the minimal cadential sets μ along with the modulator g. Then, for each minimal cadential set, if there is a quantum Q that fulfills the required properties, it is shown on the same row, thus describing a quantised modulation. Notice that, in the notation used for the modulation quantum Q and its trace $\tau = Q \cap r$, the number 0 correspons to the tonic of the departure scale s. On the other hand, cadential sets μ and pivots are indicated as degrees of r, the target scale of the modulation.

Table 7 gives complete information on how to modulate. First one should choose a target scale $r = T_p(s)$ for which there exists at least one quantised modulation. Then, choose a cadential set, if there is more than one possibility. The table provides the pivot chords that should be used for the transition progression, before confirming the new key with the cadence.

It is interesting to note that for the values of $p = 9$ and $p = 11$, there is a quantised modulation for each of all five minimal cadential sets. This maximises the number of alternative cadences to modulate to these two scales. Notice that notes 9 and 11 correspond to the degrees VI and VII, respectively, and they are the tonics of the two keys which are closest to s, if distance is measured by the *circle of sevenths*, which corresponds to a *generalised circle of fifths* in $\mathbb{Z}_{20}$.

Table 6. Key signatures for $\mathbb{Z}_{20}$.

Key		S	i	g	n	a	t	u	r	e
$C=0$										
$I=11$	$H^\#=10$									
$D=2$	$H^\#=10$	$C^\#=1$								
$J=13$	$H^\#=10$	$C^\#=1$	$I^\#=12$							
$E=4$	$H^\#=10$	$C^\#=1$	$I^\#=12$	$D^\#=3$						
$K=15$	$H^\#=10$	$C^\#=1$	$I^\#=12$	$D^\#=3$	$J^\#=14$					
$F=6$	$H^\#=10$	$C^\#=1$	$I^\#=12$	$D^\#=3$	$J^\#=14$	$E^\#=5$				
$A=17$	$H^\#=10$	$C^\#=1$	$I^\#=12$	$D^\#=3$	$J^\#=14$	$E^\#=5$	$K^\#=16$			
$G=8$	$H^\#=10$	$C^\#=1$	$I^\#=12$	$D^\#=3$	$J^\#=14$	$E^\#=5$	$K^\#=16$	$F^\#=7$		
$B=19$	$H^\#=10$	$C^\#=1$	$I^\#=12$	$D^\#=3$	$J^\#=14$	$E^\#=5$	$K^\#=16$	$F^\#=7$	$A^\#=18$	
$H=10$	$H^\#=10$	$C^\#=1$	$I^\#=12$	$D^\#=3$	$J^\#=14$	$E^\#=5$	$K^\#=16$	$F^\#=7$	$A^\#=18$	$G^\#=9$
$D^b=1$	$I^b=10$	$D^b=1$	$J^b=12$	$E^b=3$	$K^b=14$	$F^b=5$	$A^b=16$	$G^b=7$	$B^b=18$	
$J^b=12$		$D^b=1$	$J^b=12$	$E^b=3$	$K^b=14$	$F^b=5$	$A^b=16$	$G^b=7$	$B^b=18$	
$E^b=3$			$J^b=12$	$E^b=3$	$K^b=14$	$F^b=5$	$A^b=16$	$G^b=7$	$B^b=18$	
$K^b=14$				$E^b=3$	$K^b=14$	$F^b=5$	$A^b=16$	$G^b=7$	$B^b=18$	
$F^b=5$					$K^b=14$	$F^b=5$	$A^b=16$	$G^b=7$	$B^b=18$	
$A^b=16$						$F^b=5$	$A^b=16$	$G^b=7$	$B^b=18$	
$G^b=7$							$A^b=16$	$G^b=7$	$B^b=18$	
$B^b=18$								$G^b=7$	$B^b=18$	
$H=9$									$B^b=18$	

Table 7. Quantised modulations in $\mathbb{Z}_{20}$.

p	μ	Q	g	Pivots	$\tau = Q \cap r$
1	(iii,v)	(0,4,5,9,10,13,14,15,16,19)	$T_9\,I_0$	(iii,v,VIII,xi$_o$)	(5,9,10,14,16,0)
2	(iii,v)	(0,1,4,6,9,10,11,13,15,17,19)	$T_{10}\,I_0$	(II,iii,v,VII,VIII,x,xi_o)	(4,6,10,11,13,15,17,19,1)
	(v,VI)	(0,1,2,8,9,10,11,13,15,17,19)	$T_{10}\,I_0$	(I,iv,v,VI,ix)	(2,8,10,11,13,15,17,19,1)
	(iii,VIII)	(1,4,6,9,11,13,15,17,19)	$T_{10}\,I_0$	(iii,VII,VIII,xi_o)	(4,6,11,13,15,17,19,1)
	(VI,VIII)	(1,2,4,6,8,9,11,13,15,17,19)	$T_{10}\,I_0$	(I,iii,iv,VI,VII,VIII,ix,xi$_o$)	(2,4,6,8,11,13,15,17,19,1)
3	(VI,VIII)	(2,3,4,7,8,9,12,13,15,16,18,19)	$T_{11}I_0$	(iii,VI,VIII,ix,xi$_o$)	(3,7,9,12,16,18,2)
4	(xi$_o$)	(3,4,8,9,13,19)	$T_{12}I_0$	(iii,VI,xi$_o$)	(4,8,13,19,3)
5	(iii, v)	(0,4,9,13,14,15,18,19)	$T_{13}I_0$	(iii,v,VIII,xi$_o$)	(9,13,14,18,0,4)
	(iii,VIII)	(0,4,9,13,14,15,18,19)	$T_{13}I_0$	(iii,v,VIII,xi$_o$)	(9,13,14,18,0,4)
6	(iii,v)	(0,1,4,5,9,10,13,14,15,19)	$T_{14}I_0$	(iii,v,VIII,xi$_o$)	(10,14,15,19,1,5)
	(v,VI)	(0,1,5,6,8,9,13,14,15,19)	$T_{14}I_0$	(II,v,VI)	(6,8,14,15,19,1,5)
	(xi$_o$)	(4,5,9,10,15,19)	$T_{14}I_0$	(VIII,xi$_o$)	(10,15,19,5)
7	(VI,VIII)	(0,2,4,6,7,8,9,11,13,15,16,19)	$T_{15}I_0$	(II,iii,v,VI,VIII,ix,x,xi$_o$)	(7,9,11,13,15,16,0,2,4,6)
8	-	-	$T_{16}I_0$	-	-
9	(iii,v)	(0,2,4,8,9,13,15,17,18,19)	$T_{17}I_0$	(I,iii,v,VI,VIII,ix,xi$_o$)	(9,13,15,17,18,0,2,4,8)
	(v,VI)	(0,2,4,8,9,13,15,17,18,19)	$T_{17}I_0$	(I,iii,v,VI,VIII,ix,xi$_o$)	(9,13,15,17,18,0,2,4,8)
	(iii,VIII)	(2,4,8,9,13,15,18,19)	$T_{17}I_0$	(iii,VI,VIII,ix,xi$_o$)	(9,13,15,18,2,4,8)
	(VI,VIII)	(2,4,8,9,13,15,18,19)	$T_{17}I_0$	(iii,VI,VIII,ix,xi$_o$)	(9,13,15,18,2,4,8)
	(xi$_o$)	(4,8,9,13,18,19)	$T_{17}I_0$	(iii,VI,xi$_o$)	(9,13,18,4,8)
10	(iii,v)	(0,3,4,5,9,13,14,15,18,19)	$T_{18}I_0$	(iii,v,VIII,xi$_o$)	(14,18,19,3,5,9)
	(VI,VIII)	(3,4,5,8,9,10,13,14,15,19)	$T_{18}I_0$	(iii,VI,VIII,xi$_o$)	(10,14,19,3,5,9)
	(iii,v)	(3,4,5,8,9,13,14,15,18,19)	T_{10}	(iii,v,VIII,xi$_o$)	(14,18,19,3,5,9)
	(VI,VIII)	(0,3,4,5,9,10,13,14,15,19)	T_{10}	(iii,VI,VIII,xi$_o$)	(10,14,19,3,5,9)
11	(iii,v)	(0,4,6,9,10,13,15,19)	$T_{19}I_0$	(II,iii,v,VIII,xi$_o$)	(13,15,19,0,4,6,10)
	(v,VI)	(0,4,6,8,9,10,11,13,15,19)	$T_{19}I_0$	(II,iii,v,VI,VIII,x,xi$_o$)	(11,13,15,19,0,4,6,8,10)
	(iii,VIII)	(0,4,6,9,10,13,15,19)	$T_{19}I_0$	(II,iii,v,VIII,xi$_o$)	(13,15,19,0,4,6,10)
	(VI,VIII)	(0,4,6,8,9,10,11,13,15,19)	$T_{19}I_0$	(II,iii,v,VI,VIII,x,xi$_o$)	(11,13,15,19,0,4,6,8,10)
	(xi$_o$)	(0,4,9,10,15,19)	$T_{19}I_0$	(v,VIII,xi$_o$)	(15,19,0,4,10)
12	-	-	T_0I_0	-	-
13	(iii,v)	(0,1,2,4,6,8,9,12,13,15,17,19)	T_1I_0	(I,II,iii,v,VI,VIII,ix,xi$_o$)	(13,15,17,19,1,2,4,6,8,12)
14	(v,VI)	(0,2,3,7,8,9,13,14,15,19)	T_2I_0	(v,VI,ix)	(14,0,2,3,7,9,13)
	(VI,VIII)	(3,4,7,8,9,13,14,15,18,19)	T_2I_0	(iii,VI,VIII,xi$_o$)	(14,18,3,7,9,13)
	(xi$_o$)	(3,4,9,13,18,19)	T_2I_0	(iii,xi$_o$)	(18,3,9,13)
15	(iii,VIII)	(4,8,9,10,13,14,15,19)	T_3I_0	(iii,VI,VIII,xi$_o$)	(15,19,4,8,10,14)
	(VI,VIII)	(4,8,9,10,13,14,15,19)	T_3I_0	(iii,VI,VIII,xi$_o$)	(15,19,4,8,10,14)
16	(xi$_o$)	(0,4,5,9,15,19)	T_4I_0	(v,VIII,xi$_o$)	(0,4,5,9,15)
17	(iii,v)	(0,1,4,5,6,9,10,12,13,15,16,19)	T_5I_0	(II,iii,v,VIII,xi$_o$)	(19,1,5,6,10,12,16)
18	(iii,v)	(0,2,4,6,7,9,11,13,15,17,19)	T_6I_0	(II,iii,iv,v,VII,VIII,x,xi$_o$)	(0,2,4,6,7,9,11,13,15,17)
	(v,VI)	(0,6,7,8,9,11,13,15,17,18,19)	T_6I_0	(II,v,VI,VII,x)	(18,0,6,7,9,11,13,15,17)
	(iii,VIII)	(2,4,7,9,11,13,15,17,19)	T_6I_0	(iii,iv,VIII,xi$_o$)	(2,4,7,9,11,13,15,17)
	(VI,VIII)	(2,4,7,8,9,11,13,15,17,18,19)	T_6I_0	(I,iii,iv,VI,VIII,ix,xi$_o$)	(18,2,4,7,9,11,13,15,17)
19	(iii,VIII)	(3,4,8,9,12,13,14,15,18,19)	T_7I_0	(iii,VI,VIII,xi$_o$)	(19,3,8,12,14,18)
	(VI,VIII)	(3,4,8,9,12,13,14,15,18,19)	T_7I_0	(iii,VI,VIII,xi$_o$)	(19,3,8,12,14,18)

References

1. Balzano, G.J.: The group-theoretic description of 12-fold and microtonal pitch systems. Comput. Music J. **4**(4), 66–84 (1980)
2. Clough, J., Myerson, G.: Variety and multiplicity in diatonic systems. J. Music Theory **29**, 249–270 (1985)
3. Mazzola, G.: Gruppen und Kategorien in der Musik. Heldermann, Berlin (1985)
4. Mazzola, G.: Geometrie der Töne Elemente der Mathematischen Musiktheorie. Birkhäuser, Basel (1990)
5. Mazzola, G.: The Topos of Music. Birkhäuser, Basel (2003)
6. Montiel, M.: Matemáticas y Música: Perspectivas a travé s del Tiempo. Undergraduate thesis. Facultad de Ciencias, UNAM, México (1996)
7. Muzzulini, D.: Musical Modulation by Symmetries. J. Music Theory **39**, 311–327 (1995)
8. Schönberg, A.: Harmonielehre (1911). Universal Edition, Viena (1996)
9. Zweifel, P.F.: Generalized diatonic and pentatonic scales: a group-theoretic aproach. Perspecti. New Music **1**, 141–161 (1996)

Almost Difference Sets in Transformational Music Theory

Robert W. Peck[✉]

Louisiana State University, Baton Rouge, LA, USA
rpeck@lsu.edu

Abstract. The combinatorial theory of difference sets has prior applications in the field of mathematical music theory. The theory of almost difference sets, however, has not received similar attention from music scholars. Nevertheless, these types of structures also have significant musical applications. For instance, the well known all-interval tetrachords of pitch-class set theory are almost difference sets. To that end, we investigate the various categories of almost difference sets (cyclic, abelian, and non-abelian) in terms of their representations in Lewinian music-transformational groups.

Keywords: Difference set · Almost difference set
Flat interval distribution · All-interval chord
Generalized Interval System

1 Introduction

The combinatorial theory of difference sets has prior applications in the field of mathematical music theory. For instance, Gamer and Wilson [1] relate various n-chords in microtonal systems to difference sets. Wild [2] generalizes this idea further to flat-interval distributions. The theory of *almost* difference sets, however, has not received similar attention from music scholars. Nevertheless, these types of structures also have significant musical applications. For instance, the well known all-interval tetrachords of pitch-class set theory [3] are almost difference sets. To that end, we investigate the various categories of almost difference sets (cyclic, non-cyclic abelian, and non-abelian) in terms of their representations in Lewinian music-transformational groups [4].

2 Mathematical Preliminaries

2.1 Difference Sets

Before proceeding to the concept of almost difference sets, let us first establish what is a difference set.

© Springer International Publishing AG 2017
O. A. Agustín-Aquino et al. (Eds.): MCM 2017, LNAI 10527, pp. 63–75, 2017.
https://doi.org/10.1007/978-3-319-71827-9_6

Definition 1. *Let G be a finite, additively notated group of order v, and let D be a k-member subset of G. D is a (v, k, λ) **difference set (DS)** in G if every non-identity element of G can be written as a difference $d_1 - d_2$ of elements $d_1, d_2, \in D$ in exactly λ ways. (In multiplicatively notated groups, we use the product $d_1 d_2^{-1}$ as an analog for the difference $d_1 - d_2$.) We call $n = k - \lambda$ the **order** of D.*

Example 1. Put $G = \mathbb{Z}/7\mathbb{Z}$, observing that G is a group of order $v = 7$. Let $D = \{0, 1, 3\}$ be a subset of G of cardinality $k = 3$. We note that every non-identity element of G can be expressed in exactly $\lambda = 1$ way as a difference of elements $d_1, d_2 \in D$.

$$1 - 0 = 1 \text{ (modulo 7)}$$
$$3 - 1 = 2$$
$$3 - 0 = 3$$
$$0 - 3 = 4$$
$$1 - 3 = 5$$
$$1 - 0 = 6$$

Hence, D is a $(7, 3, 1)$ DS of order $n = 2$.

Below are some properties of DSs that have relevance to our later results.

Definition 2. *A DS is **cyclic, abelian,** or **non-abelian,** in accordance with the property of the particular group G that contains it as a subset.*

We thereby make a distinction between cyclic DSs and other, non-cyclic abelian DSs by referring to the latter merely as *abelian*, following the convention in combinatorics [5]. Further, the following property is of special significance to our musical applications below.

Definition 3. *If $\lambda = 1$, we call D a **planar difference set (PDS)**.*

For instance, D in Example 1 above is a PDS, as each non-identity element of G occurs only once as a difference of elements in D. This particular D is also cyclic, because $\mathbb{Z}/7\mathbb{Z}$ is a cyclic group.

Theorem 1. $\lambda = 1 \Longrightarrow v = n^2 + n + 1$.

Proof. Let G be a group of order v, and let $G^{\#}$ be the set of non-identity elements of G. We note that, for a DS D of size k in G, there exist exactly $k P 2 = 2\binom{k}{2}$ possible differences $g - h$, where $g, h \in D$ and $g \neq h$. If $\lambda = 1$, then each non-identity element of G appears once and only once as a difference of elements in D; therefore, $G^{\#}$ is also of size $k P 2$. As $n = k - 1$ (using $n = k - \lambda$ and $\lambda = 1$) and $k P 2 = k(k - 1)$, we observe that $k(k - 1) = (k - 1)^2 + (k - 1) = n^2 + n$. Adding back the identity element gives the order of G as $v = n^2 + n + 1$. $\qquad \square$

Theorem 2. *If D is a PDS, then G cannot contain an involution.*

Proof. For the condition $\lambda = 1$ to hold, there must exist some $d_1, d_2 \in D$ for every non-identity element $g \in G$ such that $d_1 d_2^{-1} = g$, but for which

$$d_2 d_1^{-1} \neq g.$$

If, however, $g = g^{-1}$, then $d_2 d_1^{-1} = (d_1 d_2^{-1})^{-1} = g^{-1} = g$, contradicting the above statement. $\square$

DSs may relate to one another in various structural ways.

Definition 4. *Let D_1 be a DS in a group G_1, and let D_2 be a DS in a group G_2. Then, we say that D_1 is **isomorphic** to D_2 if they share the same (v, k, λ) parameters. Further, we say that D_1 is **equivalent** to D_2 if they share the same (v, k, λ) parameters and if G_1 is related to G_2 by a group isomorphism.*[1]

It is possible that two non-equivalent DSs D_1 and D_2 can be isomorphic to one another while also not sharing the same property from Definition 2. For example, one finds a $(21, 5, 1)$ DS in $\mathbb{Z}/21\mathbb{Z}$ and another $(21, 5, 1)$ DS in $\mathbb{Z}/7\mathbb{Z} \rtimes \mathbb{Z}/3\mathbb{Z}$. Whereas these DSs have different properties—one is cyclic and the other is non-abelian—they are isomorphic because they share the same (v, k, λ) parameters.

Definition 5. *Let D_1 and D_2 be DSs in an additive group G. Then, D_2 is a G-**translate** of D_1 if $D_2 = \{d + g \mid d \in D_1\}$ for some $g \in G$. (We use $D_2 = \{dg \mid d \in D_1\}$ in multiplicative groups.)*

Any two DSs D_1 and D_2 that are G-translates of one another possess the same (v, k, λ) parameters. Therefore, as G is trivially isomorphic (as a group) to itseslf, D_1 and D_2 are equivalent (as DSs).

2.2   Almost Difference Sets

An almost difference set is a related combinatorial structure.

Definition 6. *An **almost difference set (ADS)** is a subset $D = (v, k, \lambda, t)$ of G, where v and k are the same as in Definition 1; t non-identity elements of G appear exactly λ times as differences $g - h$ of elements $g, h \in D$; and the remaining $v - 1 - t$ non-identity elements of G appear $\lambda + 1$ times as differences. (Again, in multiplicatively notated groups, we use the product gh^{-1} as an analog for the difference $g - h$.) Similar to DSs, we give the **order** of D as $n = k - \lambda$.*

---

[1] In the field of combintorics, the use of the term *isomorphism* in connection with difference sets derives from the use of the same term in the theory of balanced incomplete block designs (BIBDs), as a (v, k, λ) difference set D is equivalent to a symmetric (v, k, λ)-design [5, Theorem 18.6]. Two BIBDs are considered isomorphic if there exists a bijection from one design to another such that, if we rename every point in one design with its image in the other, the collection of blocks in the first is transformed into that of the second.

In essence, an ADS is a generalization of a DS, wherein the latter is an ADS with $t = 0$ or $t = v - 1$ [6]. We call ADSs with $0 < t < v - 1$ *proper* ADSs, to differentiate them from ADSs that are also DSs (improper ADSs).

Example 2. Put $G = \mathbb{Z}/6\mathbb{Z}$; G is of order $v = 6$. Let $D = \{0, 1, 3\}$ be a subset of G of cardinality $k = 3$. We note that $t = 4$ non-identity elements of G can be expressed in exactly $\lambda = 1$ way as differences of elements $d_1, d_2 \in D$, whereas the remaining $v - 1 - t = 1$ non-identity element of G can be expressed as a difference of elements $d_1, d_2 \in D$ in exactly $\lambda + 1 = 2$ ways.

$$1 - 0 = 1 \text{ (modulo 6)}$$
$$3 - 1 = 2$$
$$3 - 0 = 0 - 3 = 3$$
$$1 - 3 = 4$$
$$0 - 1 = 5$$

Therefore, D is a $(6, 3, 1, 4)$ ADS.

Definition 7. *As with DSs, an ADS D can be **cyclic**, **abelian**, or **non-abelian**. A G-translate—Dg, for some $g \in G$—of a (v, k, λ, t) ADS is also a (v, k, λ, t) ADS. Further, ADSs can be related in various ways: they may be **isomorphic** or **equivalent**, etc., as described above for DSs.*

Definition 8. *We call an ADS with $\lambda = 1$ a **planar almost difference set** (PADS).*

Recalling that ADSs are generalizations of DSs, we offer the following result concerning PADSs as a corollary to Theorem 1.

Corollary 1. $\lambda = 1 \Longrightarrow v = n^2 + n + 1 - \frac{n^2 + n - t}{2}$.

However, an important distinction that relates to our later results exists between PDSs and PADSs. Whereas Theorem 2 states that a PDS cannot contain an involution, PADSs may contain involutions, as we see in Example 2 above.

Much of the research on ADSs deals with questions of existence or with applications in various branches of engineering, including cryptography, coding theory, and CDMA communications. Whereas many open questions remain, particularly regarding existence, the following are considered basic properties [7]. The first calculates the number of differences in a ADS in two ways.

Theorem 3. *Let D be an ADS of size k in a group G of order v. Then, we observe the following:*

$$k(k - 1) = \lambda t + (\lambda + 1)(v - 1 - t).$$

The second considers the complement of an abelian ADS.

Theorem 4. *If D is (v, k, λ) ADS in an abelian group G with $|D| \leq \frac{v}{2}$, then $D' = G \setminus D$ is a $(v, v - k, v - 2k + \lambda, t)$ ADS.*

3 Musical Preliminaries

3.1 Generalized Interval Systems

The music-theoretical context we employ in this work is that presented in David Lewin's *Generalized Musical Intervals and Transformations*. Following Lewin [4, p. 26], we define a Generalized Interval System as follows.

Definition 9. *A **Generalized Interval System (GIS)** is an ordered triple (S, G, int), where the set S is the space of the GIS; $G = (G, *)$ is the group of intervals; and $int : S \times S \to G$ is the interval function, satisfying the following two conditions:*

(a) For all $r, s, t \in S$, $int(r, s) * int(s, t) = int(r, t)$,

(b) For all $s \in S$ and $g \in G$, there exists a unique $t \in S$ which lies the interval g from s. In other words, there exists a unique $t \in S$ such that $int(s, t) = g$.

In essence, a GIS consists of the simply transitive (regular) action of a group G on a set S, wherein an interval is an element of G.

Example 3. Let S be the space of twelve chromatic pitch classes as represented by the integers modulo 12, and let $G = \mathbb{Z}/12\mathbb{Z}$ be a group of intervals with an action on S. Further, let int be the interval function that maps $(s, t) \in S \times S$ to $g = t - s$ (modulo 12) in G. The reader can easily verify that this example satisfies conditions (a) and (b) of Definition 9.

3.2 Intervals and Differences

Definition 10. *An **interval** is an element $g \in G$. If g is the unique element of G for which $g : s \mapsto t$ for some $s, t \in S$ (by Definition 9), then we call $(s, t) \in S \times S$ an **occurrence** of g.*

A significant relationship exists between the concepts of intervals and differences. To facilitate their comparison, we consider the GIS formed by the simply transitive action of a group G on itself under addition (in the abelian case), or under (right) multiplication (in the non-abelian case). In such a context, G acts as both interval group and space. For any g, h in an abelian group G, $i = h - g$ is the interval that carries g to h. Here, i is construed in the same manner as the difference $d = h - g$. In the non-abelian case, however, we reckon the interval from g to h as $i = g^{-1}h$. This notation differs from the multiplicative analog for a difference that we use above in Definition 1: i.e., $d = hg^{-1}$. Therefore, we require an additional layer of structure in relating non-abelian intervals with differences.

Definition 11. *Let $(G, *)$ be a group. Then, $(G^{opp}, *')$ is the **opposite group** for G if G^{opp}, as a set, is the same as G, and if $g *' h = h * g$ for all $g, h \in G$.*

Theorem 5. *Put $\phi : G \to G$, where $\phi(g) = g^{-1}$ for all $g \in G$. Then, $\phi' : G \to G^{opp}$, where $\phi'(g) = \phi(g)$, is an isomorphism.*

Proof. First, we show that ϕ is an anti-automorphism of G. As $\phi(g) * \phi(h) = g^{-1} * h^{-1}$, and $\phi(h * g) = g^{-1} * h^{-1}$ we note that $\phi(g) * \phi(h) = \phi(h * g)$ for all $g, h \in G$. Hence, ϕ is an anti-homomorphism. Then, by the group axiom that stipulates the existence of a unique inverse for every $g \in G$, ϕ is a bijection. Therefore, ϕ is an anti-automorphism.

Next, as

$$
\begin{aligned}
\phi'(g) *' \phi'(h) &= \phi(g) *' \phi(h) && \text{(by } \phi'(g) = \phi(g)) \\
&= \phi(h) * \phi(g) && \text{(by } g *' h = h * g) \\
&= \phi(g * h) && \text{(by def. of anti-automorphism)} \\
&= \phi'(g * h), && \text{(by } \phi'(g) = \phi(g))
\end{aligned}
$$

we observe that ϕ' is an isomorphism from G to G^{opp}. $\qquad\square$

Thus, the set of intervals in G and the set of differences in G are images of one another under ϕ'. Because of this structural identity, we henceforth speak informally of intervals and differences as being equivalent, without invoking the isomorphism.

Example 4. Let the set $S = \{T_0, T_3, T_6, T_9, I_1, I_4, I_7, I_{10}\}$ of pitch-class operations be the space for our GIS. We note that S has a simply transitive action on itself under right multiplication, isomorphic to the non-abelian dihedral group of order 8; this action constitutes the interval group G for the GIS. Put $g = T_3$ and $h = I_4$. Then, $i = g^{-1}h = T_3^{-1}I_4 = I_7$ is the unique element of G that satisfies the equation $g * i = h$. That is, $i = I_7$ is the interval from g to h. (We note that $i \neq d$, where $d = hg^{-1} = I_4 T_3^{-1} = I_1$ is the multiplicative analog to the difference $h - g$.)

3.3   Total Intervallic Content and Interval Vector

We are interested in the total intervallic content of a subset $D \subseteq S$ in a GIS: a tally of the number of occurrences of each interval in G that obtains between the members of D. For that purpose, we utilize the interval vector of D.

Definition 12. *Let G be an interval group of order v with a regular action on a set S, and let $D \subseteq S$. The **interval vector** of D, $IV(D) = (i_1, i_2, ...i_v)$, is a v-member array that tallies the number of occurrences in $D \times D$ of each directed interval in G. In interval groups for which the constituent intervals are measurable (such as in a vector space), it is customary to list the coordinates of the interval vector in order of increasing intervallic size, beginning with the unison (identity) interval (see [8]).*

Example 5. Let us use the same GIS as in Example 3, which incorporates the simply transitive action of the interval group $G = \mathbb{Z}/12\mathbb{Z}$ on the set of twelve pitch-class integers $S = \{0, 1, 2, ...11\}$. The subset $D = \{0, 1, 2, 3, 4, 6, 8, 9\}$ of S has the following total interval content.

$$0 - 0 = 1 - 1 = 2 - 2 = 3 - 3 = 4 - 4 = 6 - 6 = 8 - 8 = 9 - 9 = 0$$
$$1 - 0 = 2 - 1 = 3 - 2 = 4 - 3 = 9 - 8 = 1$$
$$2 - 0 = 3 - 1 = 4 - 2 = 6 - 4 = 8 - 6 = 2$$
$$3 - 0 = 4 - 1 = 6 - 3 = 9 - 6 = 0 - 9 = 3$$
$$4 - 0 = 6 - 2 = 8 - 4 = 0 - 8 = 1 - 9 = 4$$
$$6 - 1 = 8 - 3 = 9 - 4 = 1 - 8 = 2 - 9 = 5$$
$$6 - 0 = 8 - 2 = 9 - 3 = 0 - 6 = 2 - 8 = 3 - 9 = 6$$
$$1 - 6 = 3 - 8 = 4 - 9 = 8 - 1 = 9 - 2 = 7$$
$$0 - 4 = 2 - 6 = 4 - 8 = 8 - 0 = 9 - 1 = 8$$
$$0 - 3 = 1 - 4 = 3 - 6 = 6 - 9 = 9 - 0 = 9$$
$$0 - 2 = 1 - 3 = 2 - 4 = 4 - 6 = 6 - 8 = 10$$
$$0 - 1 = 1 - 2 = 2 - 3 = 3 - 4 = 8 - 9 = 11$$

We note that D has 8 occurrences of the identity interval 0,[2] 6 occurrences of the interval 6, and 5 occurrences of each of the remaining intervals in G. Hence,

$$IV(D) = (8, 5, 5, 5, 5, 5, 6, 5, 5, 5, 5, 5),$$

where the vector's first coordinate represents the number of occurrences of interval size 0; the second, interval size 1; the third, 2; and so on.

4   PADSs in GISs

4.1   FLIDs and NFLIDs

Sets of musical objects that have flat interval distributions, or FLIDs, are DSs, as the number of occurrences of every non-identity interval in a FLID's interval vector is equal to some integer λ. FLIDs are of considerable interest in mathematical music theory; for instance, they contribute to results that incorporate the Discrete Fourier Transform [9, Sect. 4.3.3].

Similarly, sets that are ADSs have *near*-flat interval distributions (NFLIDs). The interval vector for such a set displays t coordinates that are equal to λ, and $v - 1 - t$ coordinates that equal $\lambda + 1$. For instance, the set D in Example 5 is a $(12, 8, 5, 10)$ ADS: its interval vector, $IV(D) = (8, 5, 5, 5, 5, 5, 6, 5, 5, 5, 5, 5)$, has $t = 10$ coordinates that are equal to $\lambda = 5$, and $v - 1 - t = 1$ coordinate that equals $\lambda + 1 = 6$. Like FLIDs, NFLIDs are also of music-theoretical and compositional interest. The classical example is the set of all-interval tetrachords from pitch-class set theory, the members of which appear frequently in the post-tonal repertoire and its theories. These 48 sets—members of set classes $[0, 1, 4, 6]_{12}$ and $[0, 1, 3, 7]_{12}$—share the interval vector $(4, 1, 1, 1, 1, 1, 1, 2, 1, 1, 1, 1, 1)$. Accordingly, they are $(12, 4, 1, 10)$ ADSs.

---

[2] A set D of size k will always display k occurrences of the identity interval. Typically, this is shown as the first coordinate in an interval vector.

One special property of the members of set classes $[0, 1, 4, 6]_{12}$ and $[0, 1, 3, 7]_{12}$ is that they are the smallest pitch-class sets (in this case, the only pcsets with $k = 4$) to contain at least one of every interval in their interval group $G = \mathbb{Z}/12\mathbb{Z}$, hence, their name "all-interval tetrachords." As such, they demonstrate remarkable efficiency in their compactness and intervallic completeness. This attribute is common to FLIDs and NFLIDs with $\lambda = 1$. To that end, we survey the PADSs of small order. Table 1 presents a list of the 3,800 proper and improper (i.e., those with $t = v - 1$) PADSs of order $2 \leq n \leq 6$. These sets, which appear in fourteen isomorphism classes of groups, include cyclic, abelian, and non-abelian PADSs. In the following subsections, we consider the equivalence classes of these PADSs in order of ascending values for n. The group-theoretical data was collected primarily using computer testing in GAP [10] (see the Appendix for sample code).

Table 1. PADSs of small order ($2 \leq n \leq 6$)

n	G	(v, k, λ, t)	$|\mathscr{D}|$
2	$\mathbb{Z}_6$	$(6, 3, 1, 4)$	12
2	$\mathbb{Z}_7$	$(7, 3, 1, 6)$	14
3	D_8	$(8, 4, 1, 2)$	16
3	$\mathbb{Z}_{12}$	$(12, 4, 1, 10)$	48
3	$\mathbb{Z}_{13}$	$(13, 4, 1, 12)$	52
4	SD_{16}	$(16, 5, 1, 10)$	128
4	$S_3 \times \mathbb{Z}_3$	$(18, 5, 1, 14)$	108
4	$\mathbb{Z}_7 \rtimes \mathbb{Z}_3$	$(21, 5, 1, 20)$	294
4	$\mathbb{Z}_{21}$	$(21, 5, 1, 20)$	42
5	$A_4 \times \mathbb{Z}_2$	$(24, 6, 1, 16)$	192
5	$\mathbb{Z}_2^3 \times \mathbb{Z}_3$	$(24, 6, 1, 16)$	1344
5	$\mathbb{Z}_{14} \times \mathbb{Z}_2$	$(28, 6, 1, 24)$	728
5	$\mathbb{Z}_{31}$	$(31, 6, 1, 30)$	310
6	$D_8 \circ Q_8$	$(32, 7, 1, 20)$	512

In all but one of the groups G below, the complete set $\mathscr{D}$ of PADSs consists of the orbit of any one PADS $D \subset S$ under the action of the normalizer of G in the symmetric group on S, $N_{Sym(S)}G$. In the cyclic case, this normalizer is also the affine group. In the non-cyclic abelian and non-abelian cases, the normalizer acts as an analog for the affine group. The one exceptional case occurs in Subsect. 4.5, in the discussion of $n = 5$ PADSs.

4.2 PADSs with $n = 2$

Two equivalence classes of PADSs with $n = 2$ exist: those in $\mathbb{Z}/(n^2 + n)\mathbb{Z}$ and in $\mathbb{Z}/(n^2 + n + 1)\mathbb{Z}$. As both these groups are cyclic, so too are their constituent

PADSs. The 12 PADSs in $\mathbb{Z}/6\mathbb{Z}$ have $t = 4$; hence, they are proper PADSs. In contrast, the 14 PADSs in $\mathbb{Z}/7\mathbb{Z}$ are improper, as $t = v - 1 = 6$.[3] In either group, the set $\mathscr{D}$ of all PADSs consists of the orbit of some D under the action of the relevant affine group, which is isomorphic in both cases to the dihedral group D_{2v}. In music-theoretical parlance, the members of $\mathscr{D}$ belong to two transposition classes of size v, related to one another by inversion.

A musical representation of PADSs in the former case ($v = 6$) can be found in the following GIS. Let $S = \{0, 2, 4, 6, 8, 10\}$ be the six degrees of a whole-tone scale, on which the additive group $G = 2\mathbb{Z}/12\mathbb{Z}$ has a simply transitive action. Then, the set $D = \{0, 2, 6\}$ (and its various transpositions and inversions) is a $(6, 3, 1, 4)$ PADS, as demonstrated below.

$$2 - 0 = 2 \ (\text{modulo } 6)$$
$$6 - 2 = 4$$
$$6 - 0 = 0 - 6 = 6$$
$$2 - 6 = 8$$
$$0 - 2 = 10$$

The case of $v = 7$ has a similar musical representation in the GIS of seven diatonic scale degrees, acted on by the cyclic group of diatonic transposition operators.

4.3   PADSs with $n = 3$

The equivalance classes of PADSs with $n = 3$ include two classes with similar structure to those with $n = 2$ above. We find cyclic PADSs in both $\mathbb{Z}/(n^2 + n)\mathbb{Z}$ and $\mathbb{Z}/(n^2 + n + 1)\mathbb{Z}$. Now, however, the orbits under the respective affine groups are larger, as are the affine groups themselves: $\mathbb{Z}/12\mathbb{Z}$ contains $4v = 48$ $(12, 4, 1, 10)$ proper PADSs—the all-interval tetrachords of traditional pitch-class set theory—and $\mathbb{Z}/13\mathbb{Z}$ contains $4v = 52$ $(13, 4, 1, 12)$ improper PADSs, the transposition-and-inversion classes of $\{0, 1, 4, 6\}_{13}$ and $\{0, 1, 3, 9\}_{13}$.

The case of $n = 3$ also includes the smallest non-abelian PADSs. There exist 16 $(8, 4, 1, 2)$ PADSs in D_8. A musical representation occurs in the simply transitive action of the dihedral group $G = \{T_0, T_3, T_6, T_9, I_1, I_4, I_7, I_{10}\}$ of pitch-class operations on the octatonic pitch-class set $S = \{0, 1, 3, 4, 6, 7, 9, 10\}$. Then, the set $D = \{0, 1, 4, 6\}$ is an example of a PADS for this GIS.

4.4   PADSs with $n = 4$

Whereas we find cyclic PADSs in $\mathbb{Z}/(n^2+n+1)\mathbb{Z}$ for $n = 4$, with $n \geq 4$, we cease to find cyclic PADSs in $\mathbb{Z}/(n^2+n)\mathbb{Z}$. We conjecture that the reason is related to the non-existence of perfect Golomb rulers with five or more marks.[4] In addition

---

[3] The fourteen PADSs in $\mathbb{Z}/7\mathbb{Z}$ meet the stricter definition of being PDSs.

[4] A *Golomb ruler* with k marks and length L has a different measurement between any two marks. A perfect Golomb ruler is one in which every distance from 1 to L appears once as such a difference [11].

to the 42 cyclic $(21, 5, 1, 20)$ PADSs in $\mathbb{Z}/21\mathbb{Z}$, we also find 294 non-abelian $(21, 5, 1, 20)$ PADSs in $\mathbb{Z}/7\mathbb{Z} \rtimes \mathbb{Z}/3\mathbb{Z}$.[5] As the PADSs in these sets share the same (v, k, λ, t) parameters, they possess the smallest order for non-equivalent, yet isomorphic PADSs.

We find an example of a GIS that includes these non-abelian PADSs in the set of unordered dyadic subsets of a diatonic collection, as represented by $S = \binom{\mathbb{Z}/7\mathbb{Z}}{2}$. Let $G = \langle\, x, y \mid x^7 = y^3 = 1;\ yx = x^2 y \,\rangle \cong \mathbb{Z}/7\mathbb{Z} \rtimes \mathbb{Z}/3\mathbb{Z}$ be a non-abelian group with a regular action on S, where, for all $\{s, t\} \in S$,

$$\{s, t\} \cdot x = \{s + 1, t + 1\} \text{ (modulo 7)}$$

and

$$\{s, t\} \cdot y = \{2s, 2t\} \text{ (modulo 7)}.$$

Then, the set $D = \{\, \{0, 2\}, \{0, 5\}, \{1, 5\}, \{4, 5\}, \{4, 6\} \,\}$ is a $(21, 5, 1, 20)$ PADS.

The case with $n = 4$ includes also two additional equivalence classes of non-abelian PADSs. The semidihedral group of order 16, SD_{16}, contains 128 $(16, 5, 1, 10)$ PADSs, and the direct product $S_3 \times \mathbb{Z}/3\mathbb{Z}$ contains 108 $(18, 5, 1, 12)$ PADSs.

4.5   PADSs with $n = 5$

There exist four equivalence classes of PADSs with $n = 5$: three that include proper PADSs, and one that includes improper PADSs. The one class of improper sets occurs in $\mathbb{Z}/31\mathbb{Z}$, where we find 310 cyclic $(31, 6, 1, 30)$ PADSs. A musical example of such set can be found in the integer representation of 31-tone equal temperament [13], in which the set $D = \{1, 2, 4, 9, 13, 19\}_{31}$ has a FLID with $\lambda = 1$. The three classes of proper PADSs include two that are abelian and one that is non-abelian. The non-abelian class, which includes 192 $(24, 6, 1, 16)$ PADSs, is found in $A_4 \times \mathbb{Z}_2$. A class of 1344 abelian $(24, 6, 1, 16)$ PADSs appears in $\mathbb{Z}_2^3 \times \mathbb{Z}_3$. With $n = 5$, this group possesses the smallest order of non-cyclic abelian PADSs. These two classes share the same (v, k, λ, t) parameters; hence, their PADSs are isomorphic.

The third equivalence class of proper PADSs with $n = 5$ consists of 728 $(28, 6, 1, 24)$ PADS in the abelian group $\mathbb{Z}_{14} \times \mathbb{Z}_2$. It is the one exceptional class to which we allude in Subsect. 4.1, wherein the set $\mathscr{D}$ does not consist of the orbit of a single D under the action of $N_{Sym(S)}G$. Rather, $\mathscr{D}$ is the union of three orbits: two orbits of size 336, and one orbit of size 56. As such, these three classes are GISZ-related [14].

A musical representation of these PADSs exists in the following GIS. Let S consist of the set of all dyadic seconds, thirds, sixths, and sevenths in a C-major diatonic collection; we note that S is of size 28. Let G be a group isomorphic to $\mathbb{Z}_{14} \times \mathbb{Z}_2$, generated by the following operations g, h on S. First, g is a cycle of length 14, consisting of two chains of suspensions: one of 7–6 suspensions,

---

[5] A standard result in the theory of DSs states that for every cyclic PDS with $k \equiv 2$ (modulo 3)—in this case, $k = 5$—there exists an isomorphic non-abelian PDS [12].

and one of 2–3 suspensions. We model this cycle in $\mathbb{Z}/28\mathbb{Z}$, putting the diatonic third $(C, E) = 0$, the second $(C, D) = 1$, the third $(B, D) = 2$, etc., through the second $(D, E) = 13$; and the same for sixths and sevenths, beginning with the sixth $(E, C) = 14$, and running through the seventh $(E, D) = 27$. Hence, $g = (0, 1, 2, ...13)(14, 15, 16, ...27)$. Second, h exchanges every dyadic second with the seventh that contains the same scale degrees, and the same for thirds and sixths. That is, $h = (0, 14)(1, 15)...(13, 27)$. Then, the PADSs $D_1 = \{0, 1, 3, 7, 15, 24\}$ and $D_2 = \{0, 1, 7, 10, 15, 17\}$ have different orbits of length 336 under the action of the normalizer, and $D_3 = \{0, 1, 3, 13, 20, 27\}$ has a third orbit of length 56.

4.6 PADSs with $n = 6$

A classical conjecture in the theory of DSs is that no PDSs exist for any n that is not a power of a prime; and this conjecture has been proven through $n \leq 2{,}000{,}000$ [5, Remark 18.68]. Accordingly, as $n = 6$ is not a power of a prime, we find no improper PADSs of that order. However, one equivalence class of proper PADSs exists for $n = 6$: we find 512 non-abelian $(32, 7, 1, 20)$ PADSs in the central product $D_8 \circ Q_8$.[6]

Following, we construct a GIS that contains these PADSs. S is the set of 32 trichords in the octatonic collection $\{0, 1, 3, 4, 6, 7, 9, 10\}$ that do not contain a tritone. The interval group G that has a simply transitive action on this set is the central product of a dihedral group of order 8 and a quaternion group, also of order 8. The dihedral group can be generated by the pitch-class operations T_3 and I_1 on the members of S. The quaternion group can be generated by the following two 4-cycles: Q_1 alternates the pitch-class operations $(0, 6)(4, 10)$ and $(1, 7)(3, 9)$ on the members of S; and Q_2 applies the pitch-class operation $(0, 1, 6, 7)(3, 10, 9, 4)$ to members of the transposition classes $(0, 1, 4)$ and $(0, 4, 7)$ in S, and the inverse operation $(0, 7, 6, 1)(3, 4, 9, 10)$ to the transposition classes that are the inverted forms of these trichords. Then, the set $D = \{\; \{0, 3, 7\}, \{0, 4, 7\}, \{0, 4, 9\}, \{1, 6, 10\}, \{3, 4, 7\}, \{4, 7, 9\}, \{7, 9, 10\} \;\}$ is a $(32, 7, 1, 20)$ PADS.

5 Conclusions and Future Work

In the preceding sections, we have examined how ADSs are a generalization of DSs, and how certain of the same properties that have attracted music scholars to PDSs are shared by PADSs. In particular, PADSs display maximum intervallic efficiency. Further, we have investigated several GISs that include representations of cyclic, abelian, and non-abelian PADSs. Further work remains regarding ADSs, particularly with regards to music-analytical applications (and especially those that involve ADSs with $\lambda > 2$). It is our hope that the ideas presented here serve as a departure point for some of this future work.

---

[6] This group is also known as the extra special group of order $2^{(4+1)}$ of minus type.

Appendix

The following GAP code may be used to verify the results of Sect. 4 regarding the existence of PADSs of small order.

1. Determine all isomorphism classes of groups of order v.
   ```
   G := AllGroups(v);
   ```
2. For each group G_i, define an isomorphism to a permutation representation on the set $S = \{1, ..., v\}$.
   ```
   Ii := IsomorphismPermGroup(G[i]);
   ```
3. Generate the permutation group.
   ```
   Pi := Image(Ii);
   ```
4. Output all subsets D_i of size k in S.
   ```
   D := Combinations([1 .. v],k);
   ```
5. For each subset, output all pairs of elements.
   ```
   A := Combinations(D[j],2);
   ```
6. Determine the permutation in P that takes the first element in each pair to the second.
   ```
   Rj := RepresentativeAction(P,A[j][m],A[j][n]);
   ```
7. Generate the inverse element for each of the above permutations.
   ```
   Vj := Rj^-1;
   ```
8. If the set of all Rs and Vs for any D_j covers the non-identity elements of G_i, and if some non-identity element of G_i is represented only once as an R or a V and the remaining non-identity elements are represented once or twice, then D_j is a PADS.

References

1. Gamer, C., Wilson, R.: Microtones and projective planes. In: Fauvel, J., Flood, R., Wilson, R. (eds.) Music and Mathematics: From Pythagoras to Fractals. Oxford University Press, Oxford (2003)
2. Wild, J.: Presentation at the First Biennial Meeting of the Society for Mathematics and Computation in Music, Berlin (2007)
3. Forte, A.: The Structure of Atonal Music. Yale University Press, New Haven (1973)
4. Lewin, D.: Generalized Musical Intervals and Transformations. Oxford University Press, Oxford (2007)
5. Jungnickel, D., Pott, A., Smith, K.: Difference sets. In: Colburn, C., Dinitz, J. (eds.) Handbook of Combinatorial Designs, 2nd edn., pp. 419–435. Chapman & Hall/CRC, Boca Raton (2007)
6. Nowak, K.: A survey on almost difference sets. e-print. arXiv:1409.0114v1 [math.CO] (2014)
7. Arasu, K.T., Ding, C., Helleseth, T., Martinsen, H.: Almost difference sets and their sequences with optimal autocorrelation. IEEE Trans. Inf. Theor. **47**, 2934–2943 (2001)
8. Lewin, D.: The intervallic content of a collection of notes, intervallic relations between a collection of notes and its complement: an application to Schoenberg's hexachordal pieces. J. Music Theor. **4**, 98–101 (1960)

9. Amiot, E.: Music Through Fourier Space. Springer, Cham (2016). https://doi.org/10.1007/978-3-319-45581-5

10. The GAP Group: GAP - Groups, Algorithms, and Programming, Version 4.8.6 (2016). http://www.gap-system.org

11. Shearer, J.: Difference triangle sets. In: Colburn, C., Dinitz, J. (eds.) Handbook of Combinatorial Designs, 2nd edn. Discrete Mathematics and Its Applications. Chapman & Hall/CRC, Boca Raton (2007). Rosen, K., Series Editor

12. Singer, J.: A theorem in finite projective geometry and some applications to number theory. Trans. Am. Math. Soc. **43**, 377–385 (1938)

13. Wild, J.: Genus, species and mode in Vicentino's 31-tone compositional theory. Music Theor. Online **20**(2), 1–19 (2014)

14. Lewin, D.: Conditions under which, in a commutative GIS, two 3-element sets can span the same assortment of GIS-intervals; notes on the non-commutative GIS in this connection. Intégral **11**, 37–66 (1997)

Algebra of Harmony: Transformations of Just Consonances

Marek Žabka[(⊠)]

Katedra muzikológie, Univerzita Komenského, Bratislava, Slovakia
marek.zabka@uniba.sk

Abstract. The paper focuses on mathematical aspects of harmonies in extended just intonation and their relations. The first part lays down a theoretical framework for the investigation of structural features of such harmonies. Among other aspects, it addresses symmetry, inversion, and multiplication of harmonies. The second part explores transformational relations among harmonies of the same type, while the approach is intrinsically dualistic. Riemann-Klumpenhouwer's concepts of *Schritts* and *Wechsels* are generalized for 'harmony spaces' in extended just intonation. This enables a deeper analysis of harmonic 'neighborhoods.' Finally, a graphical representation of the complete neighborhood of a harmony, called 'neighborhood network,' is presented along with several simpler and more complex examples.

Keywords: Extended just intonation · Transformation · *Schritt* ·
Wechsel

Two irreconcilable principles have shaped musical theories for a long time: purity of consonance and regularity of structure. The latter seems to have prevailed since the Baroque era. The concept of twelve tones equally spaced around the pitch circle, a regular structure *par excellence*, has enabled amazing achievements. Yet, despite its inferior role in the post-Baroque period, the former principle, the principle of (extended) just intonation, has played leading roles in other musical cultures. Ancient Greek, Renaissance, various kinds of ethnic music, or certain contemporary microtonal approaches – all have searched for just consonances. The crucial disadvantage of the consonance principle is a conceptual complexity of resulting structures, an issue to which the regular systems provide a cure, albeit at the expense of purity. This paper proposes an alternative conceptual framework. It does not compromise the principle of harmoniousness and still it yields efficient means for manipulating complex pure harmonies. Its applications are at least two-fold: it exposes interesting structural features of harmonies observed in existing music, and, more importantly, it enables explorations of new musical territories. The latter is our main motivation. The purpose of this paper is to summarize theoretical aspects as a preparation for developing specialized software for manipulating complex harmonies in extended just intonation.

© Springer International Publishing AG 2017
O. A. Agustín-Aquino et al. (Eds.): MCM 2017, LNAI 10527, pp. 76–88, 2017.
https://doi.org/10.1007/978-3-319-71827-9_7

1   Introduction

The 12-tone equal temperament overwhelmingly dominates contemporary music and theory. For many it seems an ideal solution that have been achieved through a long evolution. And very often we forget that it is an imperfect compromise that distorts the most important aspect of music: its harmoniousness. While the impure 12-tone equal temperament prevails in western post-Baroque music, most of other musical cultures prefer tone systems based on the principle of just intonation. Various oriental or ancient Greek music theories emphasize harmoniousness of musical intervals based on simple integer ratios and develop manifold tone systems in just intonation [1,3]. The same principle of just consonances underpinned music theorizing also in the West for many centuries. Until quite recently, the idea that music should be restricted to 12 tones per octave was foreign to European music. Except for instruments with fixed tuning, even today $G\sharp$ and $A\flat$ are two different tones in actual musical practice. However, the difference between enharmonic tones is not formalized in the standard tone system and is achieved through fine tuning by ear only. Since Renaissance, several theorists and musicians have tried to define explicit tone systems that would capture such distinctions. A 31-tone system is probably the most famous among them. It was proposed and used in actual practice by Vicentino [20] in the sixteenth century and independently described by seventeenth-century Dutch scientist Huyghens [8]. It was noted by several later theorists [14,21,22] and most notably by Fokker [5] who put it also into a modern practice. Henk Badings, among several other Dutch composers, applied the 31-tone system in his music.

In contrast to contemporary understanding, the principle of just intonation was part of mainstream music theory still in the nineteenth century. This can be nicely illustrated by comparing neo-Riemannian [7] and 'Riemannian' *Tonnetze*. The two-dimensional lattice representation of pitches has enjoyed considerable popularity in recent music analysis and theory, especially in the neo-Riemannian circles. The term *Tonnetz* emphasizes its derivation from the nineteenth-century German music theory. However, there is a crucial conceptual difference between modern and original Riemannian *Tonnetz*. As Cohn explicitly discusses in his seminal introduction to the neo-Riemannian 1998 special issue of the *Journal of Music Theory* [2], neo-Riemannian *Tonnetz* assumes enharmonic equivalence limiting the pitch-class space to 12 items while the nineteenth-century German theorists often used the lattice constructions to demonstrate a potentially infinite tonal space based on just intonation. For instance, Tanaka [18], Oettingen [13], and even Riemann [17] derived (different) 53-tone systems in just intonation using non-circular infinite *Tonnetz*. Related 53-tone systems were discussed also by several other theorists [5,19,21]. More recently, Ben Johnston explored a 53-tone system in 5-limit just intonation both in his writings [9] and his earlier microtonal music (e.g. String Quartet no. 2).

The most significant drawback of the principle of extended just intonation is the vastness of resulting tonal spaces. Tones in (extended) just intonation can be represented as rational numbers. The 5-limit systems (generated only by the intervals of perfect fifth and just major third) can be represented on the two-dimensional *Tonnetz* and the 7-limit systems (the interval of natural seventh is added) can be represented on a 3-dimensional lattice. If just intervals based on higher prime numbers are added we arrive to even higher dimensional spaces. The problem is to find a practicable way of navigating through such complex spaces. Johnston's 53-tone and Fokker's 31-tone systems illustrate one strategy of navigating through the 5-limit and 7-limit spaces: take a finite selection that covers most important part of the infinite space. The tonality diamond, invented by Meyer [12] and popularized by Partch [15], reflects the same strategy applied to the 11-limit space. This strategy enables building physical instruments for accessing the complex spaces of extended just intonation. However, its limitation is that the instruments are quite complex and still they make available only fragments of the infinite spaces. In his later microtonal approach, Ben Johnston has applied a different approach. His extended musical notation has enabled him to access vast parts of a space in extended just intonation. As he comments on his String Quartet no. 5: 'The music is highly modulatory, thus involving a very large number of different pitches per octave, which since they are not being used as a scale, I did not bother to count' [10].

This paper follows a similar strategy. It assumes extended just intonation (without any temperament) and does not impose a limitation through simple selection of tones. Instead, it proposes a mathematical model that provides navigation through a complex space of harmonies in extended just intonation. The main idea comes from the neo-Riemannian theory of transformations. Klumpenhouwer's [11] concepts of *Schritts* and *Wechsels* are applied to harmonies in extended just intonation.

2 Harmonies

2.1 Basic Definitions: In the Footsteps of Euler

Euler's *Tentamen novae theoriae musicae* [4] offers a simple yet powerful mathematical framework for describing harmonies in just intonation. Even after nearly three centuries, it still may serve as a source of inspiration. Some of the notions defined below, such as index and exponent, follow Euler's ideas and terminology.

Harmony is a set of positive rational numbers. Its elements are called *tones*. Furthermore, consider positive integers $x_1, \ldots, x_n$ such that $\gcd(x_1, \ldots, x_n) = 1$. Then we say that the set $\mathbf{X} = \{x_1, \ldots, x_n\}$ is a *canonic harmony*.

Lemma 1. *Consider a harmony* $H = \{\frac{a_1}{b_1}, \ldots, \frac{a_n}{b_n}\}$ *where* a_i *and* b_i *are pairs of relatively prime positive integers for all* $i = 1, \ldots, n$. *Then there exist unique rational number* $\mathsf{ind}(H)$ *and unique canonic harmony* $\mathsf{T}(H)$ *such that* $H = \mathsf{ind}(H)\mathsf{T}(H)$.

We say that $\mathsf{ind}(H)$ and $\mathsf{T}(H)$ are the *index* and the *type* of the harmony H, respectively. The proof is straightforward if we put:

$$\mathsf{ind}(H) = \frac{\gcd(a_1, \ldots, a_n)}{\mathsf{lcm}(b_1, \ldots, b_n)}, \qquad \text{(index)}$$

$$\mathsf{T}(H) = \mathsf{ind}(H)^{-1}H. \qquad \text{(type)}$$

In the next step, we introduce two more characteristics of a harmony. *Counter-index* $\mathsf{cind}(H)$ and *exponent* $\mathsf{exp}(H)$ are given by the following equations:

$$\mathsf{cind}(H) = \frac{\mathsf{lcm}(a_1,\ldots,a_n)}{\mathsf{gcd}(b_1,\ldots,b_n)}, \qquad \text{(counter-index)}$$

$$\mathsf{exp}(H) = \mathsf{cind}(H)/\mathsf{ind}(H). \qquad \text{(exponent)}$$

The last characteristic, the exponent, is crucial because it is invariant for all harmonies of the same type. The following lemma confirms this statement.

Lemma 2. *Consider a harmony H of type $\mathsf{T}(H)$. Then its exponent is the least common multiplier of the elements of its type: $\mathsf{exp}(H) = \mathsf{lcm}(\mathsf{T}(H))$. It means that all harmonies of same type share the same exponent.*

The following notation is assumed for transcriptions of rational numbers as musical tones in 7-limit just intonation. 1 corresponds to tone D. Unless necessary, octaves are not specified. If necessary, a numerical coefficient is attached: harmony $\{1,2\}$ is the interval $D_0 D_1$. Powers of $\frac{3}{2}$ are transcribed through the chain of fifths with the usual system of sharps and flats: $\frac{3}{2}$ is G, $\frac{8}{9}$ is C, $\frac{81}{64}$ is F♯ and so on. Corrections by syntonic comma $\frac{81}{80}$ are shown with an apostrophe, which is placed in lowered or raised position to show correction by the comma downwards or upwards, respectively. So DF♯, is a just major third $\{1,\frac{5}{4}\} = \{1,\frac{81}{64}\frac{80}{81}\}$. The corrections by septimal comma $\frac{64}{63}$ are denoted by the slash or backslash symbols for upward and downward corrections, respectively. Thus, DC\denotes a harmonic seventh interval.

To illustrate the basic concepts defined above, consider the C major scale in a usual just intonation: CDE,FGA,B,. It corresponds to the harmony $H = \{\frac{8}{9},1,\frac{10}{9},\frac{32}{27},\frac{4}{3},\frac{40}{27},\frac{5}{3}\}$. Its index is $\mathsf{ind}(H) = \frac{1}{27}$, counterindex $\mathsf{cind}(H) = 160$, exponent $\mathsf{exp}(H) = 4320 = 2^5 3^3 5^1$, and its type $\mathsf{T}(H) = \{24,27,30,32,36,40,45\}$.

2.2 Multiplication of Harmonies

Now we will define a binary operation on the set of harmonies. Let A and B be two harmonies. Then their multiplication is defined by the following formula:

$$AB = \{ab \mid a \in A, b \in B\}. \qquad \text{(multiplication of harmonies)}$$

This operation is associative and commutative. It also has a neutral element, which is the trivial harmony $\{1\}$. However, obviously, there are no inverse elements except for trivial harmony $\{1\}$. So the algebra of harmonies based on the operation of multiplication is a commutative monoid.

We say that a harmony H is *composite* if there are non-trivial harmonies A and B such that $H = AB$. A harmony is called *prime* if it is not composite.

Consider the following equation for pairwise distinct prime harmonies $P_1, \ldots, P_n$ and positive integers $k_1, \ldots, k_n$.

$$H = P_1^{k_1} \cdots P_n^{k_n}. \qquad \text{(prime factorization)}$$

The right side is called a *prime factorization* of harmony H. Any harmony has at least one prime factorization. However, it is not necessarily unique.

Harmonies whose prime factorization is a power of a single interval (2-element harmony) exhibit special structural features. We say that a harmony H is a *chain* generated by the interval $\{1, r\}$ if $H = q\{1, r\}^n$ for positive rational numbers q and r and a positive integer n. The usual Pythagorean scales are chains generated by the perfect fifth $\{1, 3\}$.

Prime factorization provides crucial insights into the internal structure of a harmony. Let us compare two scales whose intervallic structures are significantly different: the diatonic scale and Hungarian minor scale. (We disregard octave differences in this example. It means that factors of 2 are removed from rational numbers and such rational numbers might be interpreted as 'harmony classes.')

Prime factorization of the diatonic scale of D major is $\{1, 3, 5, 9, 15, 27, 45\} = \{1, 3\}^2\{1, 3, 5\}$. It explicitly demonstrates the structure of the scale: comprised of major triads transposed to positions given by a chain of three perfect fifths. There is no doubt that major triads and the three perfect fifths (corresponding to subdominant—tonic—dominant) are the structural basis of the scale.

This harmony also illustrates the non-uniqueness of prime factorization. Its other factorization is $\{1, 3\}\{1, 5, 9, 15\}$. The second factor is Ptolemy's tense diatonic tetrachord [1]. This factorization represents a different structural decomposition of the diatonic scale, typical for classical ancient Greek music theory: it comprises two tetrachords at the distance of perfect fifth.

Now consider the Hungarian minor scale $\{\frac{1}{15}, \frac{1}{5}, 1, 3, 5, 15\}$. Its prime factorization is $\{1, 3\}\{1, 5\}\{1, \frac{1}{15}\}$. It means that it is a multiplication of the perfect fifth, just major third, and diatonic semitone. Again, this nicely highlights the structural significance of the three intervals in the Hungarian minor scale.

2.3 Inverses and Symmetries

In this subsection we investigate harmonies constructed from inverted tones. We define two related transformations: the context-independent dual and the exponent *Wechsel*, which is contextual.

Consider a harmony $H = \{h_1, \ldots, h_n\}$. We say that the harmony $H' = \{h_1^{-1}, \ldots, h_n^{-1}\}$ is the *dual* of the harmony H. Now consider another transformation that maps harmony H to the harmony $\mathsf{W_e}(H) = \mathrm{ind}(H)\exp(H)\mathsf{T}(H)' = \mathrm{cind}(H)\mathsf{T}(H)'$. We call it *exponent Wechsel*. (The naming convention will become clearer in the context of a theory of *Schritts* and *Wechsels* discussed later.)

As an example consider the C major triad $C^+ = \{\frac{8}{9}, \frac{10}{9}, \frac{4}{3}\}$ and D major triad $D^+ = \{1, \frac{5}{4}, \frac{3}{2}\}$. Then their dual harmonies are A minor triad $\{\frac{9}{8}, \frac{9}{10}, \frac{3}{4}\}$ and G minor triad $\{1, \frac{4}{5}, \frac{2}{3}\}$. On the other hand, exponent *Wechsel* results in E'_1 minor triad $\{\frac{20}{9}, \frac{8}{3}, \frac{10}{3}\}$ and $F\sharp'_1$ minor triad $\{\frac{5}{2}, 3, \frac{15}{4}\}$. It means that the dual is a context independent inversion with the tone D in the centre of symmetry. On the other hand, the exponent *Wechsel* is a contextual transformation corresponding to the neo-Riemannian *Leittonwechsel* (for the specific case of triads).

Further we say that a harmony S is *symmetric* if $S = W_e(S)$. For a symmetric harmony $S = \{s_1, \ldots, s_n\}$ we define the *point of symmetry* $\mathsf{sym}(S)$ as the geometric mean of the contained tones:

$$\mathsf{sym}(S) = \sqrt[n]{s_1 \ldots s_n}. \qquad \text{(point of symmetry)}$$

One can observe that $\mathsf{sym}^2(S) = \mathsf{ind}(S)\mathsf{cind}(S)$.

Trivially, all one-note and two-note harmonies are symmetric. Therefore, H is least symmetric if it contains no symmetric subsets but one- and two-tone subharmonies. In that case we say that harmony H is *primitive*. The concept of primitiveness will play an important role in the second part of this paper where we investigate transformations of harmonies. There is a relation between primitive and prime harmonies.

Theorem 1. *A harmony is primitive iff all its subharmonies are prime.*

The theorem has a simple corollary. If a harmony is primitive then it is prime. The converse statement is not valid. For instance, the dominant seventh chord $\{1, 3, 5, \frac{9}{5}\}$ is prime but not primitive because of a symmetric subset $\{5, 3, \frac{9}{5}\}$ with the point of symmetry equal 3.

2.4 Transpositional Intersection

In this section we define transpositional intersection. It is an important tool when we investigate musical commonalities between two harmonies.

Consider two harmonies A and B. We define the *set of transpositional intersections* of harmony A over harmony B, denoted $\overset{A}{\cap}B$, by this formula:

$$\overset{A}{\cap}B = \{X \subseteq B \mid \exists q \in \mathbb{Q} : qA \cap B = X\}. \qquad \text{(transpositional intersections)}$$

Obviously, two different transpositions of a harmony A may have the same intersections with a harmony B, i.e. for $q_1 \neq q_2$ we may have $q_1 A \cap B = q_2 A \cap B$. Therefore, we define the *set of transpositional coefficients* for a harmony included in the set of transpositional intersections of A over B:

$$\mathsf{transp}(X \in \overset{A}{\cap}B) = \{q \in \mathbb{Q} \mid qA \cap B = X\}. \qquad \text{(transpositional coefficients)}$$

We say that A and B *primitively intersect* if $|\mathsf{transp}(X)| = 1$ for all $X \in \overset{A}{\cap} B$ (it is easy to see that this relation is symmetric).

Finally, we consider intersections of a harmony and its dual. We define the *symmetry set* $\mathcal{S}(A)$ of a harmony A:

$$\mathcal{S}(A) = \overset{A'}{\cap}A. \qquad \text{(symmetry set)}$$

The following theorem summarizes properties of the symmetry set. This set plays an important role in the description of canonic *Wechsels* investigated in the second part of the paper.

Theorem 2. *Let A be a harmony. Then the following statements hold.*

(i) *A and A' intersect primitively.*
(ii) *All elements of $\mathcal{S}(A)$ are symmetric.*
(iii) *$2|A| - 1 \leq |\mathcal{S}(A)| \leq \frac{1}{2}|A|(|A| + 1)$. The lower bound is achieved iff A is a chain, the upper bound iff A is primitive.*

3   Transformations of Harmonies

3.1   *Schritts* and *Wechsels*

Let $\mathbf{X}$ be a canonic harmony and consider the set of all harmonies of type $\mathbf{X}$ and $\mathbf{X}'$. We call it *harmony space* of type $\mathbf{X}$ and denote it $\mathcal{H}(\mathbf{X})$. Obviously, this concept is dualistic and $\mathcal{H}(\mathbf{X}) = \mathcal{H}(\mathbf{X}')$.

We define two kinds of transformations on a harmony space. Let H be a harmony from the harmony space $\mathcal{H}(\mathbf{X})$. Hence there exists a positive rational number h such that $H = h\mathbf{X}$ or $H = h\mathbf{X}'$. Furthermore, let q be any positive rational number. Then *Schritt* S_q and *Wechsel* W_q are transformations defined on $\mathcal{H}(\mathbf{X})$ by the following formulas:

$$\mathsf{S}_q(h\mathbf{X}) = hq\mathbf{X}, \qquad\qquad \mathsf{S}_q(h\mathbf{X}') = hq^{-1}\mathbf{X}', \qquad\qquad (Schritt)$$
$$\mathsf{W}_q(h\mathbf{X}) = hq\mathbf{X}', \qquad\qquad \mathsf{W}_q(h\mathbf{X}') = hq^{-1}\mathbf{X}. \qquad\qquad (Wechsel)$$

As one can observe, the definitions take inspiration from the homonymous concepts as defined by Klumpenhouwer [11] in his seminal work reformulating Hugo Riemann's theories of harmony. *Schritts* preserve harmony types and transpose dual harmonies in opposite directions. *Wechsels* change harmony types and reflect dual harmonies in opposite directions.

The system of *Schritts* and *Wechsels* is an infinite group. Let us denote it *general* SW-*group*. The trivial *Schritt* S_1 is its neutral element.

Lemma 3. *Let q and r be positive rational numbers. Then:*

$$\mathsf{S}_q^{-1} = \mathsf{S}_{q^{-1}}, \qquad\qquad \mathsf{S}_q\mathsf{S}_r = \mathsf{S}_{qr}, \qquad\qquad \mathsf{W}_q\mathsf{S}_r = \mathsf{W}_{q^{-1}r},$$
$$\mathsf{W}_q^{-1} = \mathsf{W}_q, \qquad\qquad \mathsf{S}_q\mathsf{W}_r = \mathsf{W}_{qr}, \qquad\qquad \mathsf{W}_q\mathsf{W}_r = \mathsf{S}_{q^{-1}r}.$$

3.2   Harmonic Neighborhood

Consider a canonic harmony $\mathbf{X}$ and a set of tones T. Our aim is to explore all harmonies from the harmony space $\mathcal{H}(\mathbf{X})$ that share at least one tone with T. We call the set of all such harmonies the *neighborhood* of set T and denote it $\mathcal{N}_{\mathbf{X}}(T)$. Below we consider two important cases: neighborhood of a single tone and neighborhood of a harmony belonging to $\mathcal{H}(\mathbf{X})$.

Assume that the canonic set $\mathbf{X} = \{x_1, \ldots, x_n\}$ is of cardinality n. One easily observes that the neighborhood of a single tone t contains exactly n distinct

Schritts and up to n distinct *Wechsels*: $\mathcal{N}_{\mathbf{X}}(t) = \{S_{tx_j^{-1}}(\mathbf{X}), W_{tx_j}(\mathbf{X}) \mid j = 1, \ldots, n\}$. The total number of harmonies in $\mathcal{N}_{\mathbf{X}}(s)$ may be lesser than $2n$. This happens iff the canonic set X is not symmetric.

In the next step, we explore the neighborhood of a given harmony A of type $\mathbf{X}$. Let $B \in \mathcal{H}(\mathbf{X})$ be any harmony from the harmony space of type $\mathbf{X} = \{x_1, \ldots, x_n\}$ such that B belongs to the neighborhood of A, i.e. $B \in \mathcal{N}_{\mathbf{X}}(A)$. Select an element h from the intersection of A and B, i.e. $h \in A \cap B$. Considering the types of A and B, there are four possible cases:

(i) $\mathsf{T}(A) = \mathbf{X}$ and $\mathsf{T}(B) = \mathbf{X}$. Then we have $h = \mathsf{ind}(A)x_i = \mathsf{ind}(B)x_j$ for some $i, j \in \{1, \ldots, n\}$. In this case $B = S_{x_j x_i^{-1}}(A)$.

(ii) $\mathsf{T}(A) = \mathbf{X}'$ and $\mathsf{T}(B) = \mathbf{X}'$. Then we have $h = \mathsf{cind}(A)x_i^{-1} = \mathsf{cind}(B)x_j^{-1}$ for some $i, j \in \{1, \ldots, n\}$. This also leads to $B = S_{x_j x_i^{-1}}(A)$.

(iii) $\mathsf{T}(A) = \mathbf{X}$ and $\mathsf{T}(B) = \mathbf{X}'$. Then $h = \mathsf{ind}(A)x_i = \mathsf{cind}(B)x_j^{-1}$ for some $i, j \in \{1, \ldots, n\}$, in which case $B = W_{x_j x_i}(A)$.

(iv) $\mathsf{T}(A) = \mathbf{X}'$ and $\mathsf{T}(B) = \mathbf{X}$. Then $h = \mathsf{cind}(A)x_i^{-1} = \mathsf{ind}(B)x_j$ for some $i, j \in \{1, \ldots, n\}$. This final case results again in $B = W_{x_j x_i}(A)$.

The preceding reflections lead us to the following theorem.

Theorem 3. *Let $\mathbf{X}$ be a canonic harmony. Assume a harmony A from the harmony space $\mathcal{H}(\mathbf{X})$. Then we have:*

$$\mathcal{N}_{\mathbf{X}}(A) = \{S_{xy^{-1}}(A), W_{xy}(A) \mid x, y \in \mathbf{X}\}.$$

For a given canonic set $\mathbf{X}$, we call the *Schritts* and *Wechsels* from the previous theorem *canonic* and denote $S_{\mathbf{X}}$ and $W_{\mathbf{X}}$ the sets of all canonic *Schritts* and all canonic *Wechsels*, respectively. Thus, the previous theorem states that the neighborhood of a harmony A can be achieved by applying transformations from $S_{\mathbf{X}} \cup W_{\mathbf{X}}$ to it.

For a *Wechsel* W_{xy} one can distinguish two cases: either $x = y$ or $x \neq y$. In the former case we call it a *tone Wechsel*, in the latter an *interval Wechsel*. In general, a tone *Wechsel* can equal an interval *Wechsel*. However, they are always distinct if the underlying canonic harmony is primitive.

Corollary 1. *Let $\mathbf{X}$ be a canonic harmony of cardinality n and $A \in \mathcal{H}(\mathbf{X})$. Then the neighborhood of A contains up to $n(n-1)$ non-trivial Schritts, up to $n(n-1)/2$ interval Wechsels, and exactly n tone Wechsels.*

Lemma 4. *The maximal counts for Schritts and interval Wechsels in the harmonic neighborhood $\mathcal{N}(\mathbf{X})$ from Corollary 1 are achieved iff $\mathbf{X}$ is primitive. Moreover, in such case all canonic Schritts and Wechsels are distinct.*

There is a direct relation between the symmetry set of a canonic harmony $\mathbf{X}$ and the canonic *Wechsels* of type $\mathbf{X}$. Let $S \in \mathcal{S}(\mathbf{X})$ be an element of the symmetry set of $\mathbf{X}$. We define the *symmetry Wechsel* of S acting on $\mathcal{H}(X)$:

$$W_S = W_{\mathsf{sym}^2(S)}. \qquad \text{(symmetry Wechsel)}$$

It is easy to see that the symmetry *Wechsel* is a canonic *Wechsel* of type **X**. The following theorem puts the symmetry sets and canonic *Wechsels* into an even stronger relation.

Theorem 4. *Let* **X** *be a canonic harmony. Then we have:*

$$W_{\mathbf{X}} = \{W_S \mid S \in \mathcal{S}(\mathbf{X})\}.$$

Moreover, if $S_1 \neq S_2$ *for* $S_1, S_2 \in \mathcal{S}(\mathbf{X})$ *then* $W_{S_1} \neq W_{S_2}$.

3.3 Transformational Networks of Neighborhoods

We assume a primitive canonic harmony $\mathbf{X} = \{x_1, \ldots, x_n\}$ of cardinality n. Our aim is to find a graphical representation of the neighborhood of a harmony of type **X** with an ultimate goal of defining a user interface for manipulating harmonies included in the neighborhood. If $n = 3$ the solution is well known: The generalized *Tonnetz* generated by **X** represents of the entire subgroup generated by the canonic *Schritts* and *Wechsels*. There the neighborhoods can easily be located. The *Tonnetz* can be elegantly generalized to higher cardinalities through simplicial lattices of higher dimensions. Gollin [6] explores such a structure for the case of $n = 4$ where harmonies are represented via 3-simplexes (tetrahedra). However, this approach is not applicable for our purposes: higher cardinalities lead us quickly to high-dimensional spaces that are difficult to manipulate through a simple user interface. Therefore, we follow a different route here.

First, let us focus on the simple case of the neighborhood of a single tone x_i for any $i = 1, \ldots, n$. As discussed above, $\mathcal{N}_{\mathbf{X}}(x_i)$ is given by $\{S_{x_i x_j^{-1}}, W_{x_i x_j} \mid j = 1, \ldots, n\}$. Any two harmonies from $\mathcal{N}_{\mathbf{X}}(x_i)$ share exactly one tone if they are of the same type (two *Schritts* or two *Wechsels*) and exactly (we assume primitiveness of **X**) two tones if they are of opposite types (one *Schritt* and one *Wechsel*). We will represent them on a circle with $2n$ positions (or a regular $2n$-gon). Distribute the *Wechsels* to even positions so that $W_{x_i x_1}, \ldots, W_{x_i x_n}$ are ordered clockwise. And distribute *Schritts* $S_{x_i x_1^{-1}}, \ldots, S_{x_i x_n^{-1}}$ to odd positions, also ordered clockwise. As there are n options for the mutual position of the tone *Wechsel* $W_{x_i x_i}$ and the identity *Schritt* $S_{x_i x_i^{-1}} = S_1$, there are n different constellations for this cyclic construction. Any two neighbors on such a cycle are related through a *Wechsel* and, thus, share two common tones.

Now we proceed to the construction of a transformational network for a neighborhood of a harmony $A \in \mathcal{H}(\mathbf{X})$. First, assume that the cardinality n of the underlying canonic harmony $\mathbf{X} = \{x_1, \ldots, x_n\}$ is odd. Let $A = \{a_1, \ldots, a_n\}$ where $a_i = a x_i$ for all $i = 1, \ldots, n$ or $a_i = a x_i^{-1}$ for all $i = 1, \ldots, n$. Draw n equal circles (or regular $2n$-gons) $\gamma_1, \ldots, \gamma_n$ with centers distributed regularly on an auxiliary circle of the same diameter. Thus, the n circles have a common intersection in the center of the auxiliary circle. Denote this intersection Γ. Moreover, each pair of circles has exactly one more intersection. (The last condition is not met in the case of even n and, therefore, we will address that case separately.). The n circles will represent cyclic transformational networks of neighborhoods of the n tones $a_1, \ldots, a_n$ included in A, ordered clockwise. We distribute the *Wechsels* and *Schritts* of A on the circles in the following way. Interval *Wechsel* $W_{x_i x_j}$, $i \neq j$, is located in the intersection of $\gamma_i \cap \gamma_j \setminus \{\Gamma\}$ and the tone *Wechsels* $W_{x_i x_i}$ on the circles γ_i opposite of the common intersection Γ. By now, all *Wechsels* are evenly distributed along the circles, occupying exactly n positions on each. The *Schritts* will be located on the n positions regularly alternating with the *Wechsels*. The identity *Schritt* $S_1 = S_{x_i x_i^{-1}}$ is put to the common intersection Γ. All remaining *Schritts* are assigned to the corresponding positions on corresponding circles, ordered clockwise. We will call this graph *neighborhood network*.

Fig. 1. Neighborhood network for a three-note harmony. *Left:* Neighborhood network for harmony $A = \{a, b, c\}$. *Right:* Comparison with the standard neo-Riemannian case harmony space of major and minor triads.

Figure 1 illustrates the simple case when $n = 3$. The left side of the figure shows the neighborhood network for a three-element harmony $A = \{a, b, c\}$. As mentioned above, this simple case is easily modelled by the *Tonnetz* and the *Schritts* and *Wechsels* correspond to well-known neo-Riemannian transformations. The right side of the figure provides the comparison.

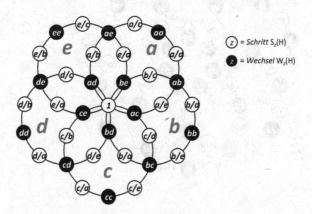

Fig. 2. Neighborhood network for a five-note harmony $\{a, b, c, d, e\}$.

Figure 2 shows the neighborhood network for harmony spaces based on five-note canonic harmonies. There is a relation between the five-note case and the famous Penrose tiling. *Schritts* and *Wechsels* from the neighborhood can be assigned to pentagons of the central part of Penrose [16] original irregular tiling.

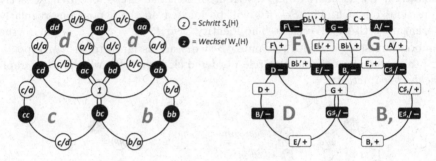

Fig. 3. Neighborhood network for a four-note harmony. *Left:* A generic case for a harmony $\{a, b, c, d\}$. *Right:* Neighborhood network of a natural seventh chord $GB,DF\backslash$.

To draw neighborhood networks for harmonies with an even number of tones we make a simple trick. For n even we consider the neighborhood network for cardinality $n + 1$. Then we remove one of the circles and the corresponding *Schritts* and *Wechsels*. The left side of Fig. 3 shows the neighborhood network obtained via such a procedure for the case $n = 4$. The right side of Fig. 3 illustrates the neighborhood network for the natural seventh chord $GB,DF\backslash$ from the harmony space $\mathcal{H}(1, 3, 5, 7)$.

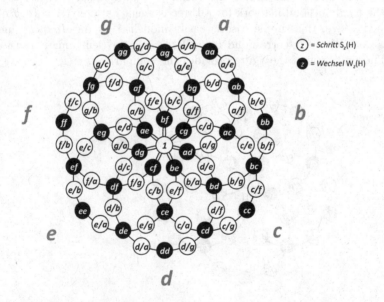

Fig. 4. Neighborhood network for a seven-note harmony.

Finally, Fig. 4 shows neighborhood networks for seven-element harmonies. Again, the neighborhood network for a six-element harmony could be derived from that network by omitting one of the circles and the corresponding *Schritts* and *Wechsels*.

Acknowledgement. This paper was supported by the scientific grant agency VEGA through the grant no. 1/0637/15.

References

1. Barker, A.: Greek Musical Writings II - Harmonic and Acoustic Theory. Cambridge University Press, Cambridge (1989)
2. Cohn, R.: Introduction to neo-Riemannian theory: a survey and a historical perspective. J. Music Theory **42**(2), 167–80 (1998)
3. Daniélou, A.: Introduction to the Study of Musical Scales. The India Society, London (1943)
4. Euler, L.: Tentamen novae theoriae musicae. St. Petersburg (1739)
5. Fokker, A.D.: Selections from the harmonic lattice of perfect fifths and major thirds containing 12, 19, 22, 31, 41 or 53 notes. In: Proceedings of Koninklijke Nederlandse Akademie van Wetenschappen, Series B, vol. 71, pp. 251–266 (1968)
6. Gollin, E.: Some aspects of three-dimensional tonnetze. J. Music Theory **42**(2), 195–206 (1998)
7. Gollin, E., Rehding, A. (eds.): The Oxford Handbook of neo-Riemannian Music Theories. Oxford University Press, New York (2011)
8. Huygens, C.: Le cycle harmonique (Rotterdam 1691), Novus cyclus harmonicus (Leiden 1724) - with Dutch and English Translations. Diapason Press, Utrecht (1986). Edited by Rasch, R
9. Johnston, B.: Scalar order as a compositional resource. Perspect. New Music **2**(2), 56–76 (1964)
10. Johnston, B.: On String quartet No. 5. In: Gilmore, B. (ed.) Maximum Clarity and Other Writings on Music, p. 203. University of Illinois Press, Illinois (2006)
11. Klumpenhouwer, H.: Some remarks on the use of Riemann transformations. Music Theory Online **0.9** (1994)
12. Meyer, M.F.: The Musician's Arithmetic. Oliver Ditson Company, Boston (1929)
13. von Oettingen, A.: Das duale Harmoniesystem. C.F.W. Siegel's Musikalienhandlung, Leipzig (1913)
14. Opelt, W.: Ueber die Natur der Musik: Ein vorläufiger Auszug aus der Bereits auf Unterzeichnung angekündigten *Algemeinen Theorie der Musik*. Hermann und Langbein, Leipzig (1834)
15. Partch, H.: Genesis of a Music: An Account of a Creative Work, Its Roots, and Its Fulfillments, 2nd edn. Da Capo Press, New York (1974)
16. Penrose, R.: The role of aesthetics in pure and applied mathematical research. Bull. Inst. Math. Appl. **10**, 266–271 (1974)
17. Riemann, H.: Katechismus der Akustik (Musikwissenschaft). Max Hesse, Leipzig (1891)
18. Tanaka, S.: Studien im Gebiete der reinen Stimmung. In: Chrysander, F., et al. (eds.) Vierteljahrsschrift für Musikwissenschaft, pp. 1–90. Breitkopf und Härtel, Leipzig (1890)

19. Thompson, T.P.: Theory and Practice of Just Intonation: With a View to the Abolition of Temperament. Effingham Wilson, London (1850)
20. Vicentino, N.: L'antica musica ridotta alla moderna prattica. Antonio Barre, Rome (1557)
21. Würschmidt, J.: Logarithmische und graphische Darstellung der musikalischen Intervalle. Zeitschrift für Physik **3**(2), 89–97 (1920)
22. Yasser, J.: A Theory of Evolving Tonality. American Library of Musicology, New York (1932)

Computer Assisted Performance

Developing Software for Dancing
Tango in *Compás*

Emmanuel Amiot[1]([✉]), Jean-Philippe Lerat[2], Bérenger Recoules[3],
and Valérie Szabo[2]

[1] LAMPS, Perpignan, France
manu.amiot@free.fr
[2] KBCOO, Nantes, France
jplerat@kbcoo.com
[3] Freelance Artist, Nantes, France
berenger.recoules@gmail.com

Abstract. Argentine Tango faces dancers with specific challenges. Since the dance is improvised, the leader is expected to follow patterns and trends in the music on the fly. While musicians have an advantage, many beginners prove unequal to the task, and are often driven to abandon. *Compass Trainer* is a piece of smartphone software intended to help dancers feel and integrate in their movements the 'Compás', the rhythmic pulse of the dance. Its development blended theoretical and down-to-earth, practical considerations. Our team had to take into account the mixed rhythmical structure – binary with a ternary component; explore signal processing techniques such as beat tracking; interview tango maestros and musicians, and build a mobile application to help our users discover the rhythmical layers of a tango track. Experimentation in Tango classes was rewardingly successful.

Keywords: Tango · Dancing · Rhythm · Beat-track
Pedagogical software · Pedagogy studies

1 Introduction

In many social dances, the strong binary pulsation helps the dancers move their feet onto the beat. However, the same is not true in Argentinian Tango. For several reasons that we tentatively explore in Sect. 2, many tango dancers are uncomfortable with the ongoing pulsation, which is as bad a translation as any for the Spanish, almost mystical term, of 'Compás'. The complexity of tango music, blending melodic lines with sometimes irregular rhythms, is a challenge for the leader. Quoting [3]:

In tango music, the melody is often given a rhythmic treatment. It is often impossible to separate them entirely, but learning to discriminate between the melody and the rhythm is one of the most important skills of all.

© Springer International Publishing AG 2017
O. A. Agustín-Aquino et al. (Eds.): MCM 2017, LNAI 10527, pp. 91–103, 2017.
https://doi.org/10.1007/978-3-319-71827-9_8

This is especially cruel for leaders, because their follower sometimes has a better perception of rhythm, with tragic effect on the harmony of the dancing couple.

One of the authors (see Sect. 3) was sensitive enough to the issue to engage into the development of a software purported to help tango dancers with their perception of the *Compás*, and provide a tool for learning to dance on the music, be it alone, in couple, at home, or during a tango class. This involved gathering a team, with computer programmer, music theorist, tango teachers, media com expert, etc., and meeting with tango musicians, maestro dancers and high-tech hardware developers. The present paper reports some of the difficulties we met, and the choices we made to solve them. The resulting application (available online) could also be used by dancer with any level, teacher, researcher on pedagogy and on the relationship between musical and dancing gestures (suggested by reviewer). The present paper begins with some general explanation of tango-specific difficulties, followed by a dancer's testimony and the goals, elaboration and testing of the software.

2   Difficulties in Dancing Tango

There are some rare, documented cases, of people having difficulties following a rhythm and coordinating their movements with it [4]. Experimental studies would rather tend to show that the ability to move in rhythm is well anchored in human behavior and is a characteristic feature of human beings.[1] Indeed most people can dance to a beat in a disco and learn easily to move in rhythm in most dances: rock, salsa, bachata, swing, rumba, waltz... Why then do so many leaders find it so difficult to guide their steps in accordance with the pace of tango music? Listening to https://www.youtube.com/watch?v=N-FyPMYjFpc[2] may hint at some of the problems involved (Fig. 1).

2.1   Practice of Argentinian Tango Dancing

Argentinian Tango is an improvised dance with a leader and a follower. Adaptation and flexibility are required: most often, the leader invites a follower for a series of three or four pieces (a *tanda*), and switches to another for the next tanda. The follower mostly fulfills (as best she can) the leader's suggestions.[3] Hence the leader is responsible for his own movements, for guiding precisely and unambiguously his partner's, and for the circulation of the couple among the

---

[1] Or sea lions. See [2].

[2] Astor Piazzolla's *Contratiempo* played by Pichuco (Anibal Troilo)'s orchestra, where the former began as a bandeonist. Many DJs shy away from Piazzolla in fear that the difficulty of his music may discourage dancers; conversely, many dancers begrudge this ban and ask for more challenging and delightful music! A frequent compromise is to have a couple of more difficult tangos (Piazzolla, Pugliese) but not before 2 a.m.

[3] We discuss common practice. Of course, experts can develop more freedom and originality.

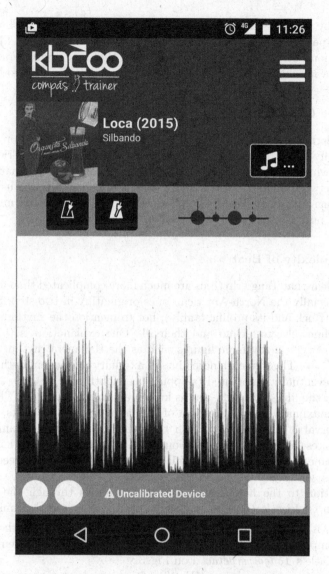

Fig. 1. Screen copy of *Compass Trainer* app

other dancers (walking anticlockwise round a circle in the ball hall and developing more complicated moves around the *rueda* when traffic allows). Best dancing follows the music closely – what this means precisely and how it is achieved are central topics in this paper. This involves stepping on the beats, or only on the strong beats, or sometimes on half-beats (contratiempo); and taking into account pauses/rests. Choosing the placement of steps presupposes perception of the different instrumental lines (which may be staccato or legato, sometimes superimposed), and mostly discerning the *Compás*, the pulsation (which may be hidden by violent syncopation or other rhythmic accidents). The following overview is perforce cursory, and counter-examples could easily be found to all of its assertions; still it purports to picture with sufficient accuracy most of the music that a *milonguero* will encounter on a typical *Milonga* (Tango ball).

2.2 Three Genres

This bewildering avalanche of difficulties is concentrated in Tango proper: there are essentially two other genres in Tango balls, Milonga and Vals Criollo (a.k.a. 'Peruano') whose beats are much more regular. The Milonga is binary (in 2/4) and fast (between 92 and 120 beats/m), Vals is of course ternary, usually with a clear strong first beat. We did test our work on some vals and milongas for comparison purposes but mostly focused on Tango.

2.3 Complexity of Beat

It is quite clear that Tango rhythms are much more complicated than most dance musics, especially the North-American ones originating in two-step, Swing and its cousins (Rock and its prolific family); not to mention the rhythmically poor Dance, Techno, electro-techno and their ilk. One explanation is the frequent occurrence of typical rhythmic figures such as the Habanera rhythm (or 'tango congo')[4], ♩♪♩. The sixteenth note hints at a ternary component in this pattern, which was eventually pinpointed (skipping the eighth note on beat 2) in many tangos from the fifties onward[5], and is known as *tresillo*, cf. Fig. 2. The latter is actually a maximally regular division of the period 8 in three parts, generated by the interval 3 in accordance with the general theory of Maximally Even Sets.[6] The presence of a ternary component is apparent on the Fourier spectrum of the Habanera rhythm, which is fairly evenly distributed between different periodicities; and even more so on the tresillo.

In addition to the habanera cell, already famous through the eponymic *Habanera* in G. Bizet's *Carmen*, another typical syncopated rhythmic motif in early tango is ♫♫♩♩ which is actually used as a marker of the genre by classical composers: see Debussy's *Une soirée dans Grenade*, Albeniz's *Tango Espagnol*, Satie's *Tango perpétuel*... on Fig. 3.[7]

It is illuminating to compare (Fig. 2) with a two-beat rock music, with its definitely binary character, and at the other end of the spectrum, with the clave,

---

[4] Spanish in origin, the Habanera is but one of the numerous ancestors of Tango music.

[5] A most famous example of the superposition of binary and ternary is Piazzolla's *Libertango*.

[6] Because $3 \times 3 = 1 \mod 8$. With an eye on the following discussion, a reference is [1]. Note that the complement set, the cinquillo 'filling the background of' tresillo, was actually used as a ternary pattern by Astor Piazzolla – just as a complement form of the Clave rhythm sometimes occurring in Cuban music.

[7] We find triolets in tangos composed by classical musicians, more than in popular Tango, cf. Ravel's *Pièce en forme de Habanera* for instance.

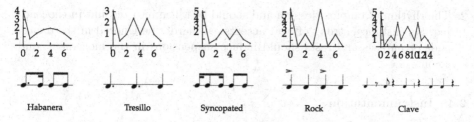

Fig. 3. A typical early-tango rhythmic cell

another complex rhythm occurring in salsa and other latin american dances. The latter is even more complex than the Habanera; however, a Salsa dancer will, just like a Rock dancer, step on the binary pulsation only: once the first beat of the ever-repeating Salsa rhythm is identified, both dancers can carry on iterating their (rhythmically speaking) simplistic four-step sequence.

Historically speaking, the density of off-beat accents (syncopes) increased with time, from none at all in the primitive *Canyengue* style (end of XIXth century), to twice a beat in late Pugliese or Piazzolla music and Tango Nuevo (1950–1980); recently the pendulum swung back to few syncopes in Electro-Tango which features more repetitive and binary rhythms.

Another typical trait is the *Arrastre*, a precipitous acceleration of a sequence of notes. It can be understood as a specific form of rubato, usually played just before a strong beat. It evolved in the extreme form of the Yum-ba with Osvaldo Pugliese's eponymic tango (1946), where the piano slaps a rumble and tumble of indistinguishable notes just before the beat.[8]

Irregularities occur also on a larger time-scale, with variations in the repetitions which usually enrich the rhythm, more or less intense rubato, rests, changes of tempo.

Undoubtedly this wealth of musical variety enhances the interest and pleasure of the listener; for the dancer however, it induces difficulty on two accounts:

1. The *perception* of the *Compás* may be difficult – this is precisely what the software *Compass Trainer* purports to remedy;

---

[8] According to Pugliese, the actual notes had little importance because of the poor quality of the instruments on which he played at the time!

2. The rhythmic complexities can and should be taken into account in the leading. For instance, strong offbeat accents can hardly be ignored in the dance, ends of phrases and rests should induce suspensions in the movement, and so on.

2.4   Instrumentation

The music ensembles used in Tango changed ceaselessly during a century or so. Astor Piazzolla even introduced percussion (including a vibraphone!) and an electric guitar in the 70's, and modern Electrotango (*Otros Aires, Gotan Project*) uses beatboxes. But most balls cling to more traditional, even traditionalist, ensembles, typically using bandoneons, violins, piano, a double bass and sometimes lyrics (tango cancion).

Most instruments can and will play either staccato or legato, the strings pizzicato or arco; sometimes alternatively (so typical of Carlos Di Sarli), sometimes in counterpoint, sometimes both in the same phrase. An expert dancer can even dance following one instrument and guide his partner to dance on another – but such dancers do not need our software.

3   The Problem

We have analyzed some of the difficulties – specific to Tango dancing – that may arise, from an academic viewpoint. From this perspective, there are no compelling reasons to invest time, money and dedication into the research and production of a piece of software remedying these difficulties. In this section, we change tack and recall the vividly painful experience of one of the authors, who was thus moved to launch the *Compass Trainer* adventure. In step with the unexpected changes of rhythm or style in Tango music, we switch from academic style to first-person writing.

A Woeful Though Common Story

My first 18 months as a tango dancer were Hell... I did suffer so much that many times I was about to give up – it took two years for tango to turn into a pleasurable activity with plenty of good music in the arms of multiple ladies. However, it was not easy to diagnose the reasons.
When discussing with experienced tango classmates, I recognized that each had a different way to live as a happy dancer, even if most of them had experienced painful starts.
Even more interesting were the interviews of those 'lost in dancelation'.[9]
I did try to get in touch with all the people giving up tango after a few months or a couple of years. Basically, we are talking about half of the

---

[9] A word forged for expressing the transformation of an ordinary individual into a dancer.

people who did start a tango course. The waste is just enormous, especially on the masculine side of the population.

Most of these men shared a similar story: tango is 'too much' for uneducated ears and feet. Among the men starting tango in the second half of their life, many come without regular practice of sport or music. Learning tango appears as an insuperable mountain to climb, generating more frustration than pleasure.

Why?

Leading in tango, which is overwhelmingly the men's part, combines four major responsibilities: the motion, the partner, the space and the music. We will not sort them by importance, but the four are crucial to becoming a happy tango dancer.

Motion is steps (forward, backward, both sides, pivots, revolving, etc.) combined with torsion between hips and chest (*disociaciòn*).

Partner means that the guider must know continuously where the partner is, his/her orientation (hip and chest), the weight partition between his/her feet, the momentum of its motion, his/her stability.

Space is the few square feet on which the couple can move (or stand). This space can vary from several feet to a few inches, depending on how crowded the dance floor is.

Music is the raw material that both dancers will transform into motion. Initially, the leader creates a motion to be followed by his partner. Happy dancers go way beyond this partition: the leader initiates a motion to be harmoniously exchanged, enriched, and playfully challenged by his partner. Tango then becomes a peer-to-peer, witty, whimsical and artsy dialog.

Going deeper into the analysis, two fundamental skills are found essential for the 'dancelation': good feet and good ears.

Good feet are the basis for the muscular tonus, required for precise motion. [...]

Good ears are the natural way to integrate music. Tango music sustains different layers (rhythm, melody, song, etc.) which are as many propositions for the dancer. The dancer is just another instrument joining the orchestra and playing with it. With experience, one can see by watching a dancer which instrument he/she follows and mimics. The beginner in Tango will usually try to follow the rhythm given by the double bass: a step on every strong beat, a pause when the instrument does not play, etc. Later on, other musical layers will be taken into account by the dancer, but the starting point is always the same: follow the bass! Easier said than done: the double-bass plays soft notes, hard to hear.[10] Most tango beginners, if asked to listen to the bass and follow it, will find the exercise a painful and frustrating challenge.

---

[10] As observed by [3] and other DJs and professionals, much of bass dynamics is lost through (old) recordings and cheap audio systems – not to mention the gut strings used of old for the bass. This is also true of most harmonics (above 5–7 kHz), but this is less of an issue for the perception of rhythm.

Taking all this in consideration, I decided it was high time someone provided these unhappy creatures with simple means to improve their feet and ears, in order to reduce their frustration and enhance their experience and learning of tango. Thinking on it, I came up with specifications for a smartphone application that would highlight the double bass, as a reference time-line to help dancers practice and progress.

A team was then gathered and financed, comprising programmers, dancers, tango teachers, music theorist, covering the required wild field of expertise. We uncovered a wealth of ideas and tracks for development. We had to choose which were to be expanded and which had to wait for the software to reach economic sustainability. What we endeavored to do, and what we eventually managed, is the topic of next section.

4　Compass Trainer: What, Why, How

4.1　Goal and Means

As outlined in the preceding section, the project started with a large spectrum of possible objectives:

– Extraction (in realtime) of relevant rhythmic information.
– Extraction (for post-processing) of musical information: recognition of rhythmic cells, segmentation in phrases and sub-phrases, by timbre, by character, by frequency (treble/medium/bass), by loudness (melody, accompaniment).
– Possibility of selection (filtering) by instrument, tempo, etc.
– Providing several superposable informations: beat-tracks by instruments, strong beats, weaker beats, climaxes, ends of phrases, staccatos, legatos.
– Analysis of finer aspects of the step of an expert tango dancer (changes of repartition of pressure on ball, toes, sides) for calibration. Analysis and recording of the steps of the dancer(s) in real time and correction against the recorded values.
– Establishing a playlist of copyright-free tango musics that could be analysed and used in the app.

The team took several months examining what was actually feasible, or desirable, leading to several reformulations of objectives. Many algorithms for beat-tracking were tested, diverse renderings of the rhythmic information (which had to be suitable for a phone's screen) were tried, most discarded. At this stage, with wider, deeper, and less shaky foundations for the project, we decided to enlist some more advice.

4.2　Consulting Experts

At several stages we fruitfully asked for expert advice maestros, musicians, teachers, and the eventual end-users, ordinary milongueros. Tango musicians (from the band *Silbando*) for instance confirmed the need for enhanced beats for the dancers, explaining how they played rhythm differently in a ball or a concert: less rubato, stronger accents especially in piano and bass. To quote the leader Chloë Pfeiffer, "in a ball we have to hear what we are playing above the din"!

We chose to abandon for the time being the question of feet tonicity and precision (along with precise measurements of position and movement of dancers, which are not yet supported by smartphone technology), but some experiments were made and data saved for future research, with captors on the sole and recordings of maestros dancing with coordinated video, abstraction of the mechanics of dancers (software-reduced to articulated skeletons), the beat track and sole pressure variations on 17 points per foot.

Though this data does not directly reflect on the current state of the app, it is all but anecdotic: for instance, video recording compared with beat tracks showed that *those maestros who were most vocal about the imperious necessity of 'following the double bass'... did not!*

4.3   Abstracting the Relevant Rhythmic Information

As sketched above, we had occasion to explore several techniques for MIR (Music Information Retrieval), constituting an ecosystem of open-source apps dedicated to the annotation of audio files with visual and/or audio markers produced in sync with the music.

The first layer is obtaining data about the beats or onsets of the piece of music. We needed absolute temporal measurements (down to milliseconds precision) for positioning the piece's metric against it, specifying any musical element's position in a bar and relatively to the different beats/sub-beats in the bar.

Among other open-source solutions, we chose Sonic Visualiser (see Fig. 4 below). Developed by Queen Mary's University in London, Sonic Visualiser is dedicated to the facilitation of analysis and visualisation of musical tracks. Units for signal processing are available as plugins in VAMP format, allowing analysis, edition and exportation of processed data.

In the simplest cases (when the rhythmic character is more robust and clearcut, for instance in Gotan Project's *Epoca*), the plugins dedicated to bar and beat tracking in the standard "Queen Mary Plugin Set" provided quite honorable results.

However, when the music exhibits large tempo variations (rubati or slowing down to fermata), or features a singer, most beat detection algorithms are off their rocker. In the first case, we found that some recent algorithms based on pre-trained neural networks [10,11] do a fair job of following the players' *rubati*, essentially extrapolating from the memory of the preceding beats – a fair job, but not perfect. Nonetheless, many traditional tangos, with old but decent recordings (1920–1950), could be processed entirely automatically (for instance Di Sarli's *La Morocha*).

Optimizing the algorithms' efficiency did require some external knowledge and user-parametrization.[11] In summary, the algorithmic detection provides a substantial reduction of the analysis time (bearing in mind that we aimed at building, in time, a library of hundreds of annotated music pieces) but some (qualified) human intervention, annotation and correction is still required, especially when musicians distort the metric a lot.

The second layer was implemented as Ruby scripts [14]. It remedied some defects in the outputs of the first layer, and formatted all the data extracted from the sources and their processing. For instance, all audio was normalized to .wav format. Besides, a data layer describing some formal musical information (e.g., bars and beats) was added, abstracting this information independently of the fluctuations of tempo. Here is a sample of the format retained for encoding a single beat:

```
"abs_beat": 3.0,
"bar": 1.0,
"beat": 3.0,
"bpb": 4.0,
"timestamp": 0.35736961371046183
```

This represents the third beat since the beginning of the piece (abs_beat), it occurs as the third beat of the first bar, the time-signature for the whole piece being four quarter notes per bar (4/4); finally, this beat strikes at 0.357... seconds from the start.

This Ruby scripts library allows to insert other beat tracks, currently either a track processed from the extraction of the double-bass line (mostly by frequency filtering and beat tracking) or user-made claps inserted into the rhythmic model of the piece. Each clap is encoded as a floating number in order to respect the musicians' freedom in departing from the strict meter of the model. It is also possible to quantize them, in order to stick the events exactly to the metric. No less than nine such different beat-tracks appear as vertical colored lines on Fig. 4.

A movie trailer can be viewed and heard at http://www.kbcoo.net/.

4.4   User Interface

In practice we have several rhythmic lines, that we can switch off or on at will. In the public online version 1.0, the app can provide the dancer with a metronomic line, or just the strong beats, or the half-beats, or the double-bass track, or any combination thereof. This fulfills one of our aims, allowing different choices on the same music for practice and pedagogical purposes. The idea of using the phone's buzzer was considered but left aside, at least for the time

---

[11] Letting the computer believe that a waltz has a binary time signature leads to hilarious results. Though automatic detection of a ternary signature has been possible for some time (cf. [9] for instance), we found it simpler to just add the information manually.

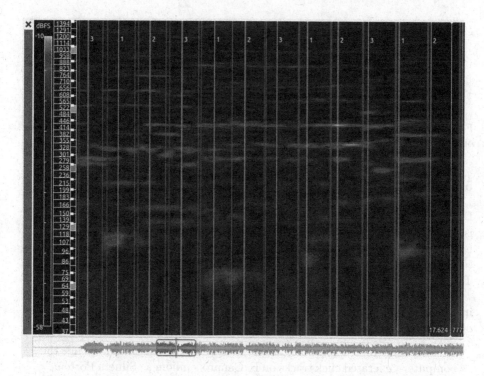

Fig. 4. A passage of d'Arienzo's "El Aeroplano" viewed in Sonic Visualiser

Fig. 5. Full spectrogram for Manuel Jovès's "Loca" (Color figure online)

being. We have preferred to focus on sonic and visual signalization. Indeed, a substantial part of our research was devoted to the visualization of the music, focusing on the perception of rhythm but also the energy variations (both in a physical and aesthetical sense: accents, accelerations, dramatic fermatas...) of a piece. Visual information is absolutely necessary for the dancer to anticipate was is going to happen. We tried numerous prototypes in order to visualize energy. Since there was too much information on a full spectrogram (see Fig. 5, where frequency ranges are already abstracted as superposed color bands), we retained a presentation in frequency bands roughly following the repartition of the instruments into bass, accompaniment and melody, producing a reduced spectrogram with three different frequency range levels coded in three colors, and vertical width proportional to their energy (Fig. 6). This visualization still appears complex at first glance, but one soon gets used to it and appreciates the wealth of useful information that it provides.

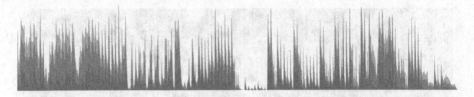

Fig. 6. Reduction of "Loca" to the energies of three spectral bandwidths (Color figure online)

5   Conclusion

Despite all its fascinating incursions into theoretical territory, *Compass Trainer* is a commercial project available for iOS and Android. This entailed some simplifications (mentioned above) of the data proffered to the end-user, and of course to beta-testing in actual Tango classes and showing the software to professional Tango masters.

To sum it up, *Compass Trainer* appeared as a very efficient tool for its intended purpose, when used by an experienced teacher. Quoting the field notes of one of the first tests:

> This is my first experimentation with *Compass Trainer*, using the computer-generated click tracks on F. Canaro's milonga "Silueta Porteña". In this class, there are three different levels of practitioners [including complete beginners in Milonga style].
> After listening once to the music, dancers are invited to walk solo on it; most are unable to follow anything like the strong beats or Compás. The teacher claps her hands to help them understand when to step [or transfer weight on the other foot]. Then *Compass Trainer* is launched:
> (1) First with the Compás and media-tempo beat tracks. All students are now able to walk in Compass, almost instantly.
> (2) Then with the double bass beat-track. With some preliminary work (listening to diverse characteristic Milonga rhythms), students quickly identify the typical patterns and begin to play (solo) with them. The use of this track is immediate and enjoyable.
> Overall, using *Compass Trainer* is a huge bonus: after some time dancing on the augmented musics, every student was able to dance several milongas correctly in phase with the strong beats, without even needing a dry run of the first bars. No need for the teacher to clap the beats anymore, freeing time for individual fine-tuning instead. The only drawback was that the will to test fully the software perturbed the flow of the class – this just has to be thought of ahead of time.

We also probed professional circles – cautiously, since we wanted to establish cooperation, not competition, and had to make clear that home practice with the software could *complete* tango classes, not replace them. So far the maestro's reactions were quite positive.

In this project, we studied the cutting-edge research on beat and meter detection and selected what suited our specific needs, fulfilling our primary goal of providing the listener with a usable rendition of several layers of rhythmic content present in the tango music *as perceived by a dancer*.[12] Interestingly however, we did require at least one state-of-the-art module making use of pre-trained Neural Networks for the detection of the *Compás* proper, mimicking both the musical culture of an educated listener and the predictive hearing as triggered by the first beats of a piece, just as a human milonguero does, gently balancing himself and his partner from side to side for a few bars before embarking on the first step of the dance.

References

1. Amiot, E.: Music Through Fourier Space. CMS. Springer, Switzerland (2016). https://doi.org/10.1007/978-3-319-45581-5
2. Cook, P., Rouse, A., Wilson, M., Reichmuth, C.: A California Sea Lion (Zalophus californianus) can keep the beat: motor entrainment to rhythmic auditory stimuli in a non vocal mimic. J. Comp. Psychol. **127**, 412–427 (2013)
3. Lavocah, M.: Tango Stories: Musical Secrets, 2nd edn. Milonga Press (2014)
4. Palmer, C., Lidji, P., Peretz, I.: Losing the beat: deficits in temporal coordination. Philos. Trans. Roy. Soc. B Biol. Sci. **369**, 1658 (2014)
5. Davies, M., Brossier, P., Fast implementations for perceptual tempo extraction. http://www.music-ir.org/evaluation/mirex-results/articles/tempo/davies.pdf
6. Stark, A.M., Davies, M.E.P, Plumbley, M.D.: Real-time beat-synchronous analysis of musical audio. In: Proceedings of 12th International Conference on Digital Audio Effects (DAFx) (2009)
7. Davies, M.E.P., Plumbley, M.D.: Context-dependent beat tracking of musical audio. IEEE Trans. Audio Speech Lang. Process. **15**(3), 1009–1020 (2007)
8. Davies, M.E.P, Plumbley, M.D.: Beat tracking with a two state model. In: Proceedings of the IEEE International Conference on Acoustics, Speech and Signal Processing, ICASSP 2005, 3 Philadelphia, USA, 19–23 March 2005, pp. 241–244 (2005)
9. Gouyon, F., Dixon, S.: A review of automatic rhythm description systems. Comput. Music J. **29**, 34–54 (2005)
10. https://github.com/CPJKU/madmom
11. Eyben, F., Böck, S., Schuller, B., Graves, A.: Universal onset detection with bidirectional long short-term memory neural networks. In: Proceedings of the 11th International Conference on Music Information Retrieval (ISMIR), August 2010, pp. 589–594 (2010)
12. Böck, S., Schedl, S.: Enhanced beat tracking with context-aware neural networks. http://citeseerx.ist.psu.edu/viewdoc/download?doi=10.1.1.227.9109&rep=rep1&type=pdf
13. Foote, J., Cooper, M.: Visualizing musical structure and rhythm via self-similarity. http://musicweb.ucsd.edu/~sdubnov/CATbox/Reader/FXPAL-PR-01-152.pdf
14. Open-source language. https://www.ruby-lang.org/en/

---

[12] For instance, pattern-recognition of specific rhythm cells will probably be used in later versions. It will help discriminate between quarter and eighth-notes, 2/4 or 4/4 for instance.

Using Inharmonic Strings in Musical Instruments

Kevin Hobby[1]($\boxtimes$), William A. Sethares[2]($\boxtimes$), and Zhenyu Zhang[2]($\boxtimes$)

[1] Synchratron, 645 S. Englewood Ave., Evansville, IN 47714, USA
synchratron@gmail.com
[2] Department of Electrical and Computer Engineering,
University of Wisconsin – Madison, Madison, WI, USA
{sethares,zzhang546}@wisc.edu

Abstract. Uniform strings have a harmonic sound; nonuniform strings have an inharmonic sound. This paper experiments with musical instruments based on nonuniform/inharmonic strings. Given a precise description of the string, its spectrum can be calculated using standard techniques. Dissonance curves are used to motivate specific choices of spectrum. A particular inharmonic string consisting of three segments (two equal unwound segments surrounding a thicker wound portion) is used in the construction of the *hyperpiano*. A second experiment designs a string with overtones that lie on steps of the 10-tone equal tempered scale. The strings are sampled, and digital (software) versions of the instruments are made available along with a call for composers interested in writing for these new instruments.

1 Ideal and Non-ideal Strings

An ideal string vibrates in a periodic fashion and the overtones of the spectrum are located at exact multiples of the fundamental period, as required by the one-dimensional linear wave equation [6]. When the string deviates from uniformity, the overtones depart from the harmonic relationship and the sound becomes inharmonic. There is only one way to be uniform, but there are many ways to be nonuniform; there is only one way to be harmonic, but many ways to be inharmonic.

In a "prepared piano," weights and other objects are placed in contact with the string, giving it a nonuniform density and a sound that can be described as bell-like, metallic, or gong-like. Such preparations tend to lack detailed control. Our recent work [4] explores one possible musical instrument and system, the *hyperpiano*, which is based on a particular inharmonic string consisting of three connected components. The design approach is illustrated in Fig. 1(a) where a nonuniform string is conceptualized as consisting of a sequence of connected segments, each of which is uniform. By carefully controlling the string segments, a large variety of inharmonic effects may be achieved.

The invention of new musical instruments and ways to tune and play them has a long history [2,12] and continues to the present, though modern approaches

O. A. Agustín-Aquino et al. (Eds.): MCM 2017, LNAI 10527, pp. 104–116, 2017.
https://doi.org/10.1007/978-3-319-71827-9_9

Fig. 1. A nonuniform string can be thought of as a sequence of connected uniform strings, shown schematically in (a) with three segments characterized by their mass densities ν_1, ν_2 and ν_3 and lengths. A typical round-wound string (b), as commonly used for guitar and piano strings, has a solid metal core with mass density μ_1. The winding (plus inner core) may be characterized as a uniform string with mass density μ_2.

are often based on digital rather than analog sound production [11]. From an acoustical point of view, the idea of designing an instrument based on an inharmonic sound contrasts with the more common approach of beginning with an inharmonic vibrating element and trying to make it more harmonic.

While there is no conceptual difficulty in imagining a string with an arbitrary density profile, it is not easy to fabricate strings with oddly varying contours. For certain specific densities, and for a small number of segments, it is possible to exploit the structure of commonly manufactured strings. A wound string, as shown schematically in Fig. 1(b), consists of a core metal wire with mass density μ_2 surrounded by a second wire wrapped around the outside (shown with a density of μ_1). Stripping away the winding from a portion of the string effectively creates a segmented string that is readily available from commercial sources, and this is how the strings of the hyperpiano are made.

The completed hyperpiano (see Fig. 2) is discussed in some detail in Sect. 2, and its musical system, based on the hyperoctave, is outlined. An inherent problem for any new musical instrument is to find composers to write for it and performers to play it; we address this issue by making digital simulations (software-based sample playback modules) of the hyperpiano available for download (http://sethares.engr.wisc.edu/papers/hyperInstruments.html) [19].

Fig. 2. The completed hyperpiano and a closeup of the nonuniform strings. The instrument can be seen and heard "in action" in Video Example 1 (http://sethares.engr.wisc.edu/papers/hyperOctave.html). The nonuniform strings of the hyperpiano have inharmonic strings, the first five overtones occur at ratios (1., 1.79, 2.84, 4.02, 5.08).

Given an inharmonic string, one way to characterize the quality of the result-ing musical system is to draw the dissonance curve. This provides a way of locat-ing the intervals that are maximally consonant and hence provides a candidate tuning for the instrument. Indeed, several examples of such systems are shown in [16], but most are based on electronic sound synthesis. The use of inharmonic strings provides a physical analog. Section 3 uses the specification obtained from dissonance curves to design a nonuniform segmented string, and we verify that the overtones match the locations of 10-tet scale steps. In order to expose com-posers and musicians to this new musical system, we present the design, the string, and a software emulation that can be easily downloaded.

2   The Hyperpiano

In [4], using the techniques of [5], we simulated a large number of different inharmonic strings of the form of Fig. 1(a), each with three segments. Visualizing the dissonance curves was useful since it allowed a rapid overview of the behavior of a given design over all possible intervals. Eventually, we chose a particular design in which each string has the (unwound, wound, unwound) lengths of $\ell_1 = 12\%$, $\ell_2 = 9.6\%$, and $\ell_3 = 78.4\%$, with densities $\nu_1 = \nu_3 = 0.00722$ and $\nu_2 = 0.0276$ kg/m. These strings have the spectrum shown in Fig. 2.

The psychoacoustic work of R. Plomp and W.J.M. Levelt [14] provides a basis on which to build a measure of sensory dissonance. In their experiments, Plomp and Levelt asked volunteers to rate the perceived dissonance or rough-ness of pairs of pure sine waves. In general, the dissonance is minimum at unity, increases rapidly to its maximum somewhere near one quarter of the critical bandwidth, and then decreases steadily back towards zero. When considering timbres that are more complex than pure sine waves, dissonance can be calcu-lated by summing up all the dissonances of all the partials, and weighting them according to their relative amplitudes. For harmonic timbres, this leads to curves having local minima (intervals of local maximum consonance) at small integer ratios (as in Fig. 3(a)), which occur near many of the steps of the 12-tone equal tempered scale. Similar curves can be drawn for inharmonic timbres [15], though the points of local consonance are generally unrelated to the steps and intervals of the 12-tone equal tempered scale.

The dissonance curve for the strings of the hyperpiano is shown in Fig. 3(b). This curve mimics the shape of a harmonic dissonance curve, except that it is stretched out over two octaves (instead of one). Thus the corresponding *hyper-octave system* is built on a tuning that has its unit of repetition at the double octave (making it analogous to the Bohlen-Pierce scale [1,8], which has its unit of repetition at the interval 3:1). Figure 3(b) can be thought of as a sensory map of the $\sqrt[12]{4}$ hyperoctave, which labels each hypermajor scale step between the unison (0 cents) and hyperoctave (2400 cents). In descending order of con-sonance, the minima formed by coinciding partials are the perfect hyperfifth, major hypersixth, perfect hyperfourth, major hypersecond, minor hyperthird,

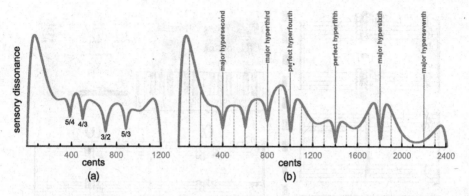

Fig. 3. The dissonance curve [16] is a plot of the summed dissonances of all the sinusoidal overtones. (a) The left curve assumes harmonic sounds with five equal partials. As shown in [14], such a curve has minima at many of the simple integer ratios. The dissonance curve for the first five partials of a hyperoctave nonuniform string are shown in (b).

and major hyperthird. In comparing these tempered intervals to the nonuniform string dissonance curve, the largest deviation is only 2 cents, suggesting an inharmonic analogy to just intonation.

2.1   The Hyperoctave System

Terhardt [18] writes "It may not only be possible but even promising to invent new tonal systems... based on the overtone structure." The hyperoctave scale is based on the overtone structure of a sound, in this case, a string with a specific nonuniform geometry. Like the Bohlen–Pierce scale, it can be played on acoustic instruments, as demonstrated by the hyperpiano. This section delves into the musical possibilities of the system with a focus on its tonal possibilities. More specifically, the goal is to consider the hyperoctave system in terms of Cope's three crucial characteristics: key, consonance and dissonance (or relaxation and tension), and hierarchical relationships [3].

As shown in Fig. 3(b), most of the scale steps fall on or near local minima, indicating that the hyperoctave scale consists mainly of consonances. Unfortunately, this also means that there is not a large degree of contrast between possible consonances and dissonances, and this may limit the ability to adequately implement Cope's second requirement. On the other hand, there may be other ways to obtain harmonic tension and release patterns.

Figure 4 outlines the available notes and the manner of notation for the hyperoctave system. The lowercase v preceding the staff designates it as a hyperstaff. Each string of the hyperpiano was recorded (these raw recordings are also available [19]). A stand-alone sample playback module was written in Max/MSP [10] to make it easier for a composer to explore hyperpiano music via a MIDI keyboard, and the interface is shown in the top right of Fig. 4.

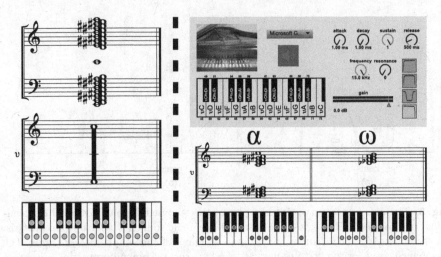

Fig. 4. (Left) A traditional grand staff displays every note in the range of the hyper-piano and the same set of pitches displayed on a grand hyperstaff via the simplified chromatic tone-cluster notation of Henry Cowell. The keyboard located below depicts how these notes are arranged in relation to a MIDI keyboard. (Top right) One of the software sample-playback devices available on the paper's website [19]. This is the Max for Live interface which integrates easily into Ableton Live. (Bottom right) The α and ω sets parallel the two whole tone scales in 12-tet.

2.2 First Compositions: Giovanni Dettori

Italian composer Giovanni Dettori's initial reaction to the system (evidently due to its overabundance of consonant intervals) was to exclaim, "Everything that I play sounds 'motionless' after a few seconds." His observation, in this regard, synchronizes well with Piston's [13] conviction that, "It cannot be too strongly emphasized that the essential quality of dissonance is its sense of movement." But as Dettori persisted in tentative compositional exploration combined with attentive listening, he began to discover ways to add harmonic motion. And upon the completion of his first piece, *Improvviso for Hyperpiano* [19], Dettori wrote:

> I feel that the more I play around with hyperpiano, the more I feel com-
> fortable. Through chromatic writing, I don't have that feeling of 'static
> soundscape' that I had at the beginning. For example, if I insist on a set
> of pitches for a while then moving the same set chromatically (upwards or
> downwards by parallel motion or both by contrary motion) I have a feeling
> of modulation, or at least of pretty strong harmonic change. Also, listening
> to my *Improvviso* piece I don't feel it lacks modulation. This is to say that
> my 'fears' about the limitations of the system were probably more 'cul-
> tural' (related to decades of listening and writing habits) than perceptive.

The preliminary pitch-set technique described above was not used systemati-cally, and it was still in the process of taking shape while Dettori was composing his *Improvviso*. In fact, he later wrote, "I found traces of it after I was sketching the music." The basic technique involves dividing the hyperchromatic scale into two transpositions of whole-hypertone scales, and normalizing one transposition before moving to the other transposition (to add a sense of tension) which then resolves back to the original normalized transposition (to add a sense of release). Although each transposition is composed of traditional augmented chords, due to the novel context and the inharmonic spectrum, Dettori continues, "they don't have the dominant function we usually assign to augmented chords." These two transpositions are depicted in Fig. 4. W.A. Mathieu [9] has referred to the two whole-tone scale transpositions in 12-tet as the α and ω sets, and these names can also be applied to the two whole-hypertone scale transpositions.

Of course, the *Improvviso* is more complex than bouncing back and forth between the α and ω sets; many parallel logics converge that are difficult to describe completely. For instance, tension can also be achieved by breaking an established harmonic rhythm (and so breaking psychological expectations). Dettori found that the pitch-set approach not only helped with inducing harmonic tension–release patterns, it also provided him with a framework of hierarchical order and it helped him to not feel "lost" in the hyperoctave soundscape. To Dettori's ear, the tonic chord consists of a pitch-set plus a specific bass note. For instance, a piece could be written in the key of $\alpha \upsilon C$, $\omega \upsilon C$, $\alpha \upsilon C\sharp$, etc. In this notation, the α or ω designates the pitch-set, while the pitch class designates the bass note. Taken together, these designate the key. Dettori prefers to think of other chords from the tonic pitch-set (i.e., with a different note in the bass) as relative tonic chords. Dettori perceives two dominant chords; they both are based on the pitch-set antithetical to the one on which the tonic is built, but one of them has an ascending leading tone in the bass, while the other one has a descending leading tone in the bass; in this respect, the leading tones are located a hypersemitone (i.e., 200 cents) above or below the tonic bass note. And when a cadence ends on a relative tonic chord, Dettori perceives less conclusiveness than when it ends on the tonic chord.

Table 1 provides a summary with links to the existing repertoire of music that adventurous composers (to whom we are grateful) have chosen to create using the hyperoctave system. Together, these works provide a fairly diverse taste of the possibilities.

2.3 Call for Compositions

"After creating a new scale, how can one quickly find out what it is good for?" asked Mathews, Pierce, and Roberts [7]. "Are there listening tests and laboratory studies that can precede the long slow process of trying to compose significant music with the new scale?" While it is certainly possible to try and "think through" many of the issues that arise with any new scale system, likely the answer to this question is that it takes time, and the ear needs to become accustomed to the new sounds. Our approach is to provide software versions of the hyperpiano (such as in Fig. 4) so that composers can easily play, listen, contemplate, and acclimate. We would like to encourage participation in what we

Table 1. Compositions and improvisations for hyperpiano

G. Dettori	Improvviso for Hyperpiano (http://sethares.engr.wisc.edu/Sounds/hyperOctaveSongs/Improvviso.wav)
	Miniature Variations, for Hyperpiano (http://sethares.engr.wisc.edu/Sounds/hyperOctaveSongs/MiniatureVariations.wav)
P. Eisenhauer	Arroyo (http://sethares.engr.wisc.edu/Sounds/hyperOctaveSongs/Arroyo.wav)
H. Straub	Gon-Tanz (http://sethares.engr.wisc.edu/Sounds/hyperOctaveSongs/Gon-Tanz.wav)
B. Hamilton	Hyperthing (http://sethares.engr.wisc.edu/Sounds/hyperOctaveSongs/Hyperthing.wav)
M. Tristan	Mandala No. 1 (http://sethares.engr.wisc.edu/Sounds/hyperOctaveSongs/MandalaNo1.wav)
	Temple Bell Sketch (http://sethares.engr.wisc.edu/Sounds/hyperOctaveSongs/TempleBellSketch.wav)
	Palimpsest (http://sethares.engr.wisc.edu/Sounds/hyperOctaveSongs/Palimpsest.wav)
	Ayutthaya Rhapsody (http://sethares.engr.wisc.edu/Sounds/hyperOctaveSongs/AyutthayaRhapsody.wav)
	Siamese Cat (http://sethares.engr.wisc.edu/Sounds/hyperOctaveSongs/SiameseCat.wav)
C. Devizia	Blue Rorqual (http://sethares.engr.wisc.edu/Sounds/hyperOctaveSongs/BlueRorqual.wav)
J.-P. Kervinen	Ten Two-Part HyperInventions (http://sethares.engr.wisc.edu/Sounds/hyperOctaveSongs/TwoPartHyperInventions.mp4)
	HyperInvention #2 (Glitch) (http://sethares.engr.wisc.edu/Sounds/hyperOctaveSongs/HyperInventionNo2(Glitch).wav)
	HyperInvention #3 (Variation) (http://sethares.engr.wisc.edu/Sounds/hyperOctaveSongs/HyperInventionNo3(Variation).wav)
	HyperInvention #8 (Variation) (http://sethares.engr.wisc.edu/Sounds/hyperOctaveSongs/HyperInventionNo8(Variation).wav)
S. Weigel	Gold-teased Peppermint (http://sethares.engr.wisc.edu/Sounds/hyperOctaveSongs/GoldTeasedPeppermint.wav)
W.A. Sethares	HyperScarlatti (http://sethares.engr.wisc.edu/Sounds/hyperOctaveSongs/hyperScarlatti.wav)

believe is a rewarding compositional experience. Digital samples of every string of the hyperpiano, and several versions of software playback modules implementing MIDI versions of the hyperpiano can be found at the paper's website (http://sethares.engr.wisc.edu/papers/hyperInstruments.html) [19].

2.4 Extending the Hypersystem

A natural extension of the hyperoctave system is to the $\sqrt[24]{4}$ *quarter-hypertone scale*. Figure 3 indicates the steps of the hyperchromatic scale with the solid red lines, while the extended notes are drawn as dashed green. Since most of the dashed lines occur near maxima of the dissonance curve, together these provide a much higher degree of contrast between consonance and dissonance. Closer examination reveals three unfortunate aspects of this system. First, the hypermajor

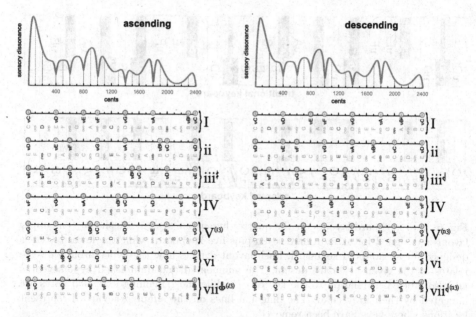

Fig. 5. Two dissonance curves for the first five partials of a hyperoctave nonuniform string are shown. The ascending hypermajor scale contains a raised leading tone, while the descending hypermajor scale contains a lowered leading tone. Below the scales are the various modes that can be derived from each hypermajor scale. Scale steps with bold red borders represent either a root, third, or fifth in relation to the various triads of a hypermajor key. Roman numerals identify the various triad types constructed on the given scales. Pitch classes which are preceded with v are hypernotes, and pitch classes which are not preceded with v are traditional notes. All of the pitch classes are structured in relation to the key of vC. Red notes represent pastel tones (see text). (Color figure online)

scale is built on the hyperchromatic scale which is an enharmonic equivalent of a traditional whole-tone scale, and each interval that the quarter-hypertone scale adds to the initial hyperchromatic scale ultimately forms a traditional whole-tone scale constructed on a different transposition. Therefore, a hypermajor key built on the hypermajor scale cannot exploit the newly generated dissonant intervals because the whole-tone scale can only be transposed two ways. Second, the perfect hyperfourth shares octave equivalence with the major hyperseventh, and thus confuses tone relations within the hypermajor scale. Finally, there is the problem of "the extra-wide leading tone" inherited from the hyperoctave system, which may cause cadences to sound less final.

Fortunately, there is an elegant solution to each of these problems: displacing the major hyperseventh. If the hyperseventh is augmented to 2300 cents when ascending and diminished to 2100 cents when descending, the scale contains every possible quarter-hypertone interval (except the hypertritone) in the hypermajor key! This allows many possibilities for dissonant intervals. It eliminates octave equivalence between the perfect hyperfourth and the major hyperseventh. And, fortuitously, it also addresses the leading tone problem.

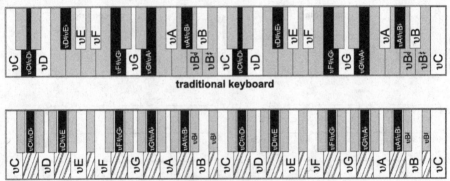

Fig. 6. Two keyboard layouts for the quarter-hypertone system are shown. The upper layout is designed for a traditional seven-plus-five keyboard, while the lower layout is designed for an adapted keyboard. Each layout depicts how the hyperchromatic scale relates to the given keyboard design. The augmented and diminished hypersevenths are also labeled in relation to the key of vC while the remaining extended notes are indicated by their gray color. The diagonal lines on the adapted keyboard represent locations where keys have been removed.

A tonal map of the quarter-hypertone system is outlined in Fig. 5. The red-colored notes (as distinguished from the gray) share octave equivalence with either the root, third, or fifth of a given triad, even though they are not part of the delineated key. For example, the fifth of the tonic triad is vG, a pitch that is traditionally identified as D. But transposing the D down an octave gives $vC\sharp$ which is foreign to the key of vC. We call such notes *pastel tones*; they are unique to hyperoctave music, and they allow for some intriguing embellishments. Pastel tones should be used cautiously, however, because they may confuse tone relations when listened to with an ear trained in octave-based music. On the other hand, they may provide a degree of ambiguity and "color."

To aid in visualization, Fig. 6 provides two quarter-hypertone keyboard layouts that complement the tonal map in Fig. 5. Collectively, the tonal map and the keyboard layouts are intended to supply enough information to propel the novice into tonal quarter-hypertone composition.

A version of the adapted keyboard in Fig. 6 was constructed by modifying a USB MIDI Controller (the M-Audio Keystation 88es). This controller was in production for many years and can generally be purchased second-hand at a reasonable price. The key arrangement on this controller has a reputation among enthusiasts as being easy to modify, and we found this to be true. Figure 7 shows a photograph of the completed controller. The gray color on some of the keys was produced by spraying on coats of primer, gray spray paint, and clear gloss polyurethane. We believe the adapted arrangement provides a more intuitive quarter-hypertone controller than the traditional keyboard arrangement in Fig. 6.

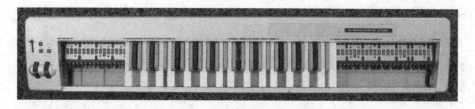

Fig. 7. The adapted keyboard arrangement of Fig. 6 embodied by a modified M-Audio Keystation 88es

3   Inharmonic Strings for 10-Tone Equal Temperament

In the familiar 12-tone equal temperament, the octave is divided into 12 equal-sounding semitones, which are in turn divided into 100 barely perceptible cents. Instead, 10-tet divides the octave into *ten* equal sounding pieces. Yet, from the orchestra to the radio, Western music overwhelmingly favors 12-tet while the existence of 10-tet is comparatively unknown. There may be an underlying reason for this lack—that harmonic tones sound out-of-tune (or dissonant) when played in 10-tet. For instance, the closest 10-tet interval to a musical fifth is 720 cents, as opposed to the 12-tet perfect fifth of 700 cents. The 10-tet fifth is likely to be heard as a sharp, out-of-tune 12-tet fifth. A full major chord is even worse. The problem is not simply that harmonic sounds are dissonant in 10-tet. In tonal music, the motion from consonance to dissonance (and back again) plays an important role. Thus the fact that most intervals in 10-tet are dissonant when using harmonic sounds makes it almost impossible to achieve the kinds of contrasts needed for tonal motion.

Using the ideas of [16], it is straightforward to design spectra for sounds that will appear consonant at the 10-tet intervals. Let $r = \sqrt[10]{2}$ and consider a sound with its first six overtones at $f^* = \{f, r^{10}f, r^{16}f, r^{20}f, r^{23}f, r^{26}f\}$. The "principle of coinciding partials" suggests that such a spectrum should have a dissonance curve with minima at many of the ratios of these partials. All of these ratios are integer powers of r, and hence form intervals that lie on steps of the 10-tet scale. Thus intervals such as the 720-cent "fifth" and the 480-cent "fourth" need not sound dissonant and out-of-tune when played with sounds that have this spectrum (even though they appear very out-of-tune when played with normal harmonic sounds).

While it may be straightforward to create electronic simulations of sounds such as f^* [17], it is less obvious how to create such sounds acoustically. This can be formulated mathematically as an optimization problem by assuming that an inharmonic string consists of n segments, each with length ℓ_i and mass density μ_i. The spectrum of that string will have overtones at $f(\ell, \mu)$ where $\ell = \{\ell_1, \ell_2, \ldots, \ell_n\}$ and $\mu = \{\mu_1, \mu_2, \ldots, \mu_n\}$. Then the goal is to minimize

$$J = \min_{\ell, \mu} ||f^* - f(\ell, \mu)|| \qquad (1)$$

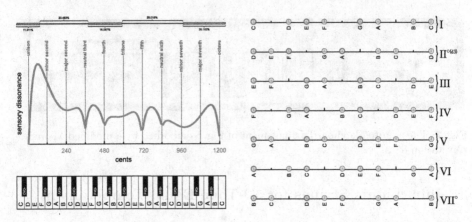

Fig. 8. The 10-tet string, its dissonance curve, a possible keyboard layout, and an overview of possible tonal functions in the 10-tet system.

in some appropriate norm. The optimization problem (1) can be solved using a gradient descent method

$$\mu(j+1) = \mu(j) - \alpha_\mu \cdot \frac{\partial J}{\partial \mu(j)} \qquad (2)$$

$$\ell(j+1) = \ell(j) - \alpha_\ell \cdot \frac{\partial J}{\partial \ell(j)}$$

where $\mu(j)$ and $\ell(j)$ are the values of the densities and lengths at iteration j, and where α_μ and α_ℓ are the algorithm stepsizes (which may be different because the units of μ and ℓ are different).

For 10-tet sounds, $f^* = f_0\{1, 2, 3.0314, 4, 4.9246, 6.0629\}$. Using $n = 5$ segments (with the constraint that there are only two different densities, for ease of construction), the optimal solution to (1) is shown in Fig. 8. We built the string (using the same technique of unwinding the wound portions as in [4]), sampled the sound, and calculated the Fourier transform to verify that the desired spectrum was achieved (the sum of absolute errors over all partials was 0.04 percent).

As with the hyperpiano in Fig. 4, we created a software sample playback module in order to enable composers and musicians to explore the instrument. An immediate response from Carlos Devizia called Ants at the Office (http://sethares.engr.wisc.edu/Sounds/hyperOctaveSongs/AntsAtTheOffice.wav) demonstrates one musical possibility. Marcus Tristan is composing an ensemble piece, *Circles of Celestial Light,* using the 10-tet string-based sampler as an "electro-acoustic" layer. Giuseppe Testa has composed two studies using the 10-tet string samples called Moten (http://sethares.engr.wisc.edu/Sounds/hyperOctaveSongs/Moten.wav) and Vinby (http://sethares.engr.wisc.edu/Sounds/hyperOctaveSongs/Vinby.wav). Figure 9 shows a possible cadence performed with this 10-tet software module in order to demonstrate the possibility of tonal music in 10-tet.

I   (IV VII°⁶) I⁶   IV   V   V⁷   I

Fig. 9. A cadence in 10-tet demonstrates how the 10-tet inharmonic string may be capable of supporting tonal structures. This cadence is played using the 10-tet string at the paper's website [19].

4   Conclusions and Acknowledgements

This paper has presented an extended analysis of the hyperoctave system, which is based on an inharmonic (nonuniform) string that forms the basis of the hyperpiano. But there are myriad possible inharmonic strings, and a design for a 10-tet string provides one example. We invite everyone to use our designs, playback modules, software, and strings to investigate inharmonic musical realms such as the hyperoctave and 10-tet. These and other resources may be found at the papers website [19].

The authors would like to thank the composers who have worked with the hyperpiano including M. Tristan, C. Devizia, P. Eisenhauer, B. Hamilton, J.-P. Kervinen, H. Straub, and S. Weigel. We would like to especially thank Giovanni Dettori for his compositions and for sharing his experiences and thoughts over the course of a long thread of emails.

References

1. Bohlen, H.: 13 tonstufen in der duodezeme. Acustica **39**, 76–86 (1978)
2. Caldersmith, G.: Designing a guitar family. Appl. Acoust. **46**(1), 3–17 (1995)
3. Cope, D.: Techniques of the Contemporary Composer, p. 12. Schirmer Books, NY (1997)
4. Hobby, K., Sethares, W.A.: Inharmonic strings and the hyperpiano. Appl. Acoust. **114**, 317–327 (2016). http://sethares.engr.wisc.edu/papers/hyperOctave.html
5. Kalotas, T.M., Lee, A.R.: The transverse modes of a string with variable mass density. Acustica **76**, 20–26 (1992)
6. Kinsler, L.E., Frey, A.R., Coppens, A.B., Sanders, J.V.: Fundamentals of Acoustics, 4th edn, pp. 22–45. Wiley, Hoboken (1999)
7. Mathews, M.V., Pierce, J.R., Reeves, A., Roberts, L.A.: Harmony and new scales. In: Sundberg, J. (ed.) Harmony and Tonality, Royal Swedish Academy of Music, Stockholm (1987)
8. Mathews, M.V., Pierce, J.R., Reeves, A., Roberts, L.A.: Theoretical and experimental explorations of the bohlen-pierce scale. J. Acoust. Soc. Am. **84**, 1214–1222 (1988)
9. Mathieu, W.A.: Harmonic Experience: Tonal Harmony from its Natural Origins to its Modern Expression, p. 487. Rochester, Inner Traditions International (1997)

10. Max/MSP. J. Acoust. Soc. Am. **115**, 2565 (2004). https://cycling74.com/. Accessed 21 Mar 2017
11. Milne, A., Sethares, W.A., Plamondon, J.: Tuning continua and keyboard layouts. J. Math. Music. **2**(1), 1–19 (2008)
12. Partch, H.: Genesis of a Music, pp. 1–544. Da Capo Press, New York (1974)
13. Piston, W.: Harmony, 5th edn. W. W. Norton & Co. Inc., New York (1987)
14. Plomp, R., Levelt, W.J.M.: Tonal consonance and critical bandwidth. J. Acoust. Soc. Am. **38**, 548–560 (1965)
15. Sethares, W.A.: Local consonance and the relationship between timbre and scale. J. Acoust. Soc. Am. **94**(3), 1218–1228 (1993)
16. Sethares, W.A.: Tuning, Timbre, Spectrum, Scale, 2nd edn, pp. 1–426. Springer, Heidelberg (2004). https://doi.org/10.1007/978-1-4471-4177-8
17. Sethares, W.A., Milne, A., Tiedje, S., Prechtl, A., Plamondon, J.: Spectral tools for dynamic tonality and audio morphing. Comput. Music J. **33**(2), 71–84 (2009)
18. Terhardt, E.: Pitch, consonance, and harmony. J. Acoust. Soc. Am. **55**(5), 1061–1069 (1974)
19. Website for "Inharmonic Strings and the Hyperpiano". http://sethares.engr.wisc.edu/papers/hyperInstruments.html. Accessed 21 Mar 2017

Real-Time Compositional Procedures
for Mediated Soloist-Ensemble Interaction:
The Comprovisador

Pedro Louzeiro[1,2,3]($\boxtimes$) (iD)

[1] Universidade de Évora, Évora, Portugal
pedrolouzeiro@gmail.com
[2] Centro de Estudos de Sociologia e Estética Musical, Lisbon, Portugal
[3] Fundação para a Ciência e a Tecnologia, Lisbon, Portugal

Abstract. Comprovisador is a real-time, networked system through which a
conductor/composer mediates the interaction between a solo improviser and an
ensemble of musicians who sight-read an animated score. The system uses
multiple computers – one host and several clients – to perform algorithmic
compositional procedures with the musical material improvised by the soloist
and to coordinate the response of the ensemble. The present paper focuses on
main aspects of the compositional algorithms used, overviewing the concept and
structure of this system as well as describing the main features of its notation
interface. Some of the real-world opportunities for development and testing that
have occurred are also reported.

Keywords: Musical improvisation · Algorithmic composition
Dynamic notation · Network musical performance · Graphical interface

1 Introduction

The development of Comprovisador aspired to join the broad concepts of improvisation and composition in a real-time environment, through machine listening, algorithmic procedures and dynamic notation. The goal was to enable soloist-ensemble
interaction expressed as a coordinated (composed) ensemble response to an improvisation. Efforts were made to allow the listener to perceive the composed material as
being originated in the soloist's improvisation. Additional levels of interactivity were
envisaged through a feedback loop – the soloist's reaction to the ensemble's response –
and through mediation – which consists in the manipulation of algorithmic parameters.
The system was designed to be flexible regarding instrumentation, accepting impro-
visers from different backgrounds and classically trained ensemble members.

1.1 Concept

In broad terms, Comprovisador is able to listen to an improvisation, decoding pitches,
intervals and durations, and to facilitate the creation of different musical responses
through algorithmic compositional procedures. Control of these procedures is

O. A. Agustín-Aquino et al. (Eds.): MCM 2017, LNAI 10527, pp. 117–131, 2017.
https://doi.org/10.1007/978-3-319-71827-9_10

performed in real-time, from a hardware terminal. The outcome of such procedures is displayed to players in the form of an animated staff-like score, viewed in a computer screen. Through wireless connectivity, it is possible to place the computers – and, therefore, the musicians – apart from each other, allowing non-standard settings.

1.2   Background

This system can be framed in four different areas of computer use in musical creation: (1) computer-assisted composition, (2) improvised music with human-machine interaction, (3) dynamic musical notation and (4) networked music performance.

Computer-assisted composition (CAC) was born with the creation of the work "Illiac Suite", by Lejaren Hiller, in 1957 [6]. It consists of a compositional practice that uses algorithmic procedures performed by a computer, typically in deferred-time.

As an example of a human-machine interaction system, we can point out a project entitled "OMax", carried out by the research team "Musical Representations" of IRCAM [4]. "OMax" consists of a computer program capable of learning, in real-time, the typical characteristics of a musician's improvisational style, as well as to play with him, in an interactive way. The main difference between Comprovisador and most systems designed for improvised interactive music performance lies in the type of output: in most of these systems, the computer interacts directly with the musician, outputting electroacoustic sounds either synthesized or sampled; in the case of Comprovisador, the computer coordinates the musical response by an ensemble of musicians who sight-read a generated score.

Since the late 90's, dynamic musical notation has been increasingly used in algorithmic real-time music systems enabling various kinds of new interactive features such as audience participation, allowing it to influence the behavior of the algorithms [8]. It has been increasingly used as a result of recent technological developments which facilitate its implementation, such as tablets, laptop computers and video projectors. Also, regarding software, advancements have been made which allow the use of staff-like notation in real-time applications. Among these we find MaxScore [10], INScore [7] and the library used in Comprovisador: Bach [2]. Still, many approaches to dynamic notation tend to use animated graphic scores [8, 12, 15] and other kinds of non-staff notation – for example, Jason Freeman used colored LED light tubes to convey pitch and dynamics information to performers, in his work "Glimmer" [8]. Such approaches have a visual level that can be in itself an aesthetic goal, since it is common to have the animated notation projected for audiences to see. Also, these approaches rely on the improvisational skills of all performers to make their own interpretation of the score whereas Comprovisador – apart from the soloist (or soloists) – requires more traditional sight-reading skills from the ensemble performers, since it's based on staff-like notation. Nonetheless, in certain situations, ensemble members may be required to improvise through the use of textual instructions. This concertino-ripieno kind of function separation – one of the key aspects of Comprovisador – seems to be uncommon among other real-time notation work.

Networked music performance is another practice that has emerged in recent decades thanks to development of computer network technologies and the creativity of musicians [11]. It consists on performance situations where a group of musicians interact over a local or wide area network (LAN or WAN). This interaction can be achieved, for example, by audio streaming, score rendering or strategies for graphical direction. In our case, audio streaming is not presently used (only LAN performances thus far). Among systems that use graphical direction strategies, we find Decibel ScorePlayer, Quintet.net and MaxScore [13]. In ScorePlayer, the main strategy consists on scrolling the score from right to left under a fixed vertical line, while the other two systems feature a fixed score and a cursor which moves horizontally. Both strategies were adopted by Comprovisador but the former was abandoned at an early stage. Instead, a new strategy was developed in which a bouncing ball is responsible for synchronizing attacks and/or conveying a pulse (see Sect. 4). Programming of the bouncing ball incorporates motion laws that convincingly translate arsis and thesis sensations.

On an aesthetic level, besides the four areas mentioned above, the development of Comprovisador has drawn inspiration from gestural languages for real-time composition and conducted group improvisation such as Walter Thompson's "Soundpainting" [17] and Lawrence Morris's "Conduction" [16]. In fact, the author's personal experience as a performer in this field has motivated the conceptualization of some of the system's features.

2 Performances

Since 2015, Comprovisador has been used in public performances in five different occasions. Each performance has been preceded by development stages and short periods of rehearsal. In "Comprovisação nº 5"[1], which took place in the foyer of the Lisbon College of Music (ESML), in January 2017, the rehearsal stage spanned over a four-month period of weekly rehearsals with an ensemble consisting of 12 ESML students (flute, oboe, alto saxophone, tenor trombone, bass trombone, tuba, electric bass, marimba, piano, two singers – soprano and mezzo – and violin) and served as a test field for ongoing development.

Hence, every public performance has served as a developmental milestone. For example, in "Comprovisação nº 4" (see Fig. 1), harmonic as well as multi-percussion specific notation were introduced, while "Comprovisação nº 5" featured singers (with real-time generated lyrics), multiple soloists and a non-standard setting (with musicians playing over 50 m apart from each other), among many system improvements.

---

[1] The performance video of "Comprovisação nº 5" is accessible through the following URL: https://youtu.be/rXNTrNzN5z0.

Fig. 1. Performance of "Comprovisação nº 4", during the SMC2016, Hamburg, 01/09/2016; João Barradas – MIDI accordion | [author] – Interface | Radar Ensemble.

3 System Structure

3.1 Hardware

To be fully operational, Comprovisador needs the following hardware equipment (see Fig. 2 (left)):

① a [number of] microphone(s) – to capture the improvisation of the soloist(s) (only necessary for non-MIDI instruments);

② an audio/MIDI interface – to convert the analog signal of the microphone(s) into digital signal and/or to input raw MIDI data;

③ a host computer – which receives the digital audio signal and/or MIDI data and where algorithmic procedures take place;

④ a control surface[2] – in which the algorithm parameters are manipulated;

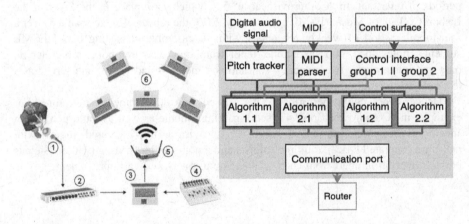

Fig. 2. Comprovisador: hardware setup (left); host application overview (right).

---

[2] Currently, the system is optimized to operate with the Novation Launch Control XL.

⑤ a wireless router – which establishes communication between computers; and

⑥ a number of client computers – to render and display the animated score to the musicians in the ensemble.

Typically, one client computer is used for every two performers. In some cases, though, it is convenient to use one computer for each performer: for instance, with large and/or multi-staff instruments like the piano. For intonation purposes, there is a feature for singers that enables them cue sounds through a set of earphones. This feature also requires one computer for every singer.

The system is fully reconfigurable regarding instrumentation of the ensemble, in regards to number, transposition and range, as will be seen in Sect. 3.2. According to our testing during rehearsals, it is compatible with Mac OSX and Windows systems. Intel Core i5 processors (or better) are recommended.

3.2 Software

Software for this system is being developed in Max 7 [14], with extensive use of Bach library [2] for its notation features, CAC tools and Max integration. The system consists of two applications: one which runs on the host computer and another which is instantiated on each of the client computers.

The host application is responsible for receiving and analyzing the input from the soloist(s), calculating the compositional procedures and responding to commands from the conductor/composer. The client application is in charge of rendering the generated score and displaying it to the musicians.

The host application consists of multiple modules (see Fig. 2 (right)), namely:

- **pitch tracker** – here, the musical notes played by the soloist are deciphered in real-time from the digital audio signal input; the object sigmund~ [3], designed by Miller Puckette, is at the heart of this module;
- **MIDI parser** – in the case of MIDI enabled instruments, a MIDI parsing module is used instead of the pitch tracker; polyphonic and multi-channel input is accepted;
- **control interface** – this module consists of two control groups containing a total of four slots for algorithms; algorithmic parameters are manipulated in real-time by the performance conductor/composer; the control interface provides graphical feedback for all commands performed on the external control surface (mirroring) and it is possible to store and recall parameter presets; it also provides information to its operator about ongoing algorithmic procedures (see Fig. 3); moreover, it features an instant message system (not visible in Fig. 3) through which messages (both predefined and written on-the-fly) can be sent to the players;
- **compositional algorithms** – there are two distinct algorithms – Harmony and Contour – instantiated in all four slots of the control groups (each slot can host any of the two algorithms); instruments can be assigned to any of these four instances, which work in parallel; each algorithm generates different musical responses (broadly, chords and melodic contours) when receiving pitch data from the pitch tracker or the MIDI parser and parametric data from the control interface; furthermore, each algorithm has two main variations; the generated musical responses

take into account idiomatic aspects of the assigned instruments such as range (and whether it is dependent on dynamics), polyphonic capabilities, etc.;

- **communication port** – generated musical data are sent via UDP or TCP protocols to client computers; data are rendered into musical notation in every client application – notation interface;

Many aspects of the host application – communication port as well as parts of algorithms and control interface – are automatically configured on startup. This is done by means of a script that looks up an instrumentation list in crossed-reference to an instrumentation dictionary, both stored in text files. The former consists of a simple list of the instruments to be used in a session while the latter contains a large set of information specific to each instrument (family, range, transposition, clef, dynamic range mapping, strings tuning, initial IP port number, etc.).

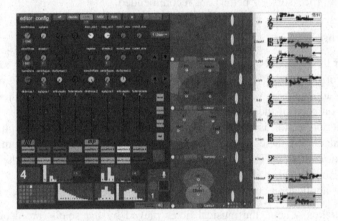

Fig. 3. Comprovisador.host: control interface.

4   Notation Interface

4.1   Overview

The notation interface was conceived in order to have one client computer for every two instruments, regardless of range or transposition of the instruments used. In some cases, it is preferable to use a single instrument per computer configuration.

Graphical objects of the interface adjust perfectly to every modern laptop computer screen, independently of the configuration used. This is achieved using JavaScript inside Max 7 to instantiate and position all graphical objects.

Comparing layouts of the two different configurations (see Fig. 4), we see that there are some advantages in single instrument layout. On one hand, multi-staves can be used, whereas on the other, both dynamics (under the staff) and direction (over the staff) bars can assume larger dimensions, ensuring a faster information detectability and better legibility, these being good principles of graphical interface design [1]. This space optimization was motivated by musicians' suggestions and was found to have a positive impact in performance.

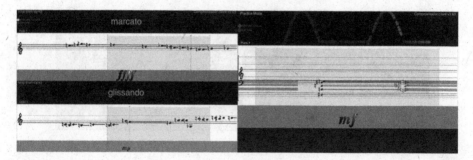

Fig. 4. Comprovisador.client: dual-instrument layout (left) vs. single-instrument layout (right). (Color figure online)

The notation objects consist on a combination of bach.roll and bach.score objects. While the former renders proportional durations, the latter renders standard rhythmic notation [2].

The dynamics bar is a colored bar over which the dynamics text is displayed at its center. Again, regarding good principles of graphical interface design, both background color and text size (3D space) change accordingly to the level of dynamics, in a reactive fashion. The color that symbolizes *pppp* is cyan and the one attributed to *ffff* is red. Any level in between will assume a proportional mixture of the two colors, maintaining the same perceived level of brightness.

Fig. 5. Comprovisador.client: dynamics bar – text size (3D space) and background color. (Color figure online)

Concerning text, whenever the level is being changed, the words *cresc.* or *dim.* appear and move forward or backward in a three-dimensional space (see Fig. 5). This feature is achieved by the use of OpenGL graphics rendering, using the Jitter object jit.gl.text3d [14].

These reactive features were highly valued by musicians who reported to being able to easily identify the dynamic level while keeping full focus on the musical notes.

Regarding the direction bar, it also features OpenGL graphics in order to render at a high frame rate (around 60 fps) a small bouncing ball which allows musicians to synchronize attacks and play in a given tempo.

Using the same OpenGL context, musical direction terms and other information are displayed. To ensure detectability, each time a new entry is displayed, it pulsates in bright white.

As demonstrated in Fig. 6, the motion described by the ball derives from a sine function with its output converted to absolute values. This has been tested against other synchronization strategies but musicians had a better response to the bouncing ball approach. Also, the fading trail has proven to have a positive impact in motion

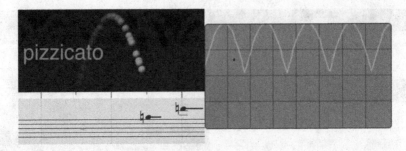

Fig. 6. Comprovisador.client: direction bar (left) vs. "folded" sine wave (right). (Color figure online)

perception. Testers reported it was easier to perceive the bouncing motion even when not looking directly at the object.

4.2 Reading Modes

There are four different reading modes or directives, corresponding to the two variations of each of the algorithms. The modes and their characteristics are:

In Sync with Green Ball – Harmony, Variation 1

- proportional notation;
- notes are written in real-time, from left to right, as they are output by the host;
- when a note's duration line stretches off the play region (fixed darker rectangular area which represents a domain of 5 s), reappears at the beginning of the same play region (see Fig. 4 (right)) – this feature replaces traditional page turning;
- likewise, notes written near the right border of the region are instantly duplicated near its left border (see Fig. 7 (left), notes E and C#);
- a reading time window of about a second and a half in duration is calculated so that the player has time to read and prepare each note on their instrument;

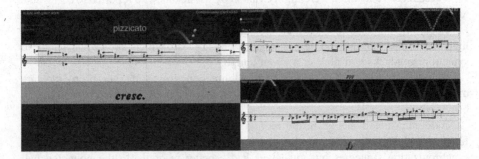

Fig. 7. Comprovisador.client: orange ball (passive gesture) and grid (underlying tempo) (left); Quantum Loop (right). (Color figure online)

- notes that have already been played are erased in order to free staff space for new notes to be written;
- the player should begin each sound precisely when the ball aligns vertically with the note and changes direction (i.e. when it "hits" the note);
- if the note is a long one, the ball will move horizontally over the note's duration line and stop at its end, disappearing, unless a new note is to be played right after it, in which case the ball bounces again, even if the previous note is still active;

In Sync with Green Ball (Grid) – Harmony, Variation 2

- the same as explained above except for the fact that there is an underlying metronomic tempo for all attacks;
- a grid representing the underlying tempo is shown in the staff (see Figs. 6 (left) and 7 (left));
- during long notes or rests, instead of moving horizontally or disappearing, the ball continues to bounce in tempo assuming instead an orange color, while the vertical amplitude of its movement is reduced – thus, simulating a conductor's passive gesture of tempo keeping [9] (see Fig. 7 (left)); whenever a response from the player is demanded (active gesture) the ball turns becomes green again and bounces higher;

Loop (Non-Sync) – Contour, Variation 1

- proportional notation;
- a melodic contour appears, all notes at once;
- the player should loop through the notes framed inside the play region which in this case can be dynamically adjusted to any arbitrary portion of the displayed melody (refer to Fig. 4 (left));
- a vertical green line (play line) cycles through the play region so to give the player an idea of the intended playing rate, although to synchronize with the line is not mandatory;
- above all, the player should not attempt to synchronize with their fellow musicians; in fact, there is an intended rate discrepancy in each client's play line in order to help avoid synchronicity between players;

Quantum Loop – Contour, Variation 2

- standard rhythmic notation;
- a quantized melodic contour appears, all notes at once, fitted in two 4/4 measures (see Fig. 7 right);
- the player should play in tempo with the green ball, which in this case aims at the beginning of every beat of a measure (instead of at every note);

- instruments assigned to the same control group (see Sect. 3.2) will always be in the same beat of the same measure and in the same tempo – hence they should play in sync;
- when two instruments are assigned to different control groups, one of three possibilities may occur: (1) both instruments are in sync; (2) both instruments are in the same tempo but in different positions within the loop (which may differ in size) or (3) the two instruments are in different tempi.

Besides graphical rendering of the notation interface, the client application also carries out some algorithmic tasks that could in theory be performed by the host application. Examples of such tasks include the quantization used in the Quantum Loop mode and the transposition necessary to all transposing instruments. The goal of this task decentralization is to unburden the host computer's CPU and to keep the wireless data traffic as lightweight as possible.

5   Algorithms

5.1   Harmony

Generally speaking, Harmony generates chords from the notes played by the soloist in his or her last musical phrase. The notes of the generated chords are automatically distributed to the assigned instruments and are written in their respective notation interfaces, under the reading directive "in sync with green ball" (grid or no grid, depending on the algorithm's variation – see Sect. 4.2).

There are two approaches on how the soloist's notes are selected and then recombined to generate chords. If the **positiveHarm** button is set via the control interface (see Fig. 8), new chords will be generated from notes that were recently **played** by the soloist. On the contrary, if **negativeHarm** is chosen, new chords will be generated from notes that were recently **avoided** by the soloist.

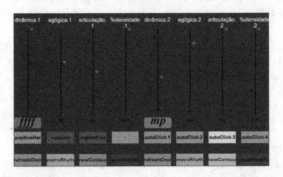

Fig. 8. Comprovisador.host: control interface; detail of the control surface mirror block – faders and buttons.

When building a chord, three different transposition modes may be used: (1) notes may be transposed one or more octaves in order to fit the range of the assigned instrument (button **equiv8a** set), (2) notes will never be transposed, meaning if a note does not fit the range of the assigned instrument, it is filtered (button **registoFixo** set) and (3) the default transposition mode (no button set). The default mode consists in finding the modulus of the soloist's phrase from its extreme notes. Thus, all transpositions obey to the latest found modulus.

Fig. 9. Comprovisador.host: control interface; detail of the control surface mirror block – knobs.

In order to obtain a smooth voice leading between chords, transposition of a note for a given instrument will always take into account the register of the previous note played by the same instrument. To ensure variety, it is possible to perform sudden changes in global register. This can be done manually by flipping the **register** knob (see Fig. 9) or automatically in reaction to the soloist's last played pitches, after a set time threshold (knob **transfmRate**).

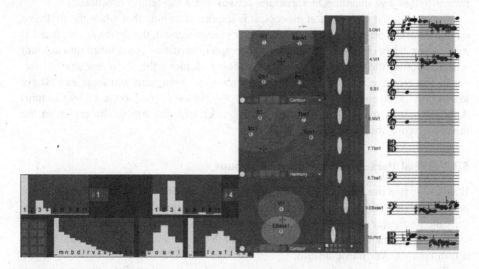

Fig. 10. Comprovisador.host: control interface – detail of the block concerning probability weights for musical durations and lyrics text (left); detail of the instrument assignment block and reference score (right).

Switching between the two variations of Harmony is done by a toggle. When it is set to **threshRhythm**, new chords are automatically triggered at the end of a phrase, after a threshold set with the fader **agogics**. When it is in **metroRhythm** state a metronome is turned on forcing all attacks to adhere to it. Durations are then calculated according to an editable set of probability weights (see Fig. 10 (left)). The rate of the metronome is set with the fader **agogics**, as well.

5.2   Contour

The algorithm Contour captures the last musical phrase of the soloist and, after procedures of truncation, filtration and transposition have taken place, writes the phrase in the notation interfaces of the assigned instruments. The written phrase is to be played in loop (non-sync – see Sect. 4.2) and may undergo transformations, such as contraction or expansion, after a few loop iterations. These transformations are always consequent to whatever the soloist meantime played.

The default modular transposition approach is also used in the Contour algorithm, when generating a new phrase or transforming an existing one.

Variation 2 of this algorithm (triggered by the **quantizeOn** toggle) consists of the quantization of the phrases, which allows musicians to play in sync. If a given phrase is quantized, its original notes remain the same (players can thus focus entirely on the rhythm, as they are familiar with the notes already). Durations are fitted in two 4/4 measures using rhythms of relatively low complexity, derived from sixteenth notes and eighth note triplets with occasional grace notes (see Fig. 7 (right)). Further melodic transformations may continue to occur in this variation after a few loop iterations.

In both variations, the fader **agogics** sets the loop rate (playing speed/ tempo). Furthermore, there is a way of synchronizing tempo between all algorithms (with metroRhythm and quantizeOn variations active) and a tap-tempo function.

In the outlined quantization process, it is important to note that while the rhythmic quantities of the original phrase are to some degree preserved, the rhythmic qualities [5] may end up fairly distorted. This is because tempo information is not taken into account when capturing the original phrase. Rather than a problem, this is an aesthetic choice.

In Fig. 9, there are two knobs worth mentioning: loop_start and loop_end. These knobs provide an easy way to manipulate the play region of the loop, used by Contour. Another way to manipulate it is to directly select with the mouse the region in the reference score shown in Fig. 10 (right).

5.3   Global Parameters and Control Groups

Besides the two main variations of each algorithm, there are several parameters that can be manipulated in real-time which result in diverse musical outcomes. Every controllable parameter can be stored in a preset which can later be recalled by the push of a single button, enabling a wide range of contrast levels in musical transitions along with a firm control over musical form.

Regarding the configuration of our control surface (see Sect. 3.1) it was necessary to come up with a layout that would be both practical and efficient. This layout would have to provide for a balanced and intuitive way to control expressive parameters in any performing context, considering all possible combinations of active algorithms. Since the control surface has 8 well-sized faders (which are ideal to operate gradual transitions between parameter states) and 16 conveniently placed buttons (which are perfect for alternating between states or triggering transformations) (see Fig. 8), the solution for this was to assign half of those controllers to control group 1 and the other half to control group 2. Therefore, faders 1 to 4 (left-hand side) control parameters named dynamics, agogics, articulation and density of group 1 while faders 5 to 8 (right-hand side) control the same named parameters of group 2.

It is important to point out that each group may control two different algorithms at the same time (hence the efficiency of this layout), although it is possible (and maybe even sensible) to activate only one of them at a time in each group. It should also be noted that musical parameters of different algorithms are in fact different procedural parameters and may behave slightly differently, despite having the same name. That being said, **dynamics** will always be dynamics, **articulation** will always control the relative length of notes, and **density** will always control the ratio of assigned instruments that will actually play. **Agogics** fader functionality is explained in Sects. 5.1 and 5.2.

Other graphical objects in the interface which are seen in Fig. 10 (left and right) facilitate the control over instrument assignment, musical durations (for Harmony's metroRhythm variation), and lyrics for singers. Most of these objects (Max objects nodes and multislider [14]) are used to control probability weights.

6 Discussion

A real-time notation system relying mainly on staff notation involves a considerable amount of failure expectation. Sight-reading is a difficult task and errors of pitch and timing are bound to occur. Thus, such a system must consider incorporating this failure factor as part of its aesthetics. For example, Nick Didkovsky's work "Zero Waste", for piano and real-time transcription algorithm, takes advantage of the failure expectation, using the mistakes of the performer and the limitations of the transcription algorithm to allow the music to evolve [8].

In Comprovisador, incorporation of the failure factor is done in several ways. In the algorithm Harmony, a reading time window is always present. In most cases, it allows the player enough time to read and prepare the note in his or her instrument, although this depends highly on many factors, ranging from note rate to type of instrument to performer's experience. The aesthetic potential of the reading time window is expressed when the soloist is able to predict the delay of the response and interact with it.

The first variation of algorithm Contour requires musicians to avoid synchronizing with each other. This is intended as a means to allow a relaxed approach by the sight-reader to the displayed melody, while creating a potentially interesting hetero-phonic texture. When activating the quantized variation, the notes of the original melody remain the same (see Sect. 5.2). This gives the performer the opportunity to focus solely on the new rhythm. From the listener's standpoint, we find it is interesting to perceive the transition from non-sync to synchronized and vice-versa.

The different approaches to note selection and transposition in chord generating (Harmony) have proven useful in creating coherent and contrasting harmonic fields.

In particular, the default modular transposition mode enables a kind of symmetry mixed with unexpectedness which we find appealing. On the other hand, the fixed register mode (no transposition) seems to be ideal to respond to soloists who play harmonic instruments (depending on their style of playing), since it provides a very coherent harmonic field – one could say reverberant or even belonging to the realm of real-time orchestration.

Using both control groups (four slots in total), any combination of the two algorithms may be activated at the same time, assigned to different instrumental groups and performing different responses which complement each other. As a simple example, some instruments may play a soft drone with manipulation in dynamics while others skim across a high-pitched melodic fragment with agogics mediation. This has the potential to create very interesting musical structures, especially if planned in advance with the help of the preset manager for mastering musical form.

Furthermore, the messenger system can be used to quickly and effortlessly send instructions to a group of musicians. For example, with two clicks it is possible to send the following: to all brass > multiphonics. Simple instructions such as these, sent simultaneously to a specified group of instruments, have a very powerful effect: the audience clearly perceives the coordinated response – and the soloist as well, thus, fostering the feedback loop.

The final version of the bouncing ball synchronization mechanism however is yet to be tested in a performance situation. The previous version used MGraphics system (JavaScript) which did not perform at a desirable frame rate. Adding to network latency and sight-reading related failure, this mechanism has not yet proved to be as effective as we have hoped. Testers who have compared both versions reported an improvement which we hope to assess in future performances.

7   Conclusions and Future Plans

We have been developing Comprovisador since 2015 and we are glad to begin to see in it signs of maturity. The performances that were carried out and the rehearsal stages that preceded them – in particular the latest stage which spanned over a four-month period – were crucial for identifying and correcting problems both on the technical and musical side. Every aspect discussed here has been tested and proven to work in a reliable fashion with the hardware available and with college level musicians.

In the short term, we plan to implement a data sequencer module which will make it possible to record all musical data generated during a performance. This will enable us an import tool for analysis. Also, in a performing context, this will allow for the use of recurrence by playing back passages that were recorded during the same performance. In the same line of thought, it will be possible to use precomposed material, based on recordings made in previous sessions. Such material may be practiced beforehand by the musicians, bringing a different aesthetic approach to the original concept, closer to open-form compositions. Nonetheless, the material will still be algorithmically composed from an improvised source and displayed as real-time notation.

References

1. User interface design (2015). https://en.wikipedia.org/wiki/User_interface_design. Accessed 20 Aug 2015
2. Agostini, A., Ghisi, D.: Bach: automated composer's helper (2010). http://www.bachproject. net. Accessed 01 Apr 2017
3. Apel, T.: Max and max for live patches and externals (1995). http://vud.org/max/. Accessed 31 Jan 2017
4. Assayag, G., Bloch, G., Chemillier, M., Lvy, B., Dubnov, S.: Omax (2015). http://repmus. ircam.fr/omax/home. Accessed 04 Apr 2017
5. Bochmann, C.: Para uma formação actualizada. Revista da Associação Portuguesa de Educação Musical, pp. 28–40 (2006). http://hdl.handle.net/10174/2915
6. Dodge, C., Jerse, T.: Computer Music: Synthesis, Composition, and Performance. Schirmer Books, New York (1997)
7. Fober, D., Daudin, C., Letz, S., Orlarey, Y.: Partitions musicales augmentées. Actes des Journées d'Informatique Musicale (2010)
8. Freeman, J.: Extreme sight-reading, mediated expression, and audience participation: real-time music notation in live performance. Comput. Music J. **32**(3), 25–41 (2008)
9. Green, E., Gibson, M.: The expressive gestures. In: The Modern Conductor, 7th edn. Pearson Prentice Hall, Upper Saddle River (2004)
10. Hajdu, G., Didkovsky, N.: MaxScore, common music notation in Max/MSP. In: Proceedings of the International Computer Music Conference (2008)
11. Hajdu, G., Didkovsky, N.: On the evolution of music notation in network music environments. Contemp. Music Rev. **28**(4–5), 395–407 (2009)
12. Hope, C., Vickery, L.: The Decibel Scoreplayer - a digital tool for reading graphic notation. In: International Conference on Technologies for Music Notation and Representation (TENOR 2015). IRCAM, Paris (2015)
13. James, S., Hope, C., Vickery, L., Wyatt, A., Carey, B., Fu, X., Hajdu, G.: Establishing connectivity between the existing networked music notation packages Quintet.net, Decibel ScorePlayer and MaxScore. In: International Conference on Technologies for Music Notation and Representation (TENOR 2017). A Coruña (2017)
14. Puckette, M., Zicarelli, D., Sussman, R., Clayton, J.K., Bernstein, J., Nevile, B., Place, T., Grosse, D., Dudas, R., Jourdan, E., Lee, M., Schabtach, A.: Max 7: Documentation. https:// docs.cycling74.com/max7/. Accessed 31 Jan 2017
15. Smith, R.R.: Animated notation dot com (2014). http://animatednotation.com
16. Stanley, T.T.: Butch Morris and the art of conduction. Ph.D. thesis, University of Maryland (2009)
17. Thompson, W.: Soundpainting: The Art of Live Composition. W. Thompson, New York (2006). https://books.google.pt/books?id=m6BMnQEACAAJ. Workbook 1

Fourier Analysis

Strange Symmetries

Emmanuel Amiot[✉]

LAMPS, Perpignan, France
manu.amiot@free.fr

Abstract. It would seem that the notion of musical inversion is one of the simplest and least mysterious: they are just run-of-the-mill symmetries around axes. However, much depends on the context and even more on the model wherein inversions are used. For instance in neo-Riemannian theory, one talks of the **local** inversion R – turning a triad into its relative –, though its actual effect on pitch-classes depends on which triad R is applied to: the connection with inversions in the circle of pcs is tenuous at best. Other models turn R into a global operation, but at the cost of the essential relation $R^2 = Id$, while still other contexts enable to embed operations on points into the more general operations on (most) pc-sets, in a natural and visual way. This paper purports to synthesize the most important situations and help understand and/or picture what an inversion really is, in its full complexity.

Keywords: Inversion · Local symmetry · Tonnetz · T/I · Homometry Spectral units · Torus of phases

1 Inversions on Circle and Tonnetz

Though inversions can be, and are, used on the whole line of pitches, the present paper will focus on pitch-classes modulo octave: pcs are modeled as integers modulo 12 and chords, scales, collections of notes as subsets of the cyclic set $\mathbf{Z}_{12}$. Most considerations throughout this paper can be applied to the more general $\mathbf{Z}_n$.

1.1 The Simplest Model

The reader is assumed to be familiar with the I_k operators defined as

$$I_k(x) = k - x \pmod{12} \qquad I_k^2 = I_k \circ I_k = \mathrm{Id}.$$

These operations generate the dihedral group[1] $\mathcal{D}_{12} = $ T/I where the product of two inversions is a transposition (mathematically speaking, a translation):

$$I_k \circ I_\ell(x) = k - (\ell - x) = x + (k - \ell) = T_{k-\ell}(x)$$

and the (maximal abelian) subgroup T of all transpositions is normal.

© Springer International Publishing AG 2017
O. A. Agustín-Aquino et al. (Eds.): MCM 2017, LNAI 10527, pp. 135–150, 2017.
https://doi.org/10.1007/978-3-319-71827-9_11

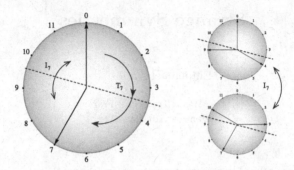

Fig. 1. Action of I_7 on a pitch-class/on a triad

As often, simplest may be best and indeed inversions are easily visualized on the cyclic (or Kremer) model of pcs.

On Fig. 1 one can see the inversion I_7 and its axis[2]; of course, inverting a whole collection of notes – here a triad – involves *two* pictures or some more complicated convention. This is hardly seen as a nuisance, since it appears to be unavoidable. However, other models like the dual Tonnetz below (or orbifolds, see [15] and Sect. 4) feature whole collections (pc-sets) as single points, suggesting alternative solutions.

1.2   The Tonnetz, Its Dual, Its Group

Of all models purporting to study relationships between *similar* chords/pc-sets, one of the most popular is the neo-Riemannian Tonnetz. On the original Tonnetz Fig. 2, points are pcs aligned in fifth and third order. Hence the triangles are all of major and minor triads. A symmetry around a side of one such triangle exchanges it for one of its three neighbors; these symmetries are the three fundamental neo-Riemannian inversions L, R, P where P, for instance, switches X minor and X major by moving the mediant.

In the context of this paper it is better to look up the *dual* Tonnetz Fig. 3, whose vertices are the triads themselves and the edges are their common pcs.

The Tonnetz has proved its worth as a powerful tool in analysis of actual music, both in describing paths of chords (or tonalities) and in encoding chord transitions. Of course the main drawback is that the PLR operations are, albeit inversions, not *constant* inversions as in the T/I group: for instance R, when considered for C major (or (0 4 7)), is the inversion I_4, turning C maj into A min (or (4 0 9)); alternatively, R in the context of G maj is I_{11}. In general, R is the inversion indexed by the sum of the pcs in the major third of the triad (similar rules apply to L and P): there is a localization operator which enables to identify which inversion on pcs

---

[1] Some authors call it $\mathcal{D}_{24}$, pinpointing its cardinality.

[2] Already the circle modelization induces this side-effect, that a center of symmetry is turned into an axis: equation $I_k(x) = x$ has two solutions not one (if k is odd, the fixed points are half-integers).

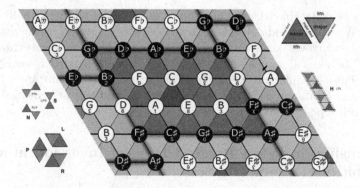

Fig. 2. The Tonnetz, a modern view (Wikipedia image)

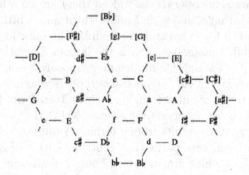

Fig. 3. The Dual Tonnetz

must be used on a given triad.[3] Mathematically, the group T/I cannot be mapped in a natural way (pc-preserving) into the group generated by the PLR operations.[4] However, this is indeed a group and – somewhat surprisingly – the PLR group is isomorphic to T/I, it is a dihedral group.[5]

---

[3] Here R, when applied to $X_{maj} = k + (0, 4, 7)$ or $X_{min} = k + (0, 3, 7)$, is I_{k+4} in the major case, and I_{10+k} in the minor case. Such maps from a local structure – say a manifold – into the linear group of its tangent space – here, its isometries – appear in other domains, theoretical physics of Fields, or pre-sheafs in Category theory. The latter have already been applied to Music Theory in [11], of course.

[4] Here we compare PLR with the left-action of T/I on triads, i.e. the image of a triad is the triad of the images of its elements. See [12] for a study of the right-action, when the set of triads is identified with the images of one triad by T/I in the context of G.I.S.

[5] Essentially because a group acting simply transitively on major/minor triads while *discerning* between both kinds – meaning there is a normal subgroup of transpositions – must be $\mathcal{D}_{12}$, though there are 48 isomorphic versions. Moreover, there are two 'good' ways to define the Tonnetz group, see [7] which pinpoints the extraordinary isomorphism between T/I (acting on pcs) and the PLR group (defined on the Tonnetz) both as subgroups of the 620,448,401,733,239,439,360,000 permutations of all 24 triads!

It may be debatable whether it is the group structure that is actually used in analysis. What is beyond doubt is that we have lost the power to *simultaneously* invert pcs and triads, since inversions have become local, i.e. contextual: what acts as P between C major and C minor, i.e. I_7, also acts as P on F$\sharp$ major or minor but as a different operation on any other triad (for instance, between D minor and B$\flat$ major, I_7 is the L operation). Should this be remedied? How? What of other pc-sets?

2   Focusing on What Really Changes: Inversion in an Orbit of Homometric Sets

One possible line of thought is to realize that inversions are special cases of *interval-content* preserving operations. Indeed there are no other trichords than maj/min triads featuring a major and minor third and a fifth, but tetrachords such as (0 1 4 6) and (0 1 3 7) are known to exhibit the same interval distribution though they have different shapes. This is the famous Z-relation, more properly called homometry [9]. Though for musicians it appeared as a byproduct of the study of the interval content $\mathbf{iv}^6$, for crystallographers who invented the notion it is better defined in terms of diffraction, i.e. Fourier Transform. Indeed diffraction is created by gaps, holes, intervals between objects (the atoms in a crystal), and the diffraction pattern results from a formula summing different sine waves (corresponding to the diverse paths that contribute to adding, or subtracting, light on a given point) which amounts to a Fourier Transform.

2.1   Homometry

Leaving crystals aside for the time being, a workmanlike discussion of homometry for musicians is the following:

1. Two objects (say subsets of $\mathbf{Z}_{12}$) are homometric if they share the same intervalic distribution.
2. The intervalic distribution **IFunc** is a convolution product. If $A \subset \mathbf{Z}_n$ for instance,

$$\mathbf{IFunc}_A(k) = \#\{x \in A \mid x - k \in A\} = \sum_{x \in \mathbf{Z}_n} \mathbf{1}_A(x)\mathbf{1}_A(x - k)$$

$$= \sum_{x \in \mathbf{Z}_n} \mathbf{1}_A(x)\mathbf{1}_{-A}(k - x) = (\mathbf{1}_A * \mathbf{1}_{-A})(k)$$

where $\mathbf{1}_A$ is the *characteristic map*[7] of set A.

---

[6] Or, more generally, of the **IFunc** of two pc-sets. Here we focus on intervals within one pc-set, i.e. **IFunc**$_A$ is essentially the interval vector $\mathbf{iv}_A$ up to definition conventions.

[7] If necessary, one can easily generalize to any distribution on $\mathbf{Z}_n$ – for instance multisets, wherein any pc can appear not only 0 or 1 time, but with any real value.

3. The Fourier transform of a pc-set A is the Fourier transform of $\mathbf{1}_A$, namely the map defined by the following values (called Fourier coefficients)

$$\mathcal{F}_A(t) = \widehat{\mathbf{1}_A}(t) = \sum_{k \in \mathbf{Z}_n} \mathbf{1}_A(k) e^{-2ikt\pi/n} = \sum_{k \in A} e^{-2ikt\pi/12}$$

in the simple case that we are studying.

4. Fourier transform turns convolution product $*$ into termwise product[8] $\times$, with the following consequence of note:

$$\widehat{\mathbf{IFunc}_A} = |\widehat{\mathbf{1}_A}|^2.$$

This value is essentially the light observed at a given point on a diffraction figure, vindicating the crystallographic interpretation. It is also the usual, modern definition of homometry:

Two objects are homometric if and only if their Fourier coefficients have the same magnitude.

5. In particular, two isometric objects are homometric, among them a pc-set and its inversions. The reciprocal is false, but the question of finding all discrete *non trivial* homometric orbits is still a formidable open problem, cf. [9]. We circumvent it by the subterfuge of introducing a continuous context: instead of characteristic functions of sets, taking values 0 or 1, we allow general *distributions*, i.e. maps from $\mathbf{Z}_n$ to $\mathbf{R}$ or even $\mathbf{C}$.

2.2 Transformations Between Homometric Distributions

In a way, it is obvious to find all transformations that permutate all pc-sets homometric to a given one: select those permutations of (the subsets of) $\mathbf{Z}_{12}$ which work, and only those! They form together a subgroup of all permutations, which can be found by appropriate software and described by relations. This is useful for compositional applications if one has an eye on a particular class of homometric objects, see [8].

On the other hand, this *a posteriori* approach does not allow to predict or understand the size and structure of the group involved; for instance, for $A = (0\ 1\ 4\ 6)$ in $\mathbf{Z}_{12}$, a group with 48 elements is found, which acts simply transitively on the 48 homometric tetrachords. It is better to look at $\mathbf{IFunc}_A$ and observe that A is an all-interval set, each interval occurring once and only once (except prime and tritone for obvious reasons); this entails that any affine transform of $\mathbf{Z}_{12}$, i.e. any map

$$x \mapsto ax + b \quad a \in \{1, 5, 7, 11\} \quad b \in \mathbf{Z}_{12},$$

---

[8] Meaning $\widehat{f * g} = \widehat{f} \times \widehat{g}$, i.e. the Fourier coefficients of convolution product $f * g$ are obtained by multiplying the corresponding coefficients for f and g. This wonderful feature (noticeably simplifying computation of any convolution-related operation) is essentially a characterization of discrete Fourier transform, cf. [1], Theorem 1.11.

since it permutates the intervalic distribution in general, will preserve it in this case; and hence these 48 operations coincide with the affine group modulo 12. This case is also gratifying in that the group operates simply transitively, a nice case of a non abelian Generalized Interval System.[9]

However, there is no universal way (in this context of permutations) to find the 'good' group acting on the orbit of all pc-sets homometric to a given one, according to

Theorem 1 (Mandereau 2011). *For $n = 8$ or $n \geq 10$, there is no subgroup G of S_n (acting both on points and on subsets of $\mathbf{Z}_n$) such that for any A subset of $\mathbf{Z}_n$, the orbit $G.A$ is equal to all pc-sets homometric with A.*

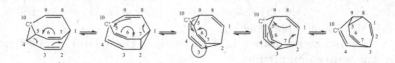

Fig. 4. Quantic transitions in the Bullvalene molecule

Nonetheless, in many situations we want to understand what the transformation exactly is, the 'magical gesture' that turns A into one of its homometric counterparts. The present paper was actually triggered by a Chemistry article [4] where the 1,209,600 states of the Bullvalene molecule *and their transitions by tunnel effect* (see Fig. 4) are described with a matricial formalism, very similar to the one we will presently introduce.[10] This formalism was originally used in [3] for the purpose of algebraically combining multisets (or scales, chords, rhythms) and is thoroughly developed in [1] where proofs and details can be found.

Definition 1. *The matrix S associated with a distribution $s : \mathbf{Z}_n \to \mathbf{C}$ is the circulating matrix whose first column is $\left(s(0), s(1) \ldots s(n-1)\right)^T$. In the usual case of a pc-set $A \subset \mathbf{Z}_n$, with distribution $\mathbf{1}_A$, we denote the associated matrix as $\mathcal{A}$.*

For a compact example, here is the matrix associated with C major triad (distribution $(1\ 0\ 1\ 0\ 1\ 0\ 0)$) in a 7-note universe:

$$\begin{pmatrix} 1 & 0 & 0 & 1 & 0 & 1 & 0 \\ 0 & 1 & 0 & 0 & 1 & 0 & 1 \\ 1 & 0 & 1 & 0 & 0 & 1 & 0 \\ 0 & 1 & 0 & 1 & 0 & 0 & 1 \\ 1 & 0 & 1 & 0 & 1 & 0 & 0 \\ 0 & 1 & 0 & 1 & 0 & 1 & 0 \\ 0 & 0 & 1 & 0 & 1 & 0 & 1 \end{pmatrix}.$$

The important feature is the obvious shift from one column to the next. Mathematically speaking, the strong point of this model is that matrix multiplication has a familiar meaning:

---

[9] Notice that, in general, a chord and its inverse constitute the simplest GIS, with the dihedral group $\mathcal{D}_1$ made up of the inversion and identity.

[10] The different states are modeled as symmetrical real matrixes with same spectra, hence transitions between them are achieved by way of unitary matrixes, just like the spectral units defined below.

Proposition 1. *The matrix associated with the convolution product of two distributions $s * t$ is $\mathcal{S} \times \mathcal{T}$.*

For instance it is easily checked that the matrix for **IFunc**$_A$, the distribution of intervals, is simply $\mathcal{A}^T \times \mathcal{A}$. Moreover, since all these circulating matrixes are polynomials in the matrix $\mathcal{J}$ associated with $j = (0, 1, 0 \ldots 0)$, they are all diagonalisable in the same linear basis $(\omega_0 \ldots \omega_{n-1})$, where $\omega_k = (1, e^{2ik\pi/n}, \ldots e^{2ikm\pi/n}, \ldots)^T$ is eigenvector for $\mathcal{J}$ with eigenvalue $e^{-2ikm\pi/n}$. The passing matrix Ω whose columns are the ω_k's is usually normalized by $1/\sqrt{n}$, making it unitary. In this basis it is classically checked that

Proposition 2. *Any circulating matrix $\mathcal{S}$ is diagonalisable in $(\omega_0 \ldots \omega_{n-1})$. The eigenvalues of $\mathcal{S}$ are the Fourier coefficients of distribution s.*

It follows

Theorem 2. *A and B are homometric if and only if there exists a unitary circulating matrix $\mathcal{U}$ such that $\mathcal{A} = \mathcal{U} \times \mathcal{B}$.*

This means that $\mathcal{U}$ lies in the same sub-algebra of circulating matrixes, but that its eigenvalues have magnitude 1 (so that A and B have Fourier coefficients with the same magnitude). Equivalently, one can consider the associated distribution u and state $\mathbf{1}_A = u * \mathbf{1}_B$.

Definition 2. *Such a matrix (or distribution) is called a spectral unit. The set* **SU** *of all spectral units is a multiplicative abelian group.*

By diagonalisation, this group is isomorphic with the group of diagonal matrixes whose eigenvalues lie on the unit circle, hence it is topologically equivalent to the torus $\mathbf{T}^n$.

Example 1. *From (0 1 4 6) to (0 1 3 7), the spectral unit is*

$$u = (\frac{1}{4}, \frac{1}{4}, 0, \frac{1}{4}, -\frac{1}{4}, -\frac{1}{2}, \frac{1}{4}, \frac{1}{4}, 0, \frac{1}{4}, -\frac{1}{4}, \frac{1}{2}).$$

The "unit" thing is clearly visible on its eigenvalues, whose magnitude is always 1, namely

$$\widehat{u} = \left(1, e^{\frac{i\pi}{6}}, e^{-\frac{i\pi}{3}}, i, 1, e^{\frac{5i\pi}{6}}, -1, e^{-\frac{5i\pi}{6}}, 1, -i, e^{\frac{i\pi}{3}}, e^{-\frac{i\pi}{6}}\right).$$

This characterization of homometry (with Fourier transform) originates in [14] (1982/84). It embodies the idea of equality of diffraction patterns, the original problem set by cristallographers.

This lengthy exposition reaches a satisfying conclusion:

Proposition 3. *For any distribution s, the homometric distributions are all the $u * s$ [or equivalently their matrixes are the $\mathcal{U} \times \mathcal{S}$] where u describes the group* **SU** *of all spectral units [$\mathcal{U}$ being its associated matrix].*

We have constructed a satisfying group whose orbits correspond exactly to homometry. Moreover, most of the time[11] the group of Spectral Units acts simply transitively on an orbit, i.e. the latter and the group form a G.I.S.

Is this then a violation of Theorem 1? Not at all: the *discrete* homometric subsets (whose distribution is a characteristic function, with values 0 or 1) are undoubtedly somewhere in the *continuous* orbit, but finding them all is no easier than before.[12]

One possible reduction is to the finite subgroup of *rational* spectral units *with finite order*: one can imagine a spectral unit as a set of several clocks, each rotating one eigenvalue by some angle. The unit has finite order if all clocks have a common period. The applet in Fig. 5 features all available values of such spectral units for $n = 12$ applied to any pc-set. The result is not always a genuine set, i.e. it may display truth values different from 0 or 1. The classification and computation of all such spectral units was achieved by the tricky Theorem 2.11 in [1].

However we are getting close to the stumbling block of this nice modelization: in this large, continuous group, some spectral units have infinite order, and notably this is the case for neo-Riemannian inversions!

2.3   Transpositions and Inversions as Spectral Units

The case of musical transpositions is straightforward: in mathematical terms, their group T is mapped to the subgroup generated by the spectral unit $j = (0, 1, 0, 0 \ldots 0)$. In another words, applying T_k (to a pc or a pc-set or any distribution) is equivalent to multiplying the appropriate circulating matrix by $\mathcal{J}^k$.[13]

Now for inversions. The general theory provides a spectral unit (and usually one only) transforming a pc-set into one of its homometric pc-sets.

Example 2. *From C major (0 4 7) to C minor (0 3 7)*[14] *the spectral unit of the inversion is*

$$p = \frac{1}{15}(7, 4, -2, 1, 7, 4, -2, 1, -8, 4, -2, 1),$$

associated with the Parallel operation P.

We begin with unexpectedly good news:

---

[11] Exceptions are pc-sets, or distributions, where one or more Fourier coefficients are nil, i.e. the matrix is singular. These sets are the famous 'Lewins's special cases' whose definition in his seminal paper [10] was so irredeemably obscure. [1], Sect. 2.2.2, shows a way round these singularities when the rank of the matrix is $n - 2$.

[12] Though maybe the strategy of exploring the continuous orbit for discrete solutions warrants further exploration.

[13] Remember that the whole algebra of circulating matrixes is made of polynomials in $\mathcal{J}$. It is deeply satisfying in a sense that in this model every single object or transformation originates in the single transposition by one semitone.

[14] Evoking the first bars of R. Strauss's *Also Sprach Zarathoustra*.

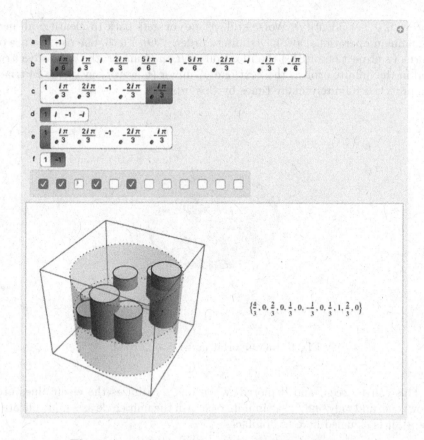

Fig. 5. An applet: transforms of 0135 by spectral units

Proposition 4. *Neo-Riemanniann transformations P, L, R in the group of spectral units are no longer local but* global, *in the sense that for any major triad X, X minor is obtained by applying the same spectral unit p as above (idem for L, R). Moreover, there is a simple, fixed relationship between all three operations:* $p = j^8 * \ell = j^3 * r$ *(meaning that ℓ, r are just circular permutations of the distribution p).*

Proof 1. *This is a corollary of the commutativity of spectral units. Say X major is C major transposed by $k : X = T_k(C)$, i.e. $x = j^k * c$ since transposition T_k corresponds with spectral unit $j^k = (0, 0 \ldots, 1, 0 \ldots)$; but since p (and similarly ℓ, r or indeed any spectral unit) commutes with j, we get*

$$p * j^k * x = j^k * (p * x),$$

meaning that X major is transformed into "C minor transposed by k", i.e. X minor. □

Now the bad news: p, ℓ, r are no longer involutions! In other words, the operation from C major to C minor is not the same as the reverse operation:

$p^2 \neq (1,0,0\ldots)$ [identity]. Worse still, p^n **never** gets back to identity: all neo-Riemannian operations now have infinite order.[15] On Fig. 6, one can picture on a torus (a projection of **SU**) the iteration of R on C major: the only other true triad in the infinite orbit is the next one, A minor (however any of the 24 triads is approached infinitely many times by this orbit).

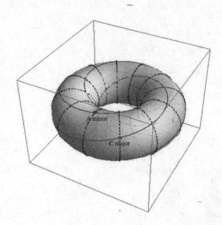

Fig. 6. Infinite orbit under R of a triad

This can be seen, and deplored, when one scrutinizes the eigenvalues of p below: one, and only one[16], has infinite order (in the unit circle as a multiplicative group); it is signaled here in boldface.

$$\widehat{p} = (1, -\frac{\sqrt{3}}{2} + \frac{i}{2}, -\frac{1}{2} + i\frac{\sqrt{3}}{2}, \boxed{\frac{3}{5} - \frac{4i}{5}}, \frac{1}{2} + i\frac{\sqrt{3}}{2}, \frac{\sqrt{3}}{2} + \frac{i}{2}, 1, \ldots)$$

Of course one could have wished for a nicer embedding of T/I (or its dual, the PLR group) in the **SU** group, but this was obviously doomed from the start since the latter is abelian and the former is not. Nonetheless, we do think that this way of interpreting P, L and R, and perhaps even the infinite subgroup[17] that they generate in the Spectral Units, should be kept in mind and has many advantages:

1. First and foremost, these operations are now global.
2. Though one needs distinguish between (say) p (from major to minor) and its inverse p^{-1} (from minor to major), this may well be a blessing in disguise:

---

[15] This stands also for compound operations, like the Slide S (exchanging F minor and E major) insofar as they exchange minor and major triads.

[16] Actually the values are repeated backwards and conjugated so that only the first 6 are featured.

[17] Its topological closure is a subgroup of **SU** (a finite union of torii with smaller dimension), whose orbit when acting on one triad contains all of them.

not all musicologists did appreciate as much as H. Ottinger or B. Riemann the notion that switching from minor to major was the same operation as its reverse.

3. Also there are numerous cases when inversions between pc-sets do have a spectral unit with finite order: for instance, from C D F♯ to B♭ C F♯ the spectral unit has order 6 – it is a square root of a major third transposition. Up to complementation, inversion and transposition this is the complete list[18]:

$$\{0,2\}, \{0,3\}, \{0,4\}, \{0,6\}, \{0,2,4\}, \{0,2,6\}, \{0,3,6\}, \{0,4,6\}, \{0,4,8\},$$
$$\{0,1,2,3\}, \{0,1,2,7\}, \{0,1,3,7\}, \{0,1,4,5\}, \{0,1,5,6\}, \{0,1,5,8\}, \{0,1,6,7\},$$
$$\{0,2,4,6\}, \{0,2,4,8\}, \{0,2,6,8\}, \{0,3,6,9\}, \{0,1,2,4,7\}, \{0,1,2,6,8\}, \{0,1,3,5,6\},$$
$$\{0,1,4,7,8\}, \{0,2,3,4,6\}, \{0,2,4,6,8\}, \{0,2,4,6,9\}, \{0,2,5,7,8\}, \{0,3,5,6,7\},$$
$$\{0,1,2,3,4,5\}, \{0,1,2,3,7,8\}, \{0,1,2,4,5,6\}, \{0,1,2,4,5,8\}, \{0,1,2,6,7,8\}, \{0,1,3,4,7,9\},$$
$$\{0,1,3,5,6,9\}, \{0,1,3,5,8,9\}, \{0,1,3,6,7,9\}, \{0,1,4,5,6,8\}, \{0,1,4,5,8,9\}, \{0,2,3,4,6,9\},$$
$$\{0,2,3,5,6,8\}, \{0,2,3,6,8,9\}, \{0,2,4,5,7,9\}, \{0,2,4,6,8,10\}.$$

Alternatively, we might desire to retain the non-commutative structure of operations and the involutivity of inversions. Nonetheless, we can improve on the T/I model, inasmuch as we can embed all (or almost all) pc-sets in the same space.

3 Remembering the Other Chords: Inversion in the Torus of Phases

3.1 The Space of Fourier Coefficients' Phases

Between homometric objects, we have seen that the magnitude of Fourier coefficients does not change. The significant dimension is the *phase* of these coefficients, i.e. the angle they make (as vectors in the complex plane) with a reference direction. Varying magnitudes between diverse pc-sets has been well studied since 2005 [13], but the study of the space of phases is much more recent [2,16–18]. Indeed, focusing on this angular dimension, which is more about harmony than shape, can also be done for non-homometric pc-sets, even with differing cardinalities. It is a very strong asset (which mysteriously escaped the author of the seminal paper [2]) that this space of phases includes the cyclic model, along with the circle of fifths, the Tonnetz and most previous models.[19] Look up on Fig. 7 the disposition of triads and of fifths, and the chromatic sequence where for instance the chromatic line (dotted, blue) appears broken but in fact is solid, coming back time and again from above after disappearing below.

So far, the most useful space of phases is defined as follows:

---

[18] Unsurprisingly, those pc-sets related to their inverse by an *involutive* spectral unit are those with a symmetry axis, like major sevenths.

[19] Quoting [18]: "... there is a different way of topologically enriching the Tonnetz that preserves the musical insights [...] and leads to a concept of harmonic distance. Such mixing of different-cardinality sets is not possible in voice-leading spaces without forfeiting their basic geometric properties".

Definition 3. *Let us write down the polar coordinates of the Fourier coefficients of a pc-set (or distribution):*

$$a_k = |a_k|\, e^{i\varphi_k}.$$

Then the pair of angles (φ_3, φ_5) lives on the Torus of Phases $T_{3,5}$.[20]

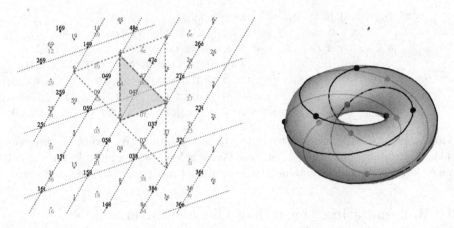

Fig. 7. Coordinates are φ_3, φ_5. Right, same torus in 3D (Color figure online)

It is a torus because both angles are defined modulo 2π: a representation of that space on a plane must be understood as glueing the left and right (and also bottom and upper) sides together, as in Fig. 7 left. A 3D immersion of such a torus is given on Fig. 7 right, with the solid chromatic line winding around it.

The pertinence of this model for musical analyses accrues every day. Still it has two drawbacks:

1. Some pc-sets (tritones, diminished sevenths) do not have coordinates in $T_{3,5}$ and cannot be represented as points here.[21]
2. Some distinct pc-sets share the same couple of coordinates.

This last may be seen, not as a bug, but as a feature: it is reasonable to identify (at least in a bidimensional model) the diatonic C major scale and its pentatonic (CDEGA). This confusion, or ambiguity, is perceptively prevalent in many

---

[20] J. Yust prefers $\varphi = 2\pi\Phi/12$ where Φ, defined modulo 12, is often an integer and is easier to compare with simple values such as those of single pcs.

[21] This is because a nil Fourier coefficient does not have a phase. One possibility is to consider that – for instance – an augmented triad has **all** values of φ_3 at the same time and can thus be represented as a vertical line. This enables modulations passing through such a chord, entering any point of the line and getting out at any other point, recalling the flexibility of these chords in Douthett's chickenwire model. See [2] for an example in Schumann's *Kinderszenen*.

dialects of Rock music for instance (guitar solos are often pentatonic, in a generally diatonic context).

Actually, those two scales share their coordinates with the single pc D, which can be explained by its role as a center of symmetry. In the torus of phases, not only do we have our cake (the diverse pc-sets positioned according to their harmonic relationships), but we can eat it, too (moving pc-sets **and** pcs by exactly the same, simple, geometric operations inherited from T/I).

3.2   The Dihedral Group Acting in $\mathcal{T}_{3,5}$

This was developed mainly by [18], see a synthesis from a mathematical angle in [1], Chap. 6.

Readers familiar with the Tonnetz will readily agree that transpositions in the torus of phases are simply translations (remembering that opposite sides are glued together so that going out to the right means going in from the left). Actually the directions of transpositions by fifths or semitones are marked on Fig. 7. J. Yust first noticed, and proved, that inversions **of any pc-set** (including single pcs) are central symmetries whose centers are the fixed points of the inversion. More precisely,

Proposition 5. *The action of I_k induces a central symmetry on the torus of phases: if pcs or pc-sets A and B are symmetrical around a center c (resp. a dyad (a b)), then their torus projections are symmetrical around the torus image of c (resp. the image of the dyad).*

See for instance on Fig. 7 how the triads (0 4 7) and (0 4 9) are exchanged by symmetry around the dyad (0 4), as are 0 and 4 – or 7 and 9. The actual operation on phase coordinates is

$$\begin{pmatrix} \varphi_3 \\ \varphi_5 \end{pmatrix} \mapsto \begin{pmatrix} \alpha - \varphi_3 \\ \beta - \varphi_5 \end{pmatrix}$$

where α, β are the sums of the angular coordinates φ_3, φ_5 of single pcs 0 and 4 (they depend on the choice of origin).

This model includes in particular the Tonnetz and its dual, and all their shared symmetries. It features a satisfying compromise: T/I acts, and it seen to act, in the same simple way on all pc-sets, regardless of their cardinalities.[22] For instance, the R transformation for C major, which the symmetry around D = 2, can be applied to the whole C major *scale*, and it is quite satisfying to find it invariant under this transformation (A minor natural).

However, some remarks are in order:

1. Because of the circular nature of the torus, caution must be exercised: lynx-eyed readers will have noticed that *another* center of symmetry should work

---

[22] There is an isomorphism between the induced left-action of T/I on subsets of $\mathbf{Z}_n$ and (a subgroup of) the dihedral group of translations/central symmetries on the torus.

in the last example, namely the single pc 8. Indeed it does, and it is the same symmetry: but drawing a line (say) through 8 from 047 to 049 involves crossing the glueing line at the top and bottom of the figure. Yust gives a nice interpretation of this choice of path in Schubert, see [18].

2. The miraculous Proposition 5 stands because the phase of the sum of *two* complex numbers *with the same magnitude* is the mean value of their phases: this involves already two different centers for the symmetry (dephased by π), which is still fine because *double* the phases are involved in the computation above, but falls apart when considering three-sets or more: triad 047 is not the center of the triangle whose vertices are 0, 4 and 7.

The embedding of T/I into the isometries of the torus of phases cannot be unduly generalized, this should not deter from using it properly.

4 Many Other Models

I have left aside a number of other possible geometric models. Some readers may have expected a discussion of orbifolds – k-uplets of pcs quotiented by some symmetry group – wherein the ordinary symmetries, such as inversions, retain their original meaning. For instance, the Moebius strip of unordered dyads (see [15]) looks very much like the pictures above of the torus of phases. Indeed, it can be embedded in it, though it lacks singles pcs, triplets and all other pc-sets. However, orbifolds of higher order are limited to pc-sets with fixed cardinalities; they are hard to visualize, and feature severe singularities which get in the way of picturing the inversions; finally, they do not readily allow a visualization of the Tonnetz of triads (in many quotients of the space of 3-pc-sets, all minor and major triads appear as one single point): in most cases, the inversions are quotiented out, which disqualifies this model from the present discussion.

A lesser known model with a nicer geometry is the 4D-Model, for all pcs and (almost all) pc-sets, described by Baroin in [5]. Its natural group of symmetries (isometries of the 4D-ambient space) is actually isomorphic to the affine group on $\mathbf{Z}_{12}$, as discussed in [6]. This brings to mind the discussion of **IFunc** and homometry above. However, despite close connexions with Fourier coefficients a_3 and a_4 for single pcs, which are essentially the complex coordinates of their representations in this 4D-space viewed as $\mathbf{C}^2$, the nice symmetry features discussed in Proposition 5 on the torus of phases vanish when pc-sets are considered in Baroin Planet 4D-model: there is simply no miracle Proposition 5 this time, and inversion around a pc or a pair of pcs just does not give the expected result for most pc-sets.

5 Conclusion

Remarkably, it appears to be impossible to retain all of the many interesting features of inversions in only one model. The simple cyclic model is cumbersome when pc-sets are introduced (as polygons), showing no structure for the Tonnetz for instance. But the latter obfuscates most pc-sets, and in it P, L and R are local operations. It was rewarding to obtain a global status for these operations in the context of spectral units structuring the space of subsets of $\mathbf{Z}_{12}$ – algebraically extended to the vector space $\mathbf{R}^{12}$ or $\mathbf{C}^{12}$, layered by $\mathbf{SU}$ in disjunct torii – but on the whole, it seems that the local nature of (say) R is too strongly rooted in the transformation to be conveniently discarded. Finally, the most satisfying model by far appears to be the torus of phases, where one can see and organize pcs and pc-sets together, and play with inversions (and transpositions) in a very visual and obvious sense, enjoying both the Tonnetz, the T/I dihedral group acting on all pc-sets, and much more, on the same 2D-picture.

Anyone interested in the natural extension of T/I to the affine group (or, say, the Schönberg group for transformations of cyclic tone-rows) would do well to scrutinize Baroin's 4D model, but such endeavors are beyond the scope of the present paper.

References

1. Amiot, E.: Music Through Fourier Space. Springer, Cham (2016)
2. Amiot, E.: The Torii of phases. In: Yust, J., Wild, J., Burgoyne, J.A. (eds.) MCM 2013. LNCS, vol. 7937, pp. 1–18. Springer, Heidelberg (2013). https://doi.org/10.1007/978-3-642-39357-0_1
3. Amiot, E., Sethares, W.: An algebra for periodic rhythms and scales. JMM **5**(3), 149–169 (2011)
4. Blaise, P., Bouamrane, R., Henrirousseau, O., Merad, N., Nafi, N.: About a molecular nuclear tunnelling effect involving a very great number of basic vibrational wave-functions-theoretical-study of bullvalene. Journal de chimie physique et de physico-chimie biologique **80**(2), 173–181 (1983). Elsevier
5. Baroin, G.: The planet-4D model: an original hypersymmetric music space based on graph theory. In: Agon, C., Andreatta, M., Assayag, G., Amiot, E., Bresson, J., Mandereau, J. (eds.) MCM 2011. LNCS, vol. 6726, pp. 326–329. Springer, Heidelberg (2011). https://doi.org/10.1007/978-3-642-21590-2_25
6. Baroin, G., Amiot, E.: Old and New Isometries between Pc sets in the Planet-4D Model, MTO 21 3, 2015 (2015). http://www.mtosmt.org/issues/mto.15.21.3/mto.15.21.3.amiot-baroin.html
7. Fiore, T., Satyendra, R.: Generalized Contextual Groups, Music Theory (2005). http://www.mtosmt.org/issues/mto.05.11.3/mto.05.11.3.fiore_satyendra.pdf
8. Jedrzejewski, F., Johnson, T.: The structure of Z-Related sets. In: Yust, J., Wild, J., Burgoyne, J.A. (eds.) MCM 2013. LNCS, vol. 7937, pp. 128–137. Springer, Heidelberg (2013). https://doi.org/10.1007/978-3-642-39357-0_10
9. Mandereau, J.L., Ghisi, D., Amiot, E., Andreatta, .M., Agon, C.: Z-relation and homometry in musical distributions. J. Math. Music **5**(2), 83–98 (2011)
10. Lewin, J.: Re: intervallic relations between two collections of notes. J. Music Theor. **3**(2), 298–301 (1959)
11. Mazzola, G.: Topos of Music. Birkhauser, Basel (1984)
12. Popoff, A.: Building generalized neo-Riemannian groups of musical transformations as extensions. J. Math. Music **7**(1), 55–72 (2013)
13. Quinn, I.: General equal-tempered harmony. Pers. New Music **44**(2–45), 114–158 (2006–2007)

14. Rosenblatt, J., Seymour, P.D.: The structure of homometric sets. SIAM J. Algebraic Discrete Meth. **3**(3), 343–350 (1982)
15. Tymoczko, D.: A Geometry of Music. Oxford University Press, Oxford (2011)
16. Yust, J.: Applications of DFT to the theory of twentieth-century harmony. In: Collins, T., Meredith, D., Volk, A. (eds.) MCM 2015. LNCS, vol. 9110, pp. 207–218. Springer, Cham (2015). https://doi.org/10.1007/978-3-319-20603-5_22
17. Yust, J.: Analysis of Twentieth-Century Music Using the Fourier Transform. Music Theory Society of New York State, Binghamton (2015)
18. Yust, J.: Special collections: renewing set theory. J. Music Theor. **60**(2), 213–262 (2016)

Interval Content vs. DFT

Emmanuel Amiot[(✉)]

LAMPS, Perpignan, France
manu.amiot@free.fr

Abstract. Several ways to appreciate the diatonicity of a pc-set can be proposed: Anatol Vierù enumerates connected fifths (or semitones, as an indicator of chromaticity), Aline Honing similarly measures 'interval categories' against prototype pc-sets [8]; numerous generalizations of the diatonic scales have been advanced, for instance John Clough and Jack Douthett 'hyperdiatonic' [5] which supersedes Ethan Agmon's model [1] and the tetrachordal structure of the usual diatonic, and many others. The present paper purports to show that magnitudes of Fourier coefficients, or 'saliency' as introduced by Ian Quinn in [9], provide better measurements of diatonicity, chromaticity, octatonicity... The latter case may help solve the controversies about the octatonic character of slavic music in the beginning of the XXth century, and generally disambiguate appreciation of hitherto mostly subjective musical characteristics.

Keywords: Diatonic · Chromatic · Octatonic · Saliency
Fourier transform · Stravinsky

1 Introduction

Tautologically, the most diatonic seven-note scale is the diatonic scale, i.e. any collection/pc-set translated from $\{0, 2, 4, 5, 7, 9, 11\}$ in $\mathbf{Z}_{12}$. Slightly less obviously, the most diatonic collection in five notes is certainly the pentatonic scale $\{0, 2, 4, 7, 9\}$. But how is one to compare, say, $\{0, 2, 3, 5, 7, 8, 11\}$, $\{0, 2, 4, 5, 7, 9\}$ or $\{0, 2, 4, 6, 7, 11\}$? The question asked here is "how can one measure (with some precise, computable definition) the *diatonic character* of a pc-set?" While we are at it, it costs nothing to ask this question while replacing 'diatonic' with 'chromatic' or 'octatonic' (other adjectives will appear subsequently). Indeed it is a vexed issue (see [11]) whether Stravinsky's music is octatonic; alternatively, it would be nice to appreciate *objectively* the evolution of chromaticity throughout Wagner's Tetralogy (with Tristan in between) and what remains of it in Parsifal – similar questions abound.

Of course several answers have been advanced. We will present some of them through a few examples, and move on to argue why the most recent one, Ian Quinn's "saliency", is the best so far.

Some knowledge of pitch classes and pitch-class sets theory is assumed, alongside with basic music theory – common scales and chords, alongside with familiarity with Western Music. More elaborate machinery will be developed in Sect. 1.2 and later.

© Springer International Publishing AG 2017
O. A. Agustín-Aquino et al. (Eds.): MCM 2017, LNAI 10527, pp. 151–166, 2017.
https://doi.org/10.1007/978-3-319-71827-9_12

1.1   Some Examples

Let us focus on four pc-sets occurring at the beginning of Stravinsky's *Rite of Spring*. The first two descending motives articulate C B G E B A i.e. the pc-set $X = \{0, 4, 7, 9, 11\}$. Then D and C♯ are added, making up $Y = \{0, 1, 2, 4, 7, 9, 11\}$; it turns into something messier with chromatic quarts in the bass, that cover the chromatic aggregate. I will complete the sample with the black-keyed motif in measures 9–12, playing C♯ F♯ D♯ with a G♯ thrown in at the end, i.e. $Z = \{1, 3, 6, 8\}$, and the new descending motif in measures 15–17 playing $T = \{0, 1, 3, 6, 7, 8, 9\}$.

Undoubtedly X can be considered diatonic. After all, it is a subset of a major scale – better, *two* major scales. There is, or was, a large current in XXth century Music Theory that focuses on **inclusion** relationships – so-called set-complex theory in American Set Theory, but also the lesser known notion of 'poor' and 'rich' modes by Anatol Vierù [12][1], an independent and fairly well contrived alternative to the previous theory. However, numerous ambiguities arise:

1. How much, *exactly*, is X diatonic? Can we *grade* it?
2. In particular, is it more or less diatonic than other 5-note pc-sets, like $\{0, 2, 4, 7, 9\}$ or $\{0, 2, 4, 5, 7\}$ which are also subsets of diatonic scales?
3. What of sets which are not *exactly included* in a diatonic mode (like Y, Z) but *almost*?

Possible answers, clinging to the set relationships of inclusion and intersection, take into account the (maximum) number of common notes between a pc-set and each and every diatonic collection; or the percentage of such common notes averaged over some common basis (the cardinality of the mode, or 7, for instance). In the chosen examples, Y shares six notes $\{0, 2, 4, 7, 9, 11\}$ with C and G major, and six others $\{1, 2, 4, 7, 9, 11\}$ with D major. On the other hand, Z is included in no less than four diatonic scales, (albeit far from the ones that 'neighbored' X or Y), so Z should be rated diatonic – but how much so, when we have so many diatonic contexts to choose from?[2] Meanwhile, T intersects three diatonic collections in five notes, five others in four notes and the remaining ones in no less than three notes. How diatonic is that? Is it actually more chromatic? Or octatonic?

I will not waste time advocating *against* the set-theoretical approach, which fails because set-theory is too poor to take into account complex musical notions[3], but rather let the more elaborate models speak for themselves.

---

[1] In short, in his theory a poor mode is a subset of several rich modes.

[2] Going to extreme cases: is a single note diatonic? What about a minor third?

[3] Among other things, it does not integrate the group structure of intervals modulo octave, not to mention subtler features. As G. Mazzolla wryly observes in the preface of [10], it is hopeless to try and apprehend the huge complexity of music with only the simplest mathematical tools – though this complexity can be reconstructed from *all* its simplifications, if one construes 'simplification' as 'forgetful functor'.

The notion of interval vector (**iv**) is more precise, and provides several illuminating informations on a pc-set.[4] Simply put (following one of the latest of D. Lewin's illuminating comments), it is the probability[5] of hearing a given interval if two pcs are chosen at random in a given pc-set. Then

$$\mathbf{iv}_X(k) = \#\{(a,b) \in X^2 \mid b - a = k\} = \#(X \cap (X + k))$$

i.e. the number of occurrences of interval k between elements of X.[6]

Since a diatonic collection has maximal value for $\mathbf{iv}(5) = \mathbf{iv}(7) = 6$ (among 7-note scales), it is natural and (important in practice) fairly elementary[7] to compute $\mathbf{iv}_X(5)$ for any pc-set X and compare it against that value.

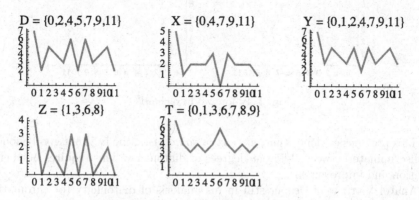

Fig. 1. **iv** for the diatonic D, X, Y, Z and T

Already **iv** provides some satisfying information (see Fig. 1):

- For X, $\mathbf{iv}(5) = 3$ is indeed the maximal coefficient; but it is far below the value for the diatonic scale, which might express the contextual ambiguity (too many different diatonic scales include X). On the other hand, $\mathbf{iv}(1) = 1$, the chromatic value, is quite small with only one semitone.
- For Y, $\mathbf{iv}(5) = 5$ is almost as large as in the case of a diatonic collection. Notice however that $\mathbf{iv}(2)$ is just as large (many whole tones) and $\mathbf{iv}(1)$ is greater than it would be for a diatonic collection.
- For Z, $\mathbf{iv}(5) = 3$ is the largest coefficient and also the maximal possible value for a 4-note scale, confirming the diatonic character despite the contextual indetermination of its many diatonic neighbors.

---

[4] The machinery involved, as we will develop below, is actually an algebra structure (with a convolution product) on the vector space of distributions, i.e. vectors describing how much of C, C♯, D and so on, are featured in a much generalized pc-set.

[5] Up to a constant.

[6] For technical reasons that will be made clear below, we do not take into account the symmetries, e.g. $\mathbf{iv}(n - k) = \mathbf{iv}(k)$ and consider $\mathbf{iv}_X$ as a vector in $\mathbf{R}^n$.

[7] Just check the number of common tones between X and $X + 5$, using the second formula in the definition above.

– Lastly, T is much more contrasted, with $\mathbf{iv}(6)$ a clear maximum[8] and other coefficients between 3 and 4.

This looks fairly close to musical perception, at least as far as diatonicity and chromaticity are concerned. However, let us take a closer look at two hexachords which share the same value for $\mathbf{iv}(5)$ (see Fig. 2): $H = \{0, 2, 4, 5, 7, 11\}$ and $H' = \{0, 1, 5, 6, 7, 8\}$. The first one, H, is a subset of C major, the second H' has only five pcs in common with C♯ and G♯ major and appears substantially more chromatic and less diatonic.[9]

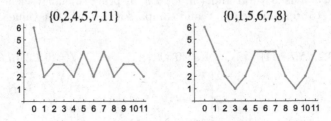

Fig. 2. iv for two hexachords

This provides evidence that, *at least in some cases*, the $\mathbf{iv}$ is not good enough to discriminate between different degrees of diatonicity. This requires both elu- cidation and improvement.

Anatol Vierù went deeper still in his analysis of diatonicity (or chromaticity), and understood the importance of *connectivity of fifths*. In a diatonic (or pentatonic) collection, we face an *uninterrupted* sequence of fifths, e.g. F C G D A E B. In H, H', there are two *broken* fifth sequences, respectively (5, 0, 7, 2), (4, 11) and (5, 0, 7), (6, 1, 8): the first collection H adheres more closely to the generating structure of the diatonic scale than H'. Hence Vierù's definition of diatonicity and chromaticity:[10]

Definition 1. *The diatonicity (resp. chromaticity) of a pc-set is the maximal number of consecutive fifths (resp. semitones) between elements of the pc-set.*

In the above example, H gets 3 and H' only 2, though the values of $\mathbf{iv}(5)$ are the same (4). Will the reader agree that the first is roughly 50% more diatonic than the second? Notice that this value is less obvious to compute than the $\mathbf{iv}$, unless one skillfully multiplies[11] the pc-set by 5 and reads the sorted result

---

[8] Actually overrated since every tritone is tallied twice.

[9] Many other examples can be devised if this one does not sound convincing to you. A more blatant one would be $\{0, 2, 7, 9\}$ vs. $\{0, 1, 7, 8\}$, both with $\mathbf{iv}(5) = 2$.

[10] *"J'ai élaboré un procédé pour mesurer le degré de diatonisme et de chromatisme d'un mode, basé sur la comparaison de la suite des quintes parfaites connexes avec la suite des demi-tons connexes à l'intérieur du même mode."* [12]; Definition 1 is more or less a translation of this.

[11] Vierù had discerned that the two notions are interchanged by multiplication by 5 (or 7) modulo 12, the classical M_5 (or M_7) operator; and offered thoughtful insights on this dichotomy as expressed by the affine group on $\mathbf{Z}_{12}$.

for chromaticity, which is a way of reading visually the value on the chain of fifths (cf. right half of Fig. 3): the first pc-set turns into $\{10, 11, 0, 1, 7, 8\}$ and the second into $\{11, 0, 1, 4, 5, 6\}$.

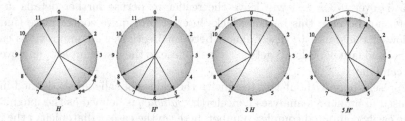

Fig. 3. Vierù's chromaticity is lesser in H than H' (left) but diatonicity stronger for H, as read on $5H$ and $5H'$ (right)

Let us cut this even finer. We would like to express that $H = \{0, 2, 4, 5, 7, 11\}$ is more diatonic than $H'' = \{0, 2, 4, 5, 7, 8\}$ (and $T = \{0, 1, 5, 6\}$ less than $T' = \{0, 3, 5, 8\}$) though the "Vierù indexes" are identical.

One possible, dual argument, would be that the *covering* chain of fifths is shorter in one case than the other: 5 0 7 2 (9) 4 11 vs 5 0 7 2 (9) 4 (11 6) 1 8 (Fig. 4). This compounds neatly the inclusion criterion, the first scale being a subset of a diatonic and not the second, but at the price of mixing two criterions and enhancing the computational complexity: should we then look up, first the lengths of connected by fifth-components, and then, in case of ex-aequo, the span of the including chain of fifths? This is getting excessively complicated.

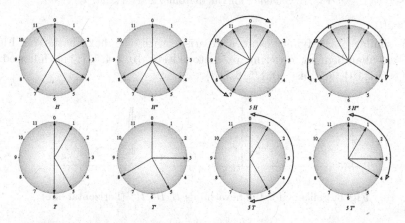

Fig. 4. Covering chain of fifths for $\{0, 2, 4, 5, 7, 11\}$, $\{0, 2, 4, 5, 7, 8\}$, $\{0, 1, 5, 6\}$ and $\{0, 3, 5, 8\}$

In [7,8], Aline Honingh endeavors to compare any pc-set with the appropriate 'prototype': for instance a hexachord will be measured against the Guidonian hexachord, a pentachord against the pentatonic, etc. For neatness, the pc-sets are first reduced to so-called 'basic-form'.[12] For instance, the two tetrachords in the last example would be compared with the prototype C D F G (numeric results depend on the choice of similarity measure), which may or may not favor 0 1 5 6   over 0 3 5 8. I will leave the reader to peruse further details in her papers, not because this measure lacks interest, but quite contrariwise (indeed it allows for instance to discriminate between Beethoven's compositions early, middle, and late periods): it gets extremely close to the last, simplest, and overall best candidate.

I present here without any technicity the values of saliency as defined in [9] and used in numerous analyses henceforth. *Saliency* is defined as the magnitude of one easily computed complex number, here (in the case of diatonicity) the fifth Fourier coefficient of a pc-set (formulas, references and properties will follow in the next section). For now, let us appreciate the values of this evaluation of diatonicity for all the above examples and some more. On Fig. 5, we can picture the magnitudes of all Fourier coefficients of the aforementioned heptachords, with the diatonic scale first. We focus on the fifth magnitude (equal to the seventh), highlighted by a dotted horizontal line, and notice that the ranking is: diatonic, Z, Y, X and T with little difference between Y and X, and a larger discrepancy with T.

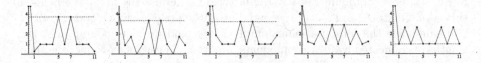

Fig. 5. Saliency for the diatonic, Z, Y, X, and T

A similarly satisfying result also arises with the hexachords on Fig. 6, with an unambiguous ordering of diatonicities: $\{0, 2, 4, 5, 7, 11\}$ followed by $\{0, 1, 5, 6, 7, 10\}$, and last $\{0, 1, 5, 6, 7, 8\}$.

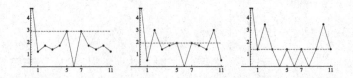

Fig. 6. Saliency for the hexachords H, H', H'' (horizontal line)

Others examples support unequivocaly this experimental evidence: that the fifth saliency corresponds very closely with the intuitive perception of diatonicity.

---

[12] In some cases this may not the best for coincidence measurements: the more compact form of a pc-set adresses its chromaticity, not its diatonicity – consider the preceding discussion where the pc-set is first transformed by M_5.

We must look into the mathematics to understand why this should be, and above all how this falls in with the competing measurements of diatonicity listed above.

1.2   Some Technical Definitions

I provide only a cursory outline; the reader of the present paper will only need to bear in mind that some easily computed[13] quantities, called Fourier coefficients, feature interesting characterizations of those pc-sets which divide the octave as evenly as possible.[14] For a very pedagogical introduction to Discrete Fourier Transform (DFT) of pc-sets, see [4]. For thorough discussion and details, see the recent reference [3] which purports to give the state of the art.

To each pc-set A considered as a subset of $\mathbf{Z}_{12}$, is associated firstly its *characteristic function*

$$1_A : x \mapsto \begin{cases} 1 & \text{if } x \in A \\ 0 & \text{if } x \notin A \end{cases} \text{ and second the Discrete Fourier Transform } \mathcal{F}_A = \widehat{1_A} \text{ of}$$

this function, the DFT of the set:

$$\mathcal{F}_A : t \mapsto \sum_{x \in A} e^{-2i\pi x t/12}.$$

This function is a sum of complex numbers of the form $e^{i\theta}$ which can all be construed as vectors $(\cos\theta, \sin\theta)$ of length 1, whose direction is given by the phase θ. The value $\mathcal{F}_A(k)$ is called the kth *Fourier coefficient*. We will mainly be concerned with its magnitude, i.e. the length of the sum of these vectors.[15]

Here is a list of elementary though useful results without proofs:

- The set A can be reconstructed from the knowledge of the Fourier coefficients $\mathcal{F}_A(k)$.
- $\mathcal{F}_A(12 - k) = \overline{\mathcal{F}_A(k)}$ (conjugate complex number).
- $\mathcal{F}_A(t) = -\mathcal{F}_{\overline{A}}(t)$ for $t \neq 0$ ($\overline{A}$ is the complement of A).
- $\mathcal{F}_A(0) = \#A$.
- $\sum |\mathcal{F}_A(k)|^2 = 12 \times \#A$.
- The Fourier transform of the (12-dimensional) interval vector $\mathbf{iv}_A$ is the square of the magnitude of $\mathcal{F}_A$:

$$\forall k \in \mathbf{Z}_{12} \quad \widehat{\mathbf{iv}_A}(k) = |\mathcal{F}_A(k)|^2. \qquad (\sharp)$$

Slightly more technical is the Huddling Lemma in [2]: in laymen's terms it states that, the closer the angles θ_k, the larger the sum $\sum_k e^{i\theta_k}$ (the vectors pull roughly in the same direction, coordinating their efforts). We will only need a simple case:

---

[13] One can compute them online at http://canonsrythmiques.free.fr/MaRecherche/styled/.

[14] Originally discovered by Quinn [9] and formally proved in excruciating detail in [2].

[15] The length of a complex number $x + iy$ is $\|(x, y)\| = |x + iy| = \sqrt{x^2 + y^2}$.

Proposition 1. *When the cardinality of A is fixed, $|\mathcal{F}_A(1)|$ reaches maximal value when the elements of A are consecutive [i.e. when A is a chromatic chunk].*

For us the most important result is

Corollary 1. *When the cardinality of A is fixed, $|\mathcal{F}_A(5)|$ reaches maximal value when the elements of A are consecutive in the chain of fifths.*

Proof. This follows from the relation $\mathcal{F}_A(5) = \mathcal{F}_{5A}(1)$, which results from $5 \times 5 = 1 \mod 12$: hence the elements of $5A$ must be consecutive, which is equivalent to the condition stated.

This is but a special case of Quinn's result:

> **Among all pc-sets with same cardinality d, the maximum magnitude for $\mathcal{F}_A(d)$ is obtained when A is a Maximally Even Set (ME set).**

ME sets admit many equivalent definitions [2,5]. We will need only to remember the most important ME sets in $\mathbf{Z}_{12}$:

1. The octatonic scale for $d = 8$.
2. The diatonic scale for $d = 7$.
3. The whole-tone scale for $d = 6$.
4. The pentatonic scale for $d = 5$.

Quinn aimed at a landscape of chords (starting from experimental knowledge) and sketched first the highest peaks. From some kind of continuity principle, it was natural to infer that the height of a chord close to a summit would still be high. Hence the definition of *saliency*, as a quality of proximity to a ME-set (that Quinn called 'prototype'):

Definition 2. *The d-saliency of a chord A is $|\mathcal{F}_A(d)|$.*

1. Among d-chords, saliency is maximal for d-ME sets.
2. Remember if convenient that $|\mathcal{F}_A(d)| = |\mathcal{F}_A(12 - d)| = |\mathcal{F}_{\overline{A}}(t)|$, hence both diatonic and (non hemitonic) pentatonic scales have maximum saliency for index 5 (namely $2 + \sqrt{3} \approx 3.73$).
3. For any (reasonable) distance on the set of pc-sets, a pc-set close to a ME set has saliency close to maximal.
4. Any pc-set (with given cardinality) distributes its saliencies according to its geometry: the sum of the squares of all saliencies is a constant. This echoes the idea in [8] that the distribution of [IC] categories throughout a piece tells of its local character.

All this provides fairly good mathematical justification, corroborated by empirical knowledge, for defining

Definition 3. – *The chromaticity of a pc-set A is $|\mathcal{F}_A(1)|$ (remembering Proposition 1).*

  – *The diatonicity of a pc-set A is* $|\mathcal{F}_A(5)|$.
  – *The octatonicity of a pc-set A is* $|\mathcal{F}_A(4)|$.

Some other values have actually been used for musical analysis: J. Yust calls 'quartal quality'[16] the magnitude $|\mathcal{F}_A(2)|$ which is, for instance, maximal among octachords for Tristan's motif pc-set $\{2, 3, 4, 5, 8, 9, 10, 11\}$; while the 'major-thirdishness' $|\mathcal{F}_A(3)|$, for want of a better term ('augmentedness'?) is maximal for an augmented triad, or for Schönberg's Napoleon hexachord $\{0, 1, 4, 5, 8, 9\}$.

    Remembering the equation $\sum |\mathcal{F}_A(k)|^2 = 12\#A$, it could be argued that the proper measure should be the *squared* magnitude – perhaps averaged by the cardinality – since the sum of all these values is a constant. Also, it is the squared value that appears in the DFT of the intervallic function. I will keep to the original definition for the present paper, but would not be surprised if the squared value were to supersede it in the future (following [17]).

2   DFT vs. iv

2.1   Theoretical Advantage

DFT is a change of (orthogonal) basis among many (polynomials, wavelets...). The major advantage[17] of expressing a (musical: pc-set, rhythm...) phenomenon in a basis of exponential functions is in the following:

Proposition 2. *The DFT exchanges convolution product* $*$ *and termwise product* $\times$. *Namely, if* f, g *are two maps from* $\mathbf{Z}_{12}$ *to* $\mathbf{C}$ *and* $\widehat{f}, \widehat{g}$ *their DFTs, then*

$$\widehat{f * g}(k) = \widehat{f}(k) \times \widehat{g}(k).$$

This is crucial because **iv** is a convolution product:

$$\mathbf{iv}_A(k) = \sum \mathbf{1}_A(t)\mathbf{1}_A(t - k) = \sum \mathbf{1}_A(t)\mathbf{1}_{-A}(k - t) = (\mathbf{1}_A * \mathbf{1}_{-A})(k)$$

and more generally, any coincidence measure or correlation (say, the number of elements of A that lie in any diatonic scale i.e. any transposition $D + k$ of $D = \{0, 2, 4, 5, 7, 9, 11\}$) can also be read on a convolution product:[18]

$$\sum \mathbf{1}_A(t)\mathbf{1}_{D+k}(t) = \sum \mathbf{1}_A(t)\mathbf{1}_D(t - k) = (\mathbf{1}_A * \mathbf{1}_{-D})(k).$$

Now the convolution product is a... convoluted operation[19] while termwise product is straightforward. Cognitively speaking, this means that complicated operations become obvious in Fourier space (i.e. computing on Fourier coefficients) and perhaps suggests that the human mind processes some equivalent of Fourier coefficients.

---

[16] In a convincing study of Ruth Crawford Seeger's *White Moon* [17].

[17] This is characteristic of DFT up to permutations: see [3], Theorem 1.11.

[18] Yust observed that conversely – by inverse DFT – the number of common tones between two pc-sets can be expressed as a sum of products of magnitudes of Fourier coefficients, pondered by cosines of the differences of phases.

[19] It has quadratic complexity, while termwise product is linear.

2.2 Multiplying Saliencies

For the sake of simplicity I present computations for diatonicity only[20], i.e. comparing a pc-set A with various transpositions of the Diatonic D and considering the fifth saliency. This is the core of the present article, making sense in a unified way of all previous diatonicity measures. We analyse first the link between coincidence and saliency. Coincidence with a prototype is a variant of Honingh's measure: $1_A * 1_B(k)$ is a high value when $A + k$ shares many common values with B. We are especially interested in the case when B is a diatonic scale, $B = D$ or $-D$ or $k - D$ etc.

Applying Proposition 2 yields immediately

$$\mathcal{F}_A(5) \times \mathcal{F}_{-D}(5) = \widehat{1_A * 1}_{-D}(5) : \qquad (\sharp)$$

the product of the (diatonic) saliencies of A and $-D$ is a Fourier coefficient of the coincidence function of A and the diatonic scale. Low values of the latter mean that bad correlation will limit the magnitude of $\mathcal{F}_A(5)$, i.e. the diatonicity of A. Conversely, when does this coincidence function $1_A * 1_{-D}$ (replaced below by $1_A * 1_D$ for simplicity's sake) exhibit a high diatonicity? On the left-hand side of equation ($\sharp$), it means simply that A is highly diatonic (large value of $|\mathcal{F}_A(5)|$). On the right-hand side, it means that the coincidence function $1_A * 1_D$

1. has at least some large values
2. **and** is 'diatonic' (large fifth Fourier coefficient).

In order to understand how the simple computation of saliency supersedes all previous notions, let us analyse this last feature, which means (in the case of diatonicity) being strongly 5-periodic: the prototype, the diatonic scale D, is a chain of fifths, meaning that $D + 5$ has $7 - 1 = 6$ common elements with D.[21] From this follows an automatic quasi-periodicity of $1_A * 1_D$ (see Fig. 7):

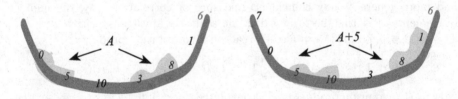

Fig. 7. Coincidence between D and A or $A + 5$ changes at most by 1

Proposition 3.

The difference between the correlations $|(1_A * 1_D)(k + 5) - (1_A * 1_D)(k)|$
is either 0 or 1.

---

[20] It would be even simpler for chromaticity (as suggested by a reviewer) but of less interest for actual analysis.

[21] One can use either 5 or 7 as generator of a chain of fifths.

Proof. These two convolution products expressed as sums share 6 common elements, plus another one than can be either 0 or 1. More precisely, setting $D = \{5m, m = 0 \ldots 6\}$ for simplicity, we get

$$(1_A * 1_D)(k) = \sum_{m=0}^{6} 1_A(k - 5m) = 1_A(k - 30) + \sum_{m=0}^{5} 1_A(k - 5m)$$

$$(1_A * 1_D)(k + 5) = \sum_{m=0}^{6} 1_A(k + 5 - 5m) = \sum_{m=0}^{6} 1_A(k - 5(m - 1))$$

$$= 1_A(k + 5) + \sum_{m=0}^{5} 1_A(k - 5m),$$

hence the two values coincide when $1_A(k + 5) = 1_A(k - 30)(= 1_A(k + 6)$ modulo 12), and differ by one if not.

How then can $\widehat{1_A * 1_D}(5)$ be as large as possible? On the one hand, the geometry of the diatonic itself partly ensures some periodicity of $1_A * 1_D$ (Proposition 3), which boosts its diatonicity. How can we further increase this periodicity?

Let for example $k = 0$ in the condition $1_A(k + 5) = 1_A(k + 6)$ just derived: we will have $1_A(5) = 1_A(6)$ when *neither F nor F♯ are* elements of A (or both), for instance when $A = \{0, 2, 4, 7, 9, 11\}$ (appropriately chiming the first notes of 'Do you know what if means'). But in order to *enlarge* the remaining sum $\sum_{m=0}^{5} 1_A(0 - 5m)$, we will need *as many elements of A as possible* in the partial chain of fifths C D E G A B (each adds 1 to the value of the convolution product). This will certainly be satisfied when A features a long *connected* subsequence of the chain of fifths.[22] We have just understood, not only how the saliency notion includes Vieru's definition, but also why it is superior: Vieru's measure is identical for H and H'' but in the latter case the elements of H are better *huddled* in the chain of fifths, providing *a larger tally of large correlation values of the convolution product* $1_H * 1_D$ (coincidence of H with the prototypical diatonic scale). Let us check this by computing some numerical values. Listing the values of the convolution products from 0 to 11 yields

$1_H * 1_D = [6, 2, 4, 3, 3, 5, 2, 5, 2, 4, 4, 2]$ and $1_{H''} * 1_D = [3, 3, 3, 3, 5, 3, 4, 3, 3, 4, 3, 5]$.

For tetrachords $T = \{0, 1, 5, 6\}$ and $T' = \{0, 3, 5, 8\}$, it is perhaps even clearer:

$1_T * 1_D = [2, 2, 2, 2, 3, 2, 3, 2, 2, 2, 2, 4]$ and $1_{T'} * 1_D = [2, 2, 3, 1, 4, 1, 3, 2, 2, 4, 0, 4]$.

Notice in the latter case how the value 4 occurs thrice in a row (in fifth order: at positions $11, 4, 9$), in agreement with the geometric constraint found above. Indeed the 5-saliency of T' is greater than T's. Similarly, H is more diatonic than H'' because of the sequence of high values (in fifth order) $\ldots 4, 5, 6, 5, 4 \ldots$

Of course, computing these correlation vectors with the diatonic would provide an effective and convincing measurement of diatonicity[23]; but as we have

---

[22] But also *almost connected chains*, like F C G A E B.

[23] As a shrewd reviewer noticed, it would also be feasible to correlate interval profiles, but our aim is to find a recipe at once simple, general and efficient.

demonstrated, the lone and straightforward value of saliency neatly subsumes the whole vector.

2.3 Inclusion and *iv*

It is redundant but perhaps useful to synthesize briefly the case of the crude inclusion as compared to saliency in the light of the above calculations. Inclusion of a pc-set inside (say) a diatonic scale is indeed a coincidence measure that can be pinpointed as one large coefficient in $1_A * 1_{-D}$ (at least one value equal to the cardinality of A, some other large values according to Proposition 3). This is but a special case of the preceding discussion, wherein it was shown that significant diatonicity depends not only on the number of coincidences but also on their grouping, or 'huddling'. The same goes for large values of $iv_A(5)$ (many fifths), which are only indicative of diatonicity when most of the fifths are neighbors in the chain.[24] The extremities of the smallest chain of fifths containing a given pc-set are of course directly related to the number of overlapping diatonic scales – i.e. tally of maximum values of the convolution product –, as foretold in Vierù's notion of 'rich modes'.

2.4 Musical Examples

To gain perspective, let us vie away from diatonicity. D. Tymoczko's thoughtful analysis of Stravinsky in [11] draws interpretation of pc-sets towards specific *classes* of scales. To his credit, he acknowledges the numerous ambiguities, criticizes fuzziness in previous analyses and avoids dogmatic pronouncements. Still, dataless statistical sentences like '... [this] scale accounts for **virtually** all of the pitches present' leave room for contestation (I highlighted the adjective). On the other hand, exact measurements of diatonicity as magnitude of $\mathcal{F}_A(5)$ – and all other saliencies – can be compared both within Stravinsky's own music, as it varies within a single piece, and from one piece to another; furthermore, this objective indicator can be applied to other composers (notably Slavic) and provide objective comparisons of their relative degrees of diatonicity, chromaticity, or octatonicity.

The interest of such comparisons warrants general and systematic research that cannot be included in this short paper. Here is but a small sample.

(1) To assess the general appreciation allowed by measurement of saliencies, I have compared all six saliencies (from chromaticity to whole-toneness) on several pieces of *The Rite of Spring* and, as an external reference, the *Dance of the Firebird*. The pieces are imported as MIDI files and a time-window of fixed width moves over it for computation of the saliencies of its pc-sets. Figure 8 simply exhibits the mean values of these saliencies.[25]

---

[24] The converse is not true: consider CDE which is undoubtedly diatonic though $iv(5) = 0$!

[25] It appears that there is little difference when the time-span of the window is expanded from 1 to 2 or even 3 s.

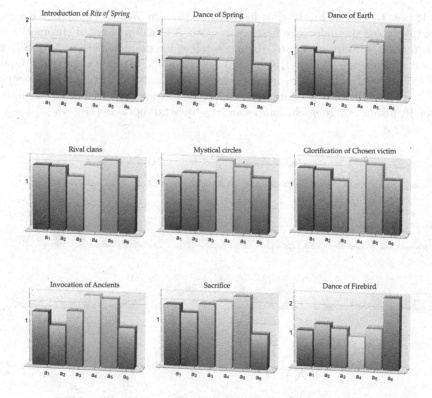

Fig. 8. Mean values of saliencies on some Stravinsky pieces

The figures show ambiguity in many pieces, which satisfyingly reflects the diversity of experts' interpretations! However, some clear-cut features do emerge:

1. Whole-tone character dominates *The Dance of the Firebird.*
2. The very first piece of *The Rite of Spring* is fairly diatonic.
3. The *Dance of Spring* is more clearly diatonic.
4. The *Dance of Earth* is mostly whole-tonish.
5. In other pieces, the balance (interplay?) between octatonic and diatonic is apparent – in line with Van der Toorn or Taruskin's analyses (as quoted in [11]).

(2) To give a feeling of the variety of these characters in the flow of the pieces, I provide some excerpts of saliencies as functions of time. On Fig. 9, following the first minute or so of the first movement of *The Rite of Spring*, the saliencies are squared (so that their sum is a constant[26]), and thus it is easily seen which character predominates in a given passage.

---

[26] Up to the cardinality of pc-sets. On these pictures, the dotted line shows the mean value of a saliency and the solid line a reference value – for a_5, say, it is the mean value found for a Mozart Sonata.

It best to look at Fig. 9 while listening to the *The Rite*'s beginning. One can practically see the indecisive first bars (motif X) flash a spurt of chromaticism (when the C♯ interferes ca. 6″) before settling for diatonicism (when the D is added to make up $Y = \{0, 1, 2, 4, 7, 9, 11\}$). Then the chromatic fourths around 15″ boost a_1; $Z = \{1, 3, 6, 8\}$ occurs between 36″ and 40″, flirting with a pentatonic i.e. largely diatonic character; finally, the last ambivalent motif T is played after 1′, a short surge of chromaticism in a 'quartal' episode (large a_2).

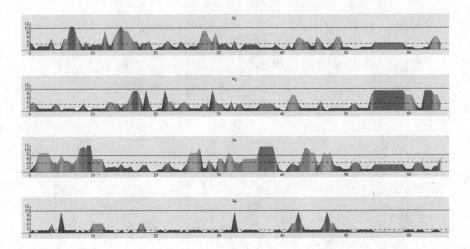

Fig. 9. Variations of saliencies in first minute of *The Rite of Spring*

This last moment exemplifies that other segmentations could, and should, be applied to music as it is perceived (as opposed to the music read on the score), for here T is clearly perceptible against the bass, though the numerical computation mixed everything together. Indeed, analyzing separate instruments, or voices, or groups, if justified on perceptual grounds, can lead to finer analyses, see examples in [11, 15], and would undoubtedly constitute an easy improvement of saliency analysis.[27]

2.5 Phase and Tonality

The (random) colors on these pictures could be adjusted to reflect the *phase* (direction of vectors) of the Fourier coefficient, which reflects a generalization of tonality (for a_5 it can be checked against the values for 12 major scales or triads, for a_6 it would be against the two whole-tone scales, etc...). Detection of the character of a passage (diatonic, octatonal etc.) can be compounded by pinpointing which (say) diatonic paradigm is involved, by computation of the phase.

---

[27] Hopefully more exhaustive analyses of saliency of Slavic music of early XXth century will soon appear, and settle once and for all the question of their octatonicity.

This is a simple way to detect tonality, and its generalizations (which whole-tone, or octatonic, scale is prevalent, etc.). More about this in [3], Chap. 6.

2.6 Possible Applications to Dodecaphonic Music

A hasty reasoning might conclude that the calculations above are meaningless in dodecaphonic music, since the Fourier coefficients of the chromatic aggregate are nil. It is not so. It is certainly true of Nicolai Obouhow's "harmonie totale"[28], but usually false in classical serial music when an appropriate time-span is used for the window of analysis, because the tone-row is often stated horizontally, not vertically; furthermore, at least in the second Viennese school, composition using the two halves (tropes) of the row are frequent. Of course a trope can be any hexachord, with distinctive saliencies, however (essentially this is Babbitt's theorem) *the saliencies of both tropes of a row are identical*. For instance, analyzing both tropes in Alban Berg's *Lyrische Suite* op. 28 and Violin Concerto op. 34 shows very strong diatonic components, see [3], p. 122. I fancy that this is a general feature of Berg's serial music (as opposed to Webern or Schönberg, say) but my ongoing computations have been impeded by the lack of available Midi files for XXth century music.

3 Conclusion

From the perspective developed here, one gets a feeling that many worthy researchers have groped for years more or less in the same direction, feeling for the right definition of diatonicity without knowing exactly where it lay. Then came Ian Quinn, and lo! the Holy Grail was there for everyone to grasp.

Not only does saliency pinpoint the character (or lack thereof) of a piece of music, the other component of the Fourier coefficients (the phase) also points its precise direction (the tonality, in the diatonic case).

Precise measurements can, at long last, supersede empirical (at best, with bevies of bored and fallible test subjects) or completely subjective (at worst, and all the more virulent for it) evaluations.

Moreover, this kind of analysis is valid for a huge repertoire, since all that was said here mostly for the diatonic character stands just as well for the 5 other characters. It is hoped that saliency diagrams, pictures and movies will be developed for many pieces of music in the very near future. Indeed, it is only a slight exaggeration to fancy deaf people enabled at last to appreciate music, simply by looking at 'Fourier clocks' ticking as the Fourier coefficients vary throughout a piece![29] It is an urgent task to develop some appropriate software for this kind of streaming analysis, picturing the Fourier flow of music on the fly.

---

[28] His chords systematically include all twelve pcs.

[29] Technically this is true since the music can be retrieved from the data of all Fourier coefficients.

References

1. Agmon, E.: A mathematical model of the diatonic system. J. Music Theor. **33**(1), 1–25 (1989)
2. Amiot, E.: David Lewin and maximally even sets. JMM **1**(3), 157–172 (2007)
3. Amiot, E.: Music Through Fourier Space. Springer, Cham (2016)
4. Callender, C.: Continuous harmonic spaces. J. Music Theor. **51**, 2 (2007)
5. Clough, J., Douthett, J.: Maximally even sets. J. Music Theor. **35**, 93–173 (1991)
6. Forte, A.: A theory of set-complexes for music. J. Music Theor. **8**, 136–184 (1964)
7. Honingh, A., Bod, R.: Clustering and classification of music by interval categories. In: Agon, C., Andreatta, M., Assayag, G., Amiot, E., Bresson, J., Mandereau, J. (eds.) MCM 2011. LNCS, vol. 6726, pp. 346–349. Springer, Heidelberg (2011). https://doi.org/10.1007/978-3-642-21590-2_30
8. Honingh, A., Bod, R.: Pitch class set categories as analysis tools for degree of tonality. In: Proceedings of ISMIR, Utrecht, Netherlands
9. Quinn, I.: General equal-tempered harmony. Pers. New Music **44**(2), 114–118 (2006). **45**(1) (2007)
10. Mazzola, G.: Topos of Music. Birkhauser, Boston (2004)
11. Tymoczko, D.: Colloquy: Stravinsky and the octatonic: octatonicism reconsidered again. Music Theor. Spect. **25**(1), 185–202 (2003)
12. Vierù, A.: Un regard rétrospectif sur la théorie des modes. The Book of Modes. Editura Muzicala, Bucarest, pp. 48 sqq (1993)
13. Yust, J.: Schubert's harmonic language and Fourier phase space. J. Music Theor. **59**, 121–181 (2015)
14. Yust, J.: Restoring the structural status of keys through DFT phase space. In: Pareyon, G., Pina-Romero, S., Agustín-Aquino, O., Lluis-Puebla, E. (eds.) The Musical-Mathematical Mind. Computational Music Science. Springer, Cham (2017). https://doi.org/10.1007/978-3-319-47337-6_32
15. Yust, J.: Applications of DFT to the theory of twentieth-century harmony. In: Collins, T., Meredith, D., Volk, A. (eds.) MCM 2015. LNCS, vol. 9110, pp. 207–218. Springer, Cham (2015). https://doi.org/10.1007/978-3-319-20603-5_22
16. Yust, J.: Analysis of twentieth-century music using the Fourier transform. Music Theory Society of New York State, Binghamton (2015)
17. Yust, J.: Special collections: renewing set theory. J. Music Theor. **60**(2), 213–262 (2016)

Probing Questions About Keys: Tonal Distributions Through the DFT

Jason Yust[(✉)]

Boston University, 855 Comm. Ave., Boston, MA 02215, USA
jyust@bu.edu

Abstract. Pitch-class distributions are central to much of the computational and psychological research on musical keys. This paper looks at pitch-class distributions through the DFT on pitch-class sets, drawing upon recent theory that has exploited this technique. Corpus-derived distributions consistently exhibit a prominence of three DFT components, f_5, f_3, and f_2, so that we might simplify tonal relationships by viewing them within two- or three-dimensional phase space utilizing just these components. More generally, this simplification, or filtering, of distributional information may be an essential feature of tonal hearing. The DFTs of probe-tone distributions reveal a subdominant bias imposed by the temporal aspect of the behavioral paradigm (as compared to corpus data). The phases of f_5, f_3, and f_2 also exhibit a special linear dependency in tonal music giving rise to the idea of a tonal index.

Keywords: Tonality · Key finding · DFT · Phase space · Probe tone

1 Introduction

Few studies in music psychology have stimulated as much interest and debate as Carol Krumhansl and Edward Kessler's 1982 article on tonal hierarchy [11]. While it is important for establishing the probe-tone technique as a behavioral correlate of the sense of key—and the consequent focus on pitch-class distributions in research on the topic—central also to its impact, one suspects, was the visualization of key relationships by deriving a toroidal space from the probe-tone data. This two-dimensional toroidal geometry of key distances, derived by applying multi-dimensional scaling (MDS) algorithms to the correlations between pitch-class distributions, was not necessary to establishing the efficacy of the probe-tone method, nor was it a necessary component of the distributional model or subsequently developed key-finding algorithms (which use correlations between distributions directly, not filtered through the two-dimensional simplification of the MDS solution). Yet the validation of common habits of thought and language relating to musical keys, spatial metaphors of distance, direction, and region, and of widely used theoretical models (such as the Schoenberg-Weber chart of regions and the Tonnetz) kindled the imaginations of a wide range of subsequent researchers.

© Springer International Publishing AG 2017
O. A. Agustín-Aquino et al. (Eds.): MCM 2017, LNAI 10527, pp. 167–179, 2017.
https://doi.org/10.1007/978-3-319-71827-9_13

Spatial models raise a number of significant questions about the nature of musical keys, many of which have been examined in the music perception and cognition literature on the topic. This paper demonstrates that the discrete Fourier transform (DFT) on pcsets can clarify these questions and in some cases suggest novel solutions.

Krumhansl ([10], pp. 99–106) noted that the spatial representation of keys in Krumhansl and Kessler 1982 could be reproduced, without recourse to MDS, by taking the third and fifth phase components of the Fourier analysis of the key profiles. For Krumhansl this theoretical reformulation of the space is primarily an expedient allowing for the plotting of various kinds of information (expert key assignments, distributional data in the music) in a fixed space. The practical problems can be overcome by clever use of computational techniques like self-organizing maps, as [12,14] have shown. But, as I will argue here, Krumhansl's simplification using the DFT is of considerable theoretical interest in its own right, especially in light of more recent applications of this same type of space [2,3,26,27]. In particular, basic mathematical properties of the DFT allow us to draw more far-reaching conclusions about this space and its significance to the nature of tonality.

Much of this research has produced different kind of pitch-class distributions that can be analyzed using the DFT on pitch-class vectors, as described by [3,13,17,25,27]. The terminology used here is taken from [25]. The entries in the DFT vector are referred to as "components" and denoted $f_0, f_1, f_2, \ldots$. They are converted to polar coordinates with magnitude $|f_n|$ and phase ϕ_n, but with phases converted to a pitch-class scale and designated $Ph_n = 2\pi(\phi_n)/12$.

2   Tonal Distributions

The large body of research that has grown out of Krumhansl's work has produced an abundant crop of tonal distributions. These come in two or three basic forms. Krumhansl and Kessler's ([11]) original distributions are probe-tone ratings from human subjects. Subsequent studies, such as [5,6,20] applied the probe-tone technique in varying contexts, or other experimental tasks that produce comparable distributional data, such as the wrong-note detection technique used by [8]. Another method that has produced many distributions for key-finding algorithms (further discussed in the next section) is to derive distributions from the frequency of occurrence of scale degrees in a corpus. Finally, other distributions (e.g. in [19,21]) are created "by hand" to optimize the performance of key-finding algorithms.

Figures 1, 2 and 3 plot distributional data from a variety of sources in three different Fourier phase spaces. (The locations of all major and minor triads and two diatonic scales are also given for reference.) Fig. 1 shows part of the Ph_3/Ph_5 space used by Krumhansl ([10]) and which is also the basis of Amiot ([2,3]) and my ([26]) continuous Tonnetz. Despite a great variety of techniques represented by corpus-derived distributions—using entire pieces with or without accounting for modulations, using just the initial or final measures, using melodies only or polyphonic textures, counting pitch-classes in different ways—all are bunched very closely together, near the tonic triad but typically with a slightly higher Ph_5, possibly reflecting a bias towards the dominant. Temperley's ("CBMS") and Sapp's hand-made distributions are close enough to these, but lie on the fringes of the pack.

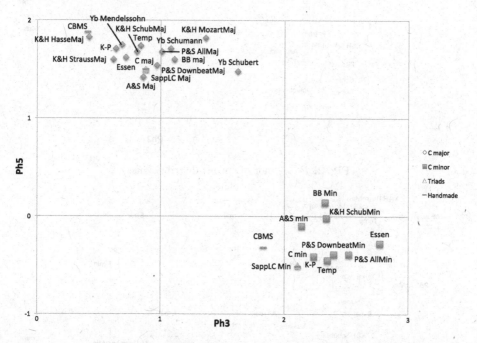

Fig. 1. $Ph_{3/5}$ plot of corpus distributions from a variety of sources (Yb: [24], K&H: [9], K-P, Essen, Temp: Kotska-Payne, Essen, and Temperley corpora from [22], BB: Bellman-Budge from [19], P&S: [16], A&S: [1])

Figure 2 plots the same data with Ph_2 replacing Ph_3. Major-key data spreads out a little more in the Ph_2 dimension, but on the whole we can reach the same conclusions. The interchangeability of Ph_2 and Ph_3 relates to a basic property of tonality, the tonal index, explored further below. Other components do not provide the same kind of essential tonal information, as the $Ph_{1/4}$ plot in Fig. 3 illustrates. Major-key data are particular unfocused in the Ph_4 dimension and the minor-key data in the Ph_1 dimension. Even where a certain amount of consistency might be found, such as the minor-key profiles in the Ph_4 dimension, it is closer to unrelated triads like B major, B minor, and D minor.

The probe-tone profiles are more variable, but reliably close to the corresponding corpus data. Figure 4 gives a variety of major-key probe-tone data reflecting a variety of experimental paradigms. Cuddy and Badertscher ("C&B") include major-triad and major scale contexts on three levels of musical background. Brown, Butler, and Jones ("BBJ") replicate these ("triad1," "scale1")

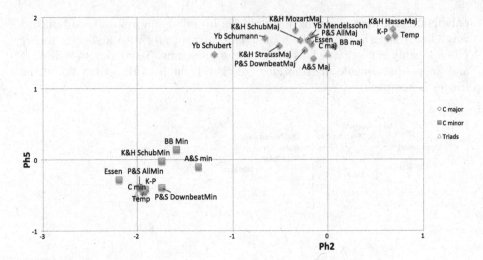

Fig. 2. $Ph_{2/5}$ plot of corpus distributions.

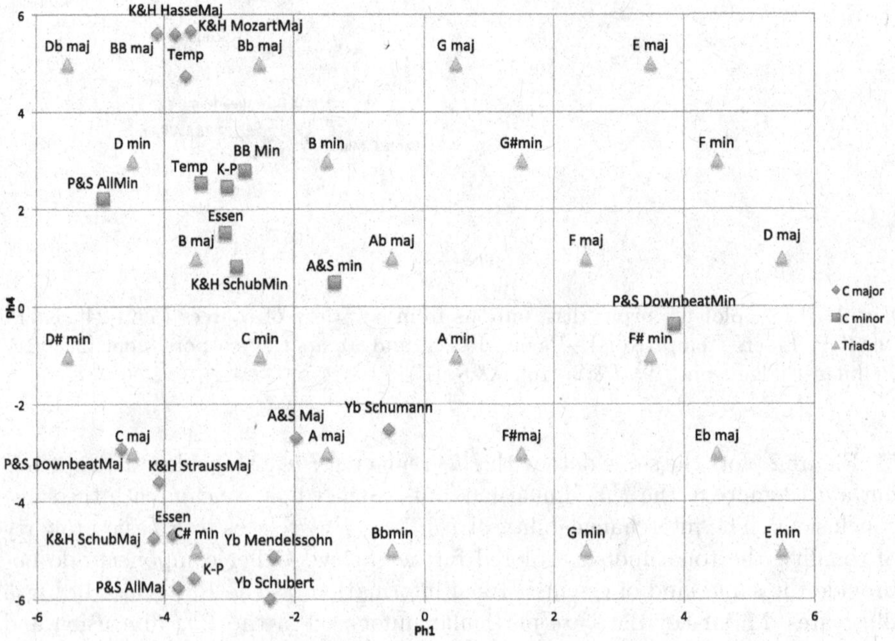

Fig. 3. $Ph_{1/4}$ plot of corpus distributions.

and also test contexts that reorder the tones of each ("triad2," "scale2"). Smith and Schmuckler randomly generate contexts using Krumhansl and Kessler's profiles weighting tones by duration ("S&S1") or frequency ("S&S2") at varying levels. Janata et al. ("J&al") use a very different method of wrong-note

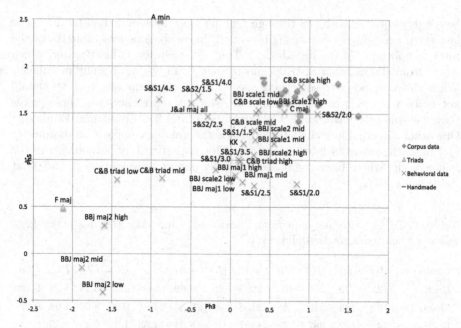

Fig. 4. $Ph_{3/5}$ plot of corpus distributions and probe tone distributions from a variety of sources (KK: [11], C&B: [6], BBJ: [5], S&S: [20], J&al: [8]).

detection. Despite such differences in experimental paradigm, these data very consistently deviate from the corpus data on the subdominant side. The difference may result from the temporal aspect of the probe-tone task: listeners evaluate, not a note merely *in* the given context, but *after* it, and motion to the left in Ph_3 (descending thirds or fifths) is much more typical of tonal music than to the right, particularly at endings and moments of resolution. Particularly striking is Brown, Butler, and Jones's reordering of the arpeggiated triad, which appears to consistently imply F major more strongly than C major.

To examine the matter more closely, let us focus on a single, fairly rich, body of corpus data collected by Prince and Schmuckler ([16]). Tables 1 and 2 show the DFTs for their data collapsed over metric position but divided by composer.[1] These data are average tonal profiles for each composer, with all pieces transposed to C major or C minor, but with no accounting for modulations. The data for Bach, Mozart, Beethoven, and Chopin represent relatively large samples (between 20,000 and 120,000 quarter notes for each data point) while those for Schubert, Liszt, Brahms, and Scriabin are smaller (1700–16,000). Despite the wide range of harmonic styles represented, one very clear conclusion can be drawn from the DFT magnitudes: With one exception, f_5 is always very large, followed by f_3, then f_2. This agrees with results from [7] whose ic5 category may

---

[1] Jon Prince generously shared this raw data through personal correspondence.

be roughly equated with $|f_5|$ through Quinn's [17] "intervallic half-truth."[2] The last three components are negligibly small, in most cases less than 1% of the total "amplitude" of the distribution. The one exception is Liszt-major, which differs from all the other distributions in that f_3 and f_5 are equally prominent. While this may point to something special in Lizst's harmonic style, we should not make too much of this distribution, since it represents only three pieces (Grande étude de Paganini 4, Liebesträume 3, and Transcendental Etude 5).[3] One other discernable stylistic difference is the greater emphasis on diatonicity (f_5) in Bach versus all later composers. This is particularly pronounced in the minor mode, where later composers typically put more weight on f_2 and f_3 at the expense of f_5.

Table 1. DFTs of corpus data from Prince and Schmuckler [16], for major keys. Squared magnitudes are multiplied by 10^4.

Composer	$|f_1|^2$	$|f_2|^2$	$|f_3|^2$	$|f_4|^2$	$|f_5|^2$	$|f_6|^2$	Ph_1	Ph_2	Ph_3	Ph_4	Ph_5	Ph_6
Bach	2	79	150	10	2095	2	9.89	0.96	0.38	3.74	1.96	6
Mozart	9	243	310	1	2158	4	9.34	11.91	0.96	8.04	1.66	0
Beethoven	4	182	287	7	1427	0	8.31	11.53	1.34	6.43	1.53	6
Schubert	7	127	337	18	1931	0	8.25	0.18	1.06	7.70	1.65	0
Chopin	10	186	357	11	1638	1	7.43	11.72	1.03	8.22	1.26	0
Brahms	2	49	224	5	1009	2	7.74	0.27	0.68	9.37	1.59	0
Liszt	5	85	402	30	394	43	6.21	0.12	0.68	9.37	1.59	0
Scriabin	11	117	352	47	2154	2	8.93	1.09	0.89	7.12	1.93	6

The generally low magnitudes of f_1, f_4, and f_6 explain another feature of the distributions: the lack of consistency in phases for these components. Phase values should become more volatile as the magnitudes approach zero where the phase becomes undefined. However, it is logically possible to expect variability in phase values for the well-represented components (f_2, f_3, and f_5). Such variation could reflect real differences between composers, who represent a wide range of styles. On the whole, however, we do not see much variability in these phases, and the data, as in Fig. 1, tends to cluster close to the Ph_2, Ph_3, and Ph_5 values

[2] There is also agreement on the minor-key data which shows smaller f_5s and correspondingly fewer ic5-category designations. The DFT data is less equivocal on the secondary features of tonality, however, which clearly relate to f_3 and f_2. This surfaces in Honingh and Bod's results in the form of ic3- or ic4-category pcsets, but it is hard to draw as clear-cut a conclusion from this aspect of their results.

[3] With the assistance of Matthew Chiu, I have recently assembled a larger data set of distributions that confirms these conclusions, including the pronounced low diatonicity of Liszt's music, especially in the minor mode where it reaches a level approximately equal to that of f_3. The tendency can also be seen in Wagner and Scriabin, but not quite as strongly as in Liszt.

Table 2. DFTs of corpus data from Prince and Schmuckler [16], for minor keys. Squared magnitudes are multiplied by 10^4.

Composer	$|f_1|^2$	$|f_2|^2$	$|f_3|^2$	$|f_4|^2$	$|f_5|^2$	$|f_6|^2$	Ph_1	Ph_2	Ph_3	Ph_4	Ph_5	Ph_6
Bach	8	69	208	49	1489	3	8.72	10.80	2.29	2.60	11.60	6
Mozart	17	195	656	10	1515	6	8.97	10.53	2.24	0.15	11.91	6
Beethoven	2	239	457	5	1150	16	6.33	10.18	2.51	1.73	11.70	6
Schubert	9	238	540	14	1815	1	9.20	10.47	2.62	11.76	11.93	0
Chopin	0	149	336	6	1002	6	6.57	10.36	2.39	2.21	11.88	6
Brahms	9	194	390	2	847	19	8.23	9.89	1.87	2.14	11.68	6
Liszt	0	254	651	1	1179	14	4.83	10.75	1.75	1.30	0.40	6
Scriabin	5	237	399	48	1524	19	10.10	10.88	1.25	0.61	11.91	0

for the tonic triads, as can be seen in Fig. 5. The only stylistic differences evident here are the greater tendency of Bach's distributions toward the diatonic scales, especially in major, and the opposite tendency of Liszt's distributions toward the parallel keys.

This striking result suggests a signal processing analogy to explain tonality as a kind of band-pass filter for pitch-class information. The tonal filter supresses certain frequencies (f_1, f_4, and f_6) while amplifying others (f_2, f_3, and f_5).

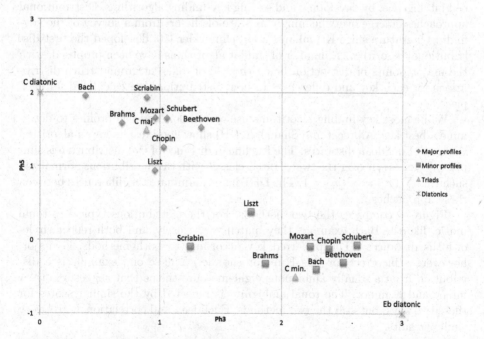

Fig. 5. $Ph_{3/5}$ plot of corpus distributions from Prince and Schmuckler [16].

As a definition of tonality, this has the advantage that it can be treated either as a property of music or as a way of hearing or interpreting music. That is, to the extent that music is tonal, it will tend to feature harmonic content that emphasizes f_2, f_3, and f_5, and tonal interpretations of music are those that filter out f_1, f_4, and f_6, possibly with disregard for a prominent status for one of those components. For instance, octatonic music (such as certain pieces by Messiaen) will have a prominent f_4, but a tonal interpretation of octatonic music will suppress this feature in order to amplify f_2, f_3, and f_5, which may be controlled by choice of subsets or emphasized notes within the given octatonic context. This means that a three-dimensional phase space, $Ph_{2/3/5}$, may be a sufficient and more stable tonal state space than the original 12-dimensional space of pitch-class distributions, since each key occupies a distinct region of $Ph_{2/3/5}$-space. However, we have also found that a two-dimensional toroidal space appears to be sufficient for distinguishing keys. This reflects an additional constraint that seems built into tonal syntax, a linear dependence between Ph_2, Ph_3, and Ph_5. This linear constraint, $Ph_2 + Ph_3 - Ph_5 \approx 0$, gives rise to a "tonality index" that will be further discussed below. Given such a linear constraint, the three-dimensional space of tonality may be projected onto any of its two-dimensional subspaces with (ideally) no essential information loss.

3   Key Finding

Many studies have approached the question of key from the standpoint of artificial intelligence, by developing and testing key-finding algorithms. Distributional approaches emerge overwhelmingly as state-of-the-art from a survey of the key-finding literature since Krumhansl and Schmuckler [10] developed the first distributional algorithm. A number of similar algorithms have been proposed, with the major points of distinction being the use of different ground truth distributions for each key and differences in how distributions are calculated for each piece.

While most key-finding algorithms use correlation between profiles to determine a best key, Albrecht and Shanahan ([1]) show good results for an algorithm that uses Euclidean distances. The Euclidean distance of two distribution is simply $\sqrt{\sum(x_i - y_i)^2}$ over the twelve pitch-classes with the distributions normalized such that $\sum(x_i) = \sum(y_i) = 1$. The DFT helps illuminate the differences between these approaches.

Figure 6 compares the two methods. For the distributions typical of tonal music, like the Bach example, they match very closely, and both reflect circle-of-fifths distances. (The preferred key according to both methods, G major, however, is incorrect for this E minor chorale.) The second example is a distribution from a tonally ambiguous eight-measure theme that suggests both A minor and E minor. The tonal ambiguity is reflected by the similar scores for these two keys, but still the two methods, Euclidean and correlational, give very similar results.

Both methods may be better understood through basic Fourier theorems. Euclidean distances remain Euclidean distances after the DFT, measured in a

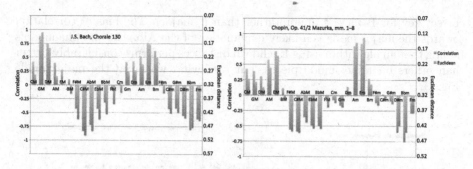

Fig. 6. Correlations and Euclidean distances comparing distributions from two tonal pieces to key profiles from Albrecht and Shanahan [1].

direct product of complex planes and scaled by $1/\sqrt{12}$, by the unitarity principle (i.e., orthogonality). Furthermore, the convolution theorem says that correlations become dot products after the DFT: $\mathcal{F}(f * g) = \mathcal{F}(f) \cdot \mathcal{F}(g)$. When the magnitudes of DFT components match—e.g. when comparing two tonal distributions with large f_2, f_3, and f_5—both measures will reflect the phase differences of the prominent components, and therefore they will tend to agree, the only difference being that correlation will be even more strongly biased towards the components that are large in both distributions (and hence will favor f_5 more strongly when comparing tonal distributions). Therefore, a simple explanation of how distributional key finding works is that the scale is selected by Ph_5 and the mode by Ph_3 or Ph_2. The same results could therefore be derived from proximity in $Ph_{3/5}$-space.

When distributions emphasize different periodicities, particularly where a DFT component is large in one distribution and close to zero in the other, the two methods respond differently. Correlation will simply supress such components (since the influence of a component is weighted by a product of magnitudes). The Euclidean measure will include a constant value that is uninfluenced by changes of phase (i.e., transposition). Therefore the range of Euclidean distances will contract more noticeably when comparing non-tonal distributions to the 24 key profiles, but the difference should not usually affect the choice of best key.

Example 7 shows two instances where the two methods do choose different keys, the first eight measures of Schubert's song "Dass sie hier gewesen" and the subject of the F♯ minor fugue from Bach's WTC I. Both distributions suppress f_2 and f_3 in favor of some non-tonal component, f_4 in Schubert's case (because of a heavily emphasized vii°⁷/ii chord) and f_1 in Bach's (because the subject is very chromatic and restricted in range). As a result, all components except for f_5 are effectively canceled out for both the correlational and Euclidean criteria. In this situation, correlation is biased towards major keys, because the major-key distribution has a slightly higher $|f_5|$. Euclidean distance chooses the key that is closest in Ph_5, which happens to be minor in both instances, whereas correlation chooses the closest major key. As a result, correlation selects the correct key for Schubert (C major as opposed to G minor) but Euclidean distance selects the correct key for Bach (F♯ minor rather than E major). The bias of correlation towards the major mode is a likely explanation of the Albrecht and Shanahan's finding that an algorithm using Euclidean distances performs considerably better than others on minor-mode pieces, but somewhat worse on the major mode (Fig. 7).

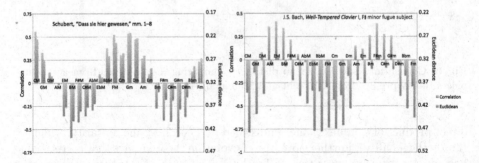

Fig. 7. Correlations and Euclidean distances comparing distributions from two tonal pieces to key profiles from Albrecht and Shanahan [1].

Studies on human subjects have been much more attentive to the influence of temporal ordering on perceptions of key than the key-finding literature. Since distributions collapse the temporal dimension, they implicitly assume that the temporal order of pitch-classes does not influence the sense of key, even though experimental studies such as [4,5,15] have amply demonstrated the importance of temporal order to key inferences. Approaches to key finding that deal with modulation by using windowed analysis, such as Temperley's ([21,22]) and Sapp's ([19]), may partially address this concern. But these only allow for the sense of key to change over time; they do not propose means by which the temporal ordering of pitch classes may influence the sense of key beyond assuming that more recently occuring pitches will have a stronger influence. More promising is Quinn's [18] approach of treating progressions as the basic elements of tonality rather than chords (built upon by White [23]). The tonal filter may provide a way of "fuzzying" the concept of chord, with progressions as characteristic kinds of motions in $Ph_{2/3/5}$ space.

4   The Tonal Index

We have observed that typical distributions in tonal music feature three prominent components, f_2, f_3, and f_5, but also that a two-dimensional space using any selection from Ph_2, Ph_3, and Ph_5, is sufficient to represent the tonal implications of a particular distribution. The reason is that typical tonal distributions seem to be constrained to keep the quantity $Ph_2 + Ph_3 - Ph_5$, the *tonal index*, close to zero. Figure 8 provides an example of how the tonal index tends to stay very consistently close to zero in the windowed analysis of a tonal piece.

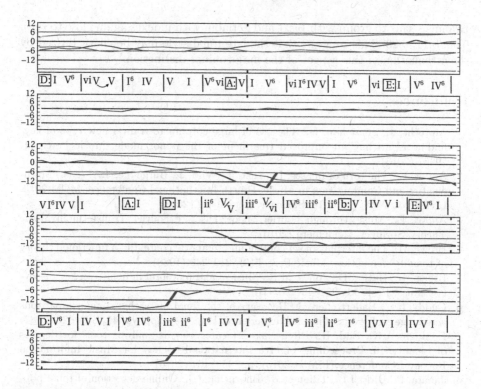

Fig. 8. Ph_2 (Green), Ph_3 (Blue), and Ph_5 (Pink) and the tonal index (blue, lower graphs) in a windowed analysis of Corelli's Violin Sonata Op. 5/1 mvt. 2, aligned with a harmonic summary of the score. (Color figure online)

The tonal index is equal to zero for certain basic, mode-neutral, pitch-class sets: unisons, perfect fifths, and diatonic scales. This is related to the mathematical fact that, for generated collections, an index of this type can only take two values, 0 or 6.[4] For major and minor triads, it is small, ± 0.62. The non-composer-specific distributions in Figs. 1, 2, 3 range from -0.20 to -1.04 for major and 0.60 to 1.05 for minor averaging -0.67 and 0.84.[5] The Prince/Schmuckler data of Fig. 5 gives averages of -0.51 and 0.41 and some evidence of historical trends. In major, the index for composers up to Brahms ranges just from -0.79 to -0.42 averaging -0.64, very close to the major triad value. The late tonal styles of Liszt and Scriabin give values much higher and closer to zero, 0.14 and 0.05. In the minor mode, Bach stands out somewhat with an index of 0.49, late-eighteenth/early-nineteenth century composers range from 0.85 to 1.16, and

---

[4] This is a consequence of Amiot's [3] Proposition 4.3, which also appears in [25] but missing a $\pm$, crucial for the recognition of 6 (or more generally π) as a possible value. This can also be extended to other inversionally symmetrical collections using Proposition 6.8 from [3]. Thanks to Emmanuel Amiot for these observations.

[5] The averaging is done in the complex plane on normalized values.

Brahms seems to group with Liszt and Scriabin with indexes again close to zero: 0.08, 0.11, 0.23. This suggests that the late tonal style may be characterized by the attenuation of this aspect of the major-minor distinction.

References

1. Albrecht, J., Shanahan, D.: The use of large corpora to train a new type of key-finding algorithm: an improved treatment of the minor mode. Music Perc. **31**, 59–67 (2013)
2. Amiot, E.: The torii of phases. In: Yust, J., Wild, J., Burgoyne, J.A. (eds.) Mathematics and Computation in Music, Fourth International Conference: MCM 2013. LNAI, vol. 7937, pp. 1–18. Springer, Heidelberg (2013)
3. Amiot, E.: Music through Fourier Space: Discrete Fourier Transform in Music Theory. Springer, Heidelberg (2016)
4. Brown, H.: The interplay of set content and temporal context in a functional theory of tonality perception. Music Perc. **5**(3), 219–249 (1988)
5. Brown, H., Butler, D., Jones, M.R.: Musical and temporal influence on key discovery. Music Perc. **11**, 371–407 (1994)
6. Cuddy, L.L., Badertscher, B.: Recovery of the tonal hierarchy: some comparisons across age and levels of musical experience. Perc. Psychophys. **41**, 609–620 (1987)
7. Honingh, A.K., Bod, R.: Pitch class set categories as analysis tools for degrees of tonality. In: Proceedings of 11th International Society for Music Information Retrieval Conference (ISMIR 2010), pp. 459–464 (2010)
8. Janata, P., Birk, J.L., Tillman, B., Bharucha, J.J.: Online detection of tonal pop-out in modulating contexts. Music Perc. **20**, 283–305 (2003)
9. Knopoff, L., Hutchinson, W.: Entropy as a measure of style: the influence of sample length. J. Music Theor. **27**, 75–97 (1983)
10. Krumhansl, C.L.: Cognitive Foundations of Musical Pitch. Oxford University Press, New York (1990)
11. Krumhansl, C.L., Kessler, E.: Tracing the dynamic changes in perceived tonal organization in a spatial representation of musical keys. Psych. Rev. **89**, 334–368 (1982)
12. Krumhansl, C.L., Toiviainen, P.: Tonal cognition. In: Peretz, I., Zatorre, R.J. (eds.) The Cognitive Neuroscience of Music (2003)
13. Lewin, D.: Special cases of the interval function between pitch-class sets X and Y. J. Music Theor. **45**, 1–29 (2001)
14. Martorell, A., Gómez, E.: Two-dimensional visual inspection of pitch-space, many time-scales and tonal uncertainty over time. In: Agon, C., Andreatta, M., Assayag, G., Amiot, E., Bresson, J., Mandereau, J. (eds.) MCM 2011. LNCS, vol. 6726, pp. 140–150. Springer, Heidelberg (2011). https://doi.org/10.1007/978-3-642-21590-2_11
15. Matsunaga, R., Abe, J.: Cues for key percpetion of a melody: pitch set alone? Music Perc. **23**, 153–164 (2005)
16. Prince, J.B., Schmuckler, M.A.: The tonal-metric hierarchy: a corpus analysis. Music Perc. **31**, 254–270 (2014)
17. Quinn, I.: General equal-tempered harmony (in two parts). Pers. New Music **45**, 4–63 (2006). 44, 114–159
18. Quinn, I.: Are pitch-class profiles really "Key for Key"? Zeitschrift der Gesellschaft für Musiktheorie **7**, 151–163 (2010)

19. Sapp, C.: Computational methods for the analysis of musical structure. Stanford University, Ph.D. dissertation (2011)
20. Smith, N.A., Schmuckler, M.A.: The percpetion of tonal structure through the differentiation and organization of pitches. J. Exp. Psych. Hum. Perc. Perf. **30**, 268–286 (2004)
21. Temperley, D.: Cognition of Basic Musical Structures. MIT Press, Cambridge (2001)
22. Temperley, D.: Music and Probability. MIT Press, Cambridge (2007)
23. White, C.W.: Some statistical properties of tonality, 1650–1900. Yale University, Ph.D. dissertation (2013)
24. Youngblood, J.E.: Style as information. J. Music Theor. **2**, 24–35 (1958)
25. Yust, J.: Applications of DFT to the theory of twentieth-century harmony. In: Collins, T., Meredith, D., Volk, A. (eds.) MCM 2015. LNCS, vol. 9110, pp. 207–218. Springer, Cham (2015). https://doi.org/10.1007/978-3-319-20603-5_22
26. Yust, J.: Schubert's harmonic language and Fourier phase space. J. Music Theor. **59**, 121–181 (2015)
27. Yust, J.: Special collections: renewing set theory. J. Musis Theor. **60**, 213–262 (2016)

Gesture Theory

Abstract Gestures: A Unifying Concept in Mathematical Music Theory

Juan Sebastián Arias(⊠)

Universidad Nacional de Colombia, Bogotá, Colombia
jsariasv1@gmail.com

Abstract. We present the notion of abstract gestures and show how it encompasses Mazzola's notions of gestures on topological spaces and topological categories, the notion of diagrams in categories, and our notion of gestures on locales. A relation to formulas is also discussed.

1 Introduction

Soon after the accomplishment of the first version of his *The Topos of Music* [9], an enterprise that achieved a topos-theoretic based framework for musicology (a theory of performance included), and that gave a very complete account of the mathematical structures present in music, Mazzola became aware of that his own activity as a free jazz pianist had little to do with the structures and procedures described in his monograph. *Gestures*, rather than formulas, were the essence of his performance. Certainly, improvisation in free jazz is mainly determined by the movements of the body's limbs, that is, by a *dancing of the body*, the classical structures of western music being secondary and auxiliary. Then a rigorous reflection on gestures is necessary, and not only in the case of musical improvisation, but in music in general, since all its power and intensity relies on its realization in bodily terms, even in the western classical tradition.

The point of departure towards a formal definition of gesture is the one given by Hugues de Saint-Victor in the chapter XII of his *De Institutione Novitiorum* [12]:

> Gestus est motus et figuratio membrorum corporis, ad omnen agendi et habendi modum.
> [Gesture is the movement and configuration of the body's limbs, towards all an action and having a modality.][1]

Based on this definition, Mazzola gives the first mathematical definition of a gesture as a *diagram* of curves in a topological space (see the Sect. 3 for the precise definition); here the diagram corresponds to the configuration of the body's limbs and the topological space corresponds to the space-time where the movement occurs. Further, this definition is generalized to topological categories in [11] to include both algebraic and topological information in gestures, and then

---

[1] Our translation.

© Springer International Publishing AG 2017
O. A. Agustín-Aquino et al. (Eds.): MCM 2017, LNAI 10527, pp. 183–200, 2017.
https://doi.org/10.1007/978-3-319-71827-9_14

to locales in [1] as a first step to define gestures on generalized notions of space. These different instances of defining gestures belong, though not so strictly, to the topological branch of the theory of gestures.

On the other hand, there is an algebraic counterpart of this. In [10, p. 39], Mazzola defines a formula in a spectroid[2] as a suitable *diagram* in this particular kind of linear category, which is the starting point to develop a mathematical framework for both the theory of nets and Lewin's transformational theory.

It is important to stress that all these different definitions rely on the notion of digraph: both gestures and formulas are morphisms of digraphs with domain a given skeleton. Moreover, following Mazzola's ideas, these instances can be regarded as attempts of reanimation of the implicit movement that the drawing of a digraph by means of arrows and nodes suggest. In Mazzola's own words [10, p. 25]:

The gesture is a morphism, where the linkage is a real movement and not only a symbolic arrow without bridging substance.

Regarding these two branches, there are two main problems. The first one deals with the search for a common universe: that is, the diamond conjecture. The second one corresponds to a gestural representation of categories in which composition of arrows can be manipulated at the level of gestural intuitions, in much the same way as the Yoneda embedding allows the representation of categories in topoi of presheaves. To a great extent, topological categories were introduced in gesture theory so as to construct a bicategory of gestures proposed by Mazzola as a first step to solve these two problems.

It is remarkable that gestures and formulas are at the core of the relation between mathematics and music. Mazzola has proposed a fundamental conceptual *adjunction*

$$\text{formulas} \xrightarrow[\text{mathematics}]{\text{music}} \text{gestures} \ ,$$

where the arrows correspond to the activities of the disciplines: mathematicians take gestures (intuitions, mental movements, analogies with reality,...) to produce formulas, musicians take formulas (scores, diagrams, musical notations,...) to produce gestures. The term adjunction refers to a relation that is more profound than a mere inversion or isomorphism, it corresponds to a true dialectic that is grasped formidably by the categorical concept of adjunction between functors. Certainly the diamond conjecture is the search for such an adjunction in precise mathematical terms.

This article is an overview of a general framework for gesture theory that could unify the definitions of gestures on several notions of space (more related to the topological branch of mathematical music theory) and the notions of formulas in spectroids and of diagrams in categories (more related to the algebraic branch

---

[2] See Sect. 7 or [4, p.29] for the definition of spectroid. Spectroids were introduced by Pierre Gabriel in representation theory of quivers or digraphs; details can be found in [4].

of mathematical music theory). In addition, this framework is flexible enough to introduce gestural ideas in other fields of mathematics given its category-theoretic nature.

The structure of this article is that of a *theme with variations*. We first present the notion of abstract gestures and then proceed to unfold different *realizations* thereof. Justifications for all statements that are not proved in this article will be found in [2].

2   Abstract Gestures

Before giving the definition of gestures we need some basic definitions and fix the notation.

Directed graphs and internal digraphs

Let G_1 be the category with two parallel arrows between two vertices $[0], [1]$ plus the identities; it can be depicted as follows:

$$id \, \circlearrowleft [0] \underset{\varepsilon_0}{\overset{\varepsilon_1}{\rightrightarrows}} [1] \circlearrowright id \, .$$

A *directed graph* (or *digraph*, for short) is a tuple $\Gamma = (A, V, t, h)$, where A, V are sets and $t, h : A \longrightarrow V$ are functions. Digraphs correspond bijectively to presheaves on the category G_1 so from now on we identify a digraph $\Gamma = (A, V, t, h)$ with its associated presheaf $\Gamma : G_1^{op} \longrightarrow \mathbf{Set}$ defined by $\Gamma([1]) = A, \Gamma([0]) = V, \Gamma(\epsilon_0) = t, \Gamma(\epsilon_1) = h$. In this way, there is a topos of digraphs, namely the Grothendieck topos[3]

$$Digraph := \mathbf{Set}^{G_1^{op}}.$$

Thus, a morphism from $\Gamma_1 = (A_1, V_1, t_1, h_1)$ to $\Gamma_2 = (A_2, V_2, t_2, h_2)$ (that is, a natural transformation) corresponds to a pair of functions (u, v), with $u : A_1 \longrightarrow A_2$ and $v : V_1 \longrightarrow V_2$, satisfying the identities

$$vt_1 = t_2 u, \ vh_1 = h_2 u.$$

Similarly, if $\mathscr{C}$ is an arbitrary category, a functor $S : G_1^{op} \longrightarrow \mathscr{C}$ can be identified with a tuple (S_1, S_2, e_0, e_1), that is, with the diagram

$$S_1 \underset{e_0}{\overset{e_1}{\rightrightarrows}} S_0$$

of morphism of $\mathscr{C}$ by putting $S_1 = S([1]), S_0 = S([0]), e_0 = S(\epsilon_0), e_1 = S(\epsilon_1)$. A tuple (S_1, S_2, e_0, e_1), where $e_0, e_1 : S_1 \longrightarrow S_0$ are morphisms of $\mathscr{C}$ is called an *internal digraph* in $\mathscr{C}$. In this way, functors $S : G_1^{op} \longrightarrow \mathscr{C}$ can be identified with internal digraphs in $\mathscr{C}$.

---

[3] Any category of presheaves on a small category is a Grothendieck topos. In fact, given a category of presheaves on a small category $\mathscr{C}$, it is a category of sheaves if we consider on $\mathscr{C}$ the *trivial topology*, whose unique covering sieve for each object of $\mathscr{C}$ is the maximal sieve.

The category of elements

Given a presheaf $P : \mathscr{C}^{op} \longrightarrow \mathbf{Set}$ on a category $\mathscr{C}$, *the category of elements of* Γ, denoted by $\int \Gamma$ is defined as follows. Its objects are pairs (C, p) where C is an object of $\mathscr{C}$ and $p \in P(C)$, and a morphism from (C', p') to (C, p) is a morphism $u : C' \longrightarrow C$ of $\mathscr{C}$ such that $P(u)(p) = p'$. Also, there is a projection functor $\pi_P : \int P \longrightarrow \mathscr{C}$ sending $u : (C', p') \longrightarrow (C, p)$ to its underlying morphism $u : C' \longrightarrow C$.

In the case when $\mathscr{C} = G_1$, note that the category $\int \Gamma$ of elements of a digraph $\Gamma = (A, V, t, h)$ can be identified with the category whose set of objects is $A \sqcup V$ and whose morphisms are the identities and the pairs of the form $(t(a), a)$ or $(h(a), a)$ where $a \in A$, domains and codomains being the first and second projections respectively. With this identification the projection $\int \Gamma \xrightarrow{\pi_\Gamma} G_1$ sends the vertices in V to $[0]$, the arrows in A to $[1]$, $(t(a), a)$ to ϵ_0, and $(h(a), a)$ to ϵ_1.

2.1   Realizations

As we will see through this article, the concept of *realization* of a digraph is closely related to that of gestures. In fact, *realization and gestures are dual concepts of each other!* (Subsect. 2.2). For simplicity, we start with realization.

Let $\mathscr{C}$ be a category with small hom-sets, $\Gamma : G_1^{op} \longrightarrow \mathbf{Set}$ a digraph, and $T : G_1 \longrightarrow \mathscr{C}$ a functor. We define *the realization of* Γ *respect to* T, denoted by $|\Gamma|_T$, as the colimit in $\mathscr{C}$ of the functor

$$\int \Gamma \xrightarrow{\pi_\Gamma} G_1 \xrightarrow{T} \mathscr{C},$$

whenever it exists.

Since Γ corresponds to a tuple (A, V, t, h) and T can be identified with a pair of morphisms $i_0, i_1 : T_0 \longrightarrow T_1$ of $\mathscr{C}$, the realization $|\Gamma|_T$ is the limit of the following diagram in $\mathscr{C}$: take a copy of T_1 for each $a \in A$, a copy of T_0 for each $x \in V$, a copy of i_0 whenever $t(a) = x$, and a copy of i_1 whenever $h(a) = x$.

If the realization $|\Gamma|_T$ exist for each digraph Γ, then there is a functor

$$|\_|_T : Digraph \longrightarrow \mathscr{C},$$

which is left adjoint[4] to the functor $\mathscr{C}(T, \_)$ that sends each object C of $\mathscr{C}$ to the digraph $\mathscr{C}(T(\_), C)$. This means that for each digraph Γ and each object of C there is a bijection

$$\mathscr{C}(|\Gamma|_T, C) \cong Digraph(\Gamma, \mathscr{C}(T(\_), C)),$$

natural in both arguments Γ and C. As we will see, this adjunction is very useful in the theory of gestures.

---

[4] See the theorem at [8, p. 47], which holds for cocomplete categories. This theorem remains valid if we only assume the existence of the colimits involved in the definition of L.

2.2   Definition

Let $\mathscr{C}$ be a category with small hom-sets. Given a digraph $\Gamma : G_1^{op} \longrightarrow \mathbf{Set}$ and a functor $S : G_1^{op} \longrightarrow \mathscr{C}$, we define *the object of $\mathscr{C}$ of gestures with skeleton Γ respect to S*, denoted by $\Gamma @ S$, as the limit of the functor

$$\left(\int \Gamma \right)^{op} \xrightarrow{\pi_\Gamma^{op}} G_1^{op} \xrightarrow{S} \mathscr{C},$$

whenever it exists.

Following this definition, since Γ corresponds to a tuple (A, V, t, h) and S can be identified with an internal digraph (S_1, S_0, e_0, e_1) in $\mathscr{C}$, the object of gestures with skeleton Γ respect to S is the limit of the following diagram in $\mathscr{C}$: take a copy of S_1 for each $a \in A$, a copy of S_0 for each $x \in V$, a copy of $e_0 : S_1 \longrightarrow S_0$ whenever $t(a) = x$, and a copy of $e_1 : S_1 \longrightarrow S_0$ whenever $h(a) = x$.

On the other hand, note that this definition is the dual of that of realization. To see this, change $\mathscr{C}$ for $\mathscr{C}^{op}$ in the definition of the realization of Γ respect to T (Subsect. 2.1). In this way, we obtain that the realization of a digraph respect to a functor $T : G_1 \longrightarrow \mathscr{C}^{op}$ (which corresponds uniquely to a functor $S : G_1^{op} \longrightarrow \mathscr{C}$, by applying $(\_)^{op}$) is to be

$$Colim \left(\int \Gamma \xrightarrow{\pi_\Gamma} G_1 \xrightarrow{T} \mathscr{C}^{op} \right) = Lim \left(\left(\int \Gamma \right)^{op} \xrightarrow{\pi_\Gamma^{op}} G_1^{op} \xrightarrow{S} \mathscr{C} \right) = \Gamma @ S.$$

So we have the following delicate and fundamental fact:

The concept of gestures is the dual of that of realization.

By dualizing the case of the realization functor, if the object of gestures $\Gamma @ S$ exists for each digraph Γ, then there is a functor

$$\_ @ S : Digraph^{op} \longrightarrow \mathscr{C},$$

which is right adjoint to the functor $\mathscr{C}(\_, S)$ that sends each object C of $\mathscr{C}$ to the digraph $\mathscr{C}(C, S(\_))$. This means that for each digraph Γ and each object C of $\mathscr{C}$ there is a bijection

$$Digraph(\Gamma, \mathscr{C}(C, S(\_))) \cong \mathscr{C}(C, \Gamma @ S),$$

natural in both arguments Γ and C.

In particular, if the category $\mathscr{C}$ has a terminal object $\mathbf{1}$, then we obtain a bijection between the set $\mathscr{C}(\mathbf{1}, \Gamma @ S)$ of points of $\Gamma @ S$ and

$$Digraph(\Gamma, \mathscr{C}(\mathbf{1}, S(\_))).$$

The digraph $\mathscr{C}(\mathbf{1}, S(\_))$ is called the *underlying digraph* of the internal digraph S.

2.3 Hypergestures

Let C be an object of $\mathscr{C}$ and $T : G_1 \longrightarrow \mathscr{C}$ a functor whose images T_0, T_1 are exponentiable in $\mathscr{C}$. We define the *internal digraph* S_C of C respect to T as the composite

$$G_1 \xrightarrow{T} \mathscr{C} \xrightarrow{C^{(\_)}} \mathscr{C},$$

which is, of course, a contravariant functor. In this case, given a digraph Γ, we write $\Gamma@C$ instead of $\Gamma@S_C$, and call it *the object of gestures with skeleton Γ and body in C*, whenever the limit exists. This construction implies that of *hypergestures*: if Γ' is another skeleton, we can construct the object $\Gamma'@\Gamma@C$, and so on, depending on the existence of suitable limits in $\mathscr{C}$.

This construction of hypergestures is the main reason for which we have defined the object of gestures $\Gamma@S$ with skeleton Γ respect to an internal digraph S. In particular, when the internal digraph is S_C we have defined the object of gestures with skeleton Γ and body in C rather than an individual gesture. Certainly, the key point of the construction of hypergetures is that $\Gamma@C$ is an object of $\mathscr{C}$ again so that can be regarded as a new body for gestures and we can iterate the construction.

2.4 Gestures from External Digraphs

The preceding construction of hypergestures relies on the existence of suitable exponentials. However, the construction of exponentials is not always available so we introduce the following notion of *external digraph* of an object. Besides, this construction allows to give the notion of *a gesture* in contrast to our preceding definition of the object of gestures.

Let $\mathscr{C}$ be a category with small hom-sets, and $T : G_1 \longrightarrow \mathscr{C}$ a functor such that the realization functor $|\_|_T$ exists. Then given an object C of $\mathscr{C}$, we define the *external digraph* s_C of C as the composite

$$G \xrightarrow{T} \mathscr{C} \xrightarrow{\mathscr{C}(\_,C)} \mathbf{Set},$$

which coincides with its underlying digraph (Subsect. 2.2) since it is a functor to **Set**. Therefore, according to Subsect. 2.2, we have a bijection between the points of $\Gamma@s_C$ (that is, its elements) and the set

$$Digraph(\Gamma, \mathscr{C}(T(\_), C)).$$

Consequently, in this case, we can define *a gesture with skeleton Γ and body in C respect to the cosimplicial object T* as a morphism

$$\delta : \Gamma \longrightarrow s_C$$

of digraphs. In this way, the set of gestures $\Gamma@s_C$ is completely determined by all the individual gestures δ, in contrast to the case of the locales of gestures, which need not be characterized by their points (see Sect. 4).

Note that, in turn, s_C coincides with the value at C of the left adjoint to the realization functor (Subsect. 2.1) and hence there is a bijection

$$\mathscr{C}(|\Gamma|_T, C) \cong Digraph(\Gamma, \mathscr{C}(T(\_), C)).$$

Thus, individual gestures with skeleton Γ and body in C correspond bijectively to morphisms from the realization $|\Gamma|_T$ to C.

2.5   An Orientation

Now we proceed to the study of the particular examples. The Fig. 1 offers an orientation for the different variations to be considered. It shows the different incarnations of the functors T and S used in the definition of gestures as well as the respective bodies of the gestures. Note that the gestures related to the columns 2–5 (left to right) come from internal digraphs of objects of the respective categories and hence yield hypergestures. This is not the case for the gestures of the column 6, where S is an external digraph $s_\mathcal{M}$. Despite this, as we have observed, it makes sense to construct individual gestures and to say that $\mathcal{M}$ is the body, but in this case, the object of gestures is not enriched as in the preceding ones. The examples from the columns 2–6 correspond to the sections 3–7 of this article, in order-preserving correspondence.

cat.	**Top**	**Loc**	**Cat(Top)**	**Cat**	**Cat$_R$**
T	$i_0, i_1 : \{*\} \to I$ endpoint inc.	$\mathcal{O}(i_0), \mathcal{O}(i_1)$	$\alpha, \beta : \mathbf{1} \to \mathbb{I}$ $\mathbb{I}$ cat. of $(I, \leq)$ $\mathbf{1}$ final cat.	$\circ\!\!\rightarrow\,\, |\circ$ $\circ\!\!\rightarrow\,\, |\circ$	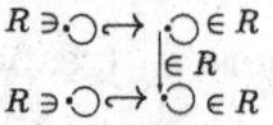 $R \ni \circ\!\!\rightarrow\,\, \circ \in R$ $|\in R$ $R \ni \circ\!\!\rightarrow\,\, \circ \in R$
body	X topological space	L locale	$\mathbb{K}$ topological category	$\mathcal{C}$ category	$\mathcal{M}$ linear category
S	S_X $e_0, e_1 : X^I \to X$ endpoint ev.	S_L $e_0, e_1 : L^{\mathcal{O}(I)} \to L$	$S_\mathbb{K}$ $e_0, e_1 : \mathbb{K}^\mathbb{I} \to \mathbb{K}$	$S_\mathcal{C}$ $dom, cod : \mathcal{C} \overset{\circ\_\circ}{\to} \mathcal{C}$	$s_\mathcal{M} = \mathbf{Cat}_R(T(\_), \mathcal{M})$ external digraph

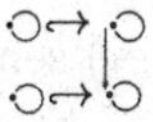

Fig. 1. Ingredients for defining gestures in different categories.

3   Gestures on Topological Spaces

Let Γ be a digraph, X a topological space, and $I = [0, 1]$ the unit *interval* in $\mathbb{R}$. In the sequel, we will denote the set of opens of the topological space X by $\mathcal{O}(X)$.

First, we construct the space X^I of *paths* in X. In fact, the space I is an exponentiable object in **Top** by Theorem [3, 5.3]: it is a *locally compact space*[5],

---

[5] A topological space X is said to be locally compact if for each point $x \in X$ and each open neighborhood U of it, there is a compact neighborhood of x contained in U. In the case when X is a Hausdorff space, this definition is equivalent to saying that each point in X has a compact neighborhood. In this way, every compact Hausdorff space is locally compact.

so $\mathscr{O}(I)$ is a *continuous lattice*[6] by Lemma [6, VII.4.2]. Furthermore, the exponential X^I is the set $\mathbf{Top}(I, X)$ of continuous maps from I to X endowed with the compact-open topology.

The internal digraph in $\mathbf{Top}$ to be considered in this instance is the *spatial digraph* $\overrightarrow{X}$ of the space X. It is the tuple (X^I, X, e_0, e_1), where e_0 and e_1 are obtained by applying the functor $X^{(-)}$ to the inclusions $i_0, i_1 : \{*\} \longrightarrow I$ of the endpoints. Note that $\overrightarrow{X}$ corresponds to the functor S_X defined in Subsect. 2.3.

In this way, since the category $\mathbf{Top}$ of all topological spaces has all small limits, following the definition in Subsect. 2.3, we have the space $\Gamma @X$ of gestures with skeleton Γ and body in X. However, in [10], Mazzola first defines a gesture as a diagram of curves in the topological space X, that is, a morphism of digraphs

$$\delta : \Gamma \longrightarrow \overrightarrow{X},$$

where $\overrightarrow{X}$ is regarded as a digraph by forgetting the topological structure. This means that the spatial digraph $\overrightarrow{X}$ can be identified with its underlying digraph (Subsect. 2.2) since topological spaces are determined by their points. In this way, the elements of $\Gamma @X$ correspond bijectively to these individual gestures δ according to our discussion of points of objects of gestures in Subsect. 2.2.

Example 1. Consider the case when $X = \mathbb{R}^2$. In this case, the spatial digraph $\overrightarrow{\mathbb{R}^2}$ of $\mathbb{R}^2$ is the tuple

$$(\mathbf{Top}(I, \mathbb{R}^2), \mathbb{R}^2, e_0, e_1),$$

where e_0 (respectively e_1) sends a continuous curve $c : I \longrightarrow \mathbb{R}^2$ to $c(0)$ (respectively $c(1)$). In this way, the digraph $\overrightarrow{\mathbb{R}^2}$ has as arrows all continuous curves in $\mathbb{R}^2$ and as vertices all points in $\mathbb{R}^2$.

Now suppose that Γ is the digraph of the Fig. 2, that is, $\Gamma = (\{a, b\}, \{x, y\}, t, h)$, where $t(a) = h(a) = t(b) = x$ and $h(b) = y$. Then a gesture $\delta : \Gamma \longrightarrow \overrightarrow{\mathbb{R}^2}$, which can be illustrated with the Fig. 2, is a pair (u, v), where $u : \{a, b\} \longrightarrow \mathbf{Top}(I, \mathbb{R}^2)$ and $v : \{x, y\} \longrightarrow \mathbb{R}^2$ are functions satisfying the conditions $u(a)(0) = u(a)(1) = u(b)(0) = v(x)$ and $u(b)(1) = v(y)$. In words, it is simply a diagram of curves that match according to the configuration of Γ.

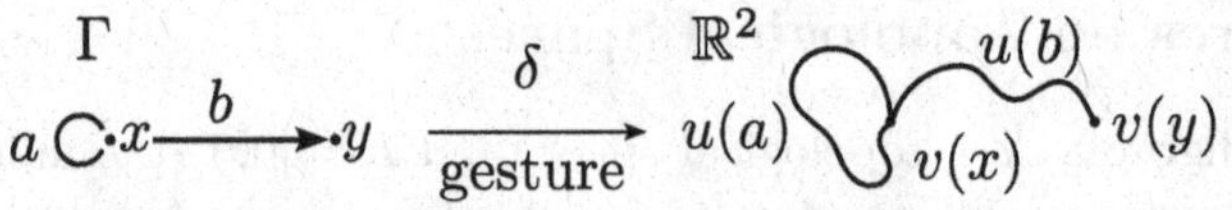

Fig. 2. A topological gesture δ.

On the other hand, the space $\Gamma @\mathbb{R}^2$ is the limit in $\mathbf{Top}$ of the diagram

$$\mathbf{Top}(I, \mathbb{R}^2) \underset{e_0}{\overset{e_1}{\rightrightarrows}} \mathbb{R}^2 \xleftarrow{e_0} \mathbf{Top}(I, \mathbb{R}^2) \xrightarrow{e_1} \mathbb{R}^2 .$$

---

[6] Or core-compact, according to the terminology in [3].

According to the construction of limits (by means of products and equalizers) in **Top**, the space $\Gamma @ \mathbb{R}^2$ is the subspace of the cartesian product (equipped with the Tychonoff topology)

$$\mathbf{Top}(I, \mathbb{R}^2) \times \mathbf{Top}(I, \mathbb{R}^2) \times \mathbb{R}^2 \times \mathbb{R}^2$$

consisting of all tuples (c_a, c_b, p_x, p_y) satisfying the conditions $c_a(0) = c_a(1) = c_b(0) = p_x$ and $c_b(1) = p_y$. Note that such a tuple is essentially the same as a gesture δ. □

Gestures and geometric realization

In the case when the functor $T : G_1 \longrightarrow \mathbf{Top}$ corresponds to the pair of inclusions $i_0, i_1 : \{*\} \longrightarrow I$ of the endpoints, the realization $|\Gamma|$ of a digraph Γ respect to T always exists since **Top** is small cocomplete and is often called the *geometric realization*[7] of Γ.

Example 2. Consider the digraph Γ of the Example 1. The geometric realization $|\Gamma|$ is the colimit in **Top** of the diagram

$$I \underset{i_1}{\overset{i_0}{\rightleftarrows}} \{*\} \xrightarrow{i_0} I \xleftarrow{i_1} \{*\} \;.$$

According to the construction of colimits (via coproducts and coequalizers) in **Top**, the geometric realization $|\Gamma|$ is the quotient of the disjoint union

$$(I \times \{a\}) \cup (I \times \{b\}) \cup \{x\} \cup \{y\}$$

by the relation $\sim$ defined by $(0,a) \sim (1,a) \sim (0,b) \sim x$ and $(1,b) \sim y$. The resulting object is illustrated in Fig. 3. In this way, an open of the quotient topology on $|\Gamma|$ corresponds to a tuple

$$(U_a, U_b, V_x, V_y),$$

where $U_a, U_b \in \mathscr{O}(I)$, $V_x \subseteq \{x\}$, and $V_y \subseteq \{y\}$ satisfying the conditions (i) $0 \in U_a$ iff $1 \in U_1$ iff $0 \in U_b$ iff $x \in V_x$ and (ii) $1 \in U_b$ iff $y \in V_y$. □

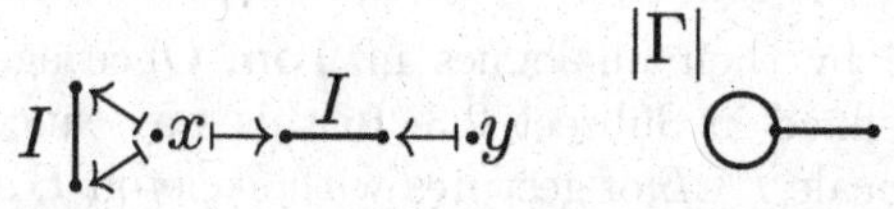

Fig. 3. The way of identifying the points of the disjoint union (left-hand) and the realization of the digraph from Fig. 2 (right-hand)

---

[7] This name is due to Milnor, who first studied the geometric realization in the context of algebraic topology, though for simplicial sets instead of digraphs. However, in [10], this object is called *spatialization*.

By the associated adjunction to the geometric realization (Subsect. 2.1), we have an isomorphism

$$\mathbf{Top}(|\Gamma|, X) \cong Digraph(\Gamma, \overrightarrow{X}),$$

natural in both arguments Γ, X. Thus, a gesture with skeleton Γ and body in X is essentially a continuous map from $|\Gamma|$ to X; for instance, note that the gesture at Fig. 2 can be interpreted as a continuos map from the geometric realization at Fig. 3 to $\mathbb{R}^2$. Moreover one may ask whether there is a homeomorphism

$$X^{|\Gamma|} \cong \Gamma@X.$$

The answer is affirmative iff Γ is a locally finite digraph[8], that is, iff $|\Gamma|$ is exponentiable in **Top**; we omit the proof here. The important point is that this result illustrates a basic problem in gesture theory: *the reduction of objects of gestures defined by the procedure in Subsect. 2.3 to exponentials*. It is important to stress that isomorphisms of the above type are not always possible; for example, if the digraph has infinitely many arrows with the same tail, the above isomorphism makes no sense. And in some respect, this is what makes topological gestures so interesting from a strictly mathematical viewpoint; if they were reducible to exponentials nothing new is to be studied.

4   Gestures on Locales

The category of *frames*, denoted by **Frm** has as objects the *complete Heyting algebras*, that is, complete lattices L satisfying the infinite distributive law $a \wedge \bigvee_{s \in S} s = \bigvee_{s \in S} a \wedge s$, for all $a \in L$ and $S \subseteq L$. The morphisms of frames are the functions that preserve finite meets including **1** and arbitrary joins including **0**. In particular these functions preserve the order. The category **Loc** of *locales* is the opposite of **Frm**. The category **Loc** is small complete and cocomplete (see [13, II.3]), the terminal object $\mathbf{2} = \{\emptyset, \{*\}\}$ being the locale of opens of the singleton.

Let $\Gamma = (A, V, t, h)$ be a digraph and L a locale. As we have already noted, the locale $\mathscr{O}(I)$ is a continuous lattice. Therefore $\mathscr{O}(I)$ is exponentiable in **Loc** (Theorem [6, VII 4.11]) and we have the *locale* $L^{\mathscr{O}(I)}$ *of paths in* L.

The *localic digraph* $\overrightarrow{L}$ of L is the tuple $(L^{\mathscr{O}(I)}, L, e_0, e_1)$ where e_0, e_1 are obtained by applying the functor $L^{(-)}$ to the endpoint inclusions $\mathscr{O}(i_0), \mathscr{O}(i_1)$: $\mathbf{2} \longrightarrow \mathscr{O}(I)$ induced by their analogues in **Top**. Once again, $\overrightarrow{L}$ corresponds to the functor S_L defined in Subsect. 2.3. In this way, since **Loc** has all small limits, we have the locale $\Gamma@L$ of gestures with skeleton Γ and body in L. This definition coincides with that given in [1].

As in the case of topological spaces, there is a realization induced by the inclusions $\mathscr{O}(i_0), \mathscr{O}(i_1) : \mathbf{2} \longrightarrow \mathscr{O}(I)$, and the realization of a digraph coincides with the locale of opens of the geometric realization in **Top**.

---

[8] A digraph is locally finite if each vertex is the tail or head of only finitely many arrows.

Example 3. Let Γ the digraph of the Example 1. The realization $|\Gamma|$ in **Loc** corresponds to the locale $\mathcal{O}(|\Gamma|)$, whose elements were already described in the Example 2. □

Also, we have a reduction to exponentials, namely an isomorphism of locales

$$L^{\mathcal{O}(|\Gamma|)} \cong \Gamma @L,$$

natural in L, for each locally finite digraph Γ.

Locales are the objects of study of the pointless topology, an approach to a great extent derived from the vision of Grothendieck of the notion of topos as a generalization of that of topological space. Locales are in some respect residues of topoi, but they exemplify transparently the spatial aspect of topoi. In first instance, locales need not be characterized by their points, and there are examples (complete boolean algebras without atoms) of locales that are non-trivial and without points at all! As a collateral effect, the objects of gestures on these complete boolean algebras are also non-trivial and with no points.

Example 4. Let $\mathcal{O}(\mathbb{R})_{\neg\neg}$ be the sublocale of $\mathcal{O}(\mathbb{R})$ induced by the double negation nucleus. The elements of $\mathcal{O}(\mathbb{R})_{\neg\neg}$ are the opens $U \in \mathcal{O}(\mathbb{R})$ for which $Int(\overline{U}) = U$. The locale $\mathcal{O}(\mathbb{R})_{\neg\neg}$ is a boolean algebra without atoms and hence has no points. In the same way, if Γ is any non-initial digraph, according to [1, Proposition 4], the space of points of $\Gamma @\mathcal{O}(\mathbb{R})_{\neg\neg}$ is homeomorphic to the space of gestures with skeleton Γ and body in the space of points of $\mathcal{O}(\mathbb{R})_{\neg\neg}$, but the latter is the empty space, and hence the space of points of $\Gamma @\mathcal{O}(\mathbb{R})_{\neg\neg}$ is empty. However, it can be shown that $\mathcal{O}(\mathbb{R})_{\neg\neg}$ is a retract of $\Gamma @\mathcal{O}(\mathbb{R})_{\neg\neg}$, and hence $\Gamma @\mathcal{O}(\mathbb{R})_{\neg\neg}$ is not a trivial locale. In particular, if Γ is the digraph $\bullet \rightarrow \bullet$, the locale $\mathcal{O}(\mathbb{R})_{\neg\neg}^{\mathcal{O}(I)} = \Gamma @\mathcal{O}(\mathbb{R})_{\neg\neg}$ of paths has no points. □

This is a fundamental example for abstract gesture theory since it shows that the notion of *an individual gesture* is insufficient if a theory of gestures on generalized spaces is desired. Besides, if we want to define a correct generalization of gestures on locales, then it is impossible to give a satisfactory definition of a gesture with skeleton Γ and body in $\mathcal{O}(\mathbb{R})_{\neg\neg}$ as a morphism of digraphs $\delta :$ $\Gamma \longrightarrow \overrightarrow{\mathcal{O}(\mathbb{R})_{\neg\neg}}$ since both the locale of paths $\mathcal{O}(\mathbb{R})_{\neg\neg}^{\mathcal{O}(I)}$ and $\mathcal{O}(\mathbb{R})_{\neg\neg}$ have no points—the object $\overrightarrow{\mathcal{O}(\mathbb{R})_{\neg\neg}}$ is not a digraph, but an internal digraph in **Loc** whose underlying digraph (Subsect. 2.2) has no vertices and no arrows!

This generalized notion of space (locales) that is concerned with notions of neighborhoods and coverings rather than points should be taken into account to model the space-time in different ways than usual. It is absolutely legitimate to ask whether the euclidean models $\mathbb{R}^n$ and their derivatives (as the interval object I), and even topological spaces, which are essentially characterized by their points, are suitable to describe processes that have to do with wraps and indecomposable movements that occur through non-atomic neighborhoods of space-time, as in the case of the human body (absolutely indecomposable in terms of points!) or the pianist's hand[9]. Probably, it is time for a new topology, closer to Grothendieck's ideas of a tame (moderate) topology and a geometry of shapes.

5   Gestures on Topological Categories

Topological categories are internal categories (see [8, V.7] or [5, B2.3.1] for the definition) in **Top**. Roughly speaking, this means that a topological category $\mathbb{K}$ is a tuple (C_1, C_0, e, d, c, m) with C_1, C_0 topological spaces of arrows and objects respectively and e, d, c, m continuous operations of unity, domain, codomain, and composition respectively. Topological categories and internal functors in **Top** (which we call *topological functors*) form a category denoted by **Cat(Top)** according to the notation in [5, B2.3.1].

Before explaining the construction of gestures, we mention two basic results on limits and exponentials of internal categories that we will need and whose justification can be found in [1].

Theorem 1. *Let $\mathscr{C}$ be a cartesian category. If $\mathbb{E} = (E_1, E_0, e', d', c', m')$ is an internal category in $\mathscr{C}$ such that E_0, E_1, and the object of composable arrows $E_2 = E_1 \times_{E_0} E_1$ are exponentiable in $\mathscr{C}$, then $\mathbb{E}$ is exponentiable in the category* **Cat**$(\mathscr{C})$ *of internal categories in $\mathscr{C}$.*

Theorem 2. *If $\mathscr{C}$ is a small complete category, then* **Cat**$(\mathscr{C})$ *is small complete.*

Let I be the unit interval in $\mathbb{R}$ and $\mathbb{I} = (E_1, E_0, e', d', c', m')$ the topological category of the poset $(I, \leq)$, that is,

(i) $(E_1, E_0) = (\{(x, y) | \ x \leq y \ \text{in} \ I\}, I)$;
(ii) $e' : E_0 \longrightarrow E_1$ is the diagonal, that is, $e'(x) = (x, x)$;
(iii) $d', c' : E_1 \longrightarrow E_0$ are the first and second projection respectively;
(iv) $E_2 = E_1 \times_{E_0} E_1 = \{((w, z), (x, y)) \in I^2 \times I^2 | \ x \leq y = w \leq z\}$, and $m' : E_2 \longrightarrow E_1$ is defined by $m'((y, z), (x, y)) = (x, z)$; and
(v) the set $E_0 = I$ has the usual topology on I, E_1 is a subspace of $I \times I$ (product topology), and E_2 is a subspace of I^4; so that e' (diagonal), d', c', m' (projections) are continuous.

To show that $\mathbb{I}$ is exponentiable in **Cat(Top)** we check the conditions of Theorem 1: in fact, E_0, E_1, E_2 are exponentiable in **Top**, that is locally compact, since they are closed subsets of some finite power of I, the latter being locally compact since finite products of locally compact spaces are locally compact.

Also, we have two endpoint inclusions into $\mathbb{I}$. In fact, note that the terminal category $\mathbf{1} = (\{*\}, \{*\}, id, id, id, !)$ is the terminal object in **Cat(Top)**. The internal functors $\alpha, \beta : \mathbf{1} \longrightarrow \mathbb{I}$ are defined by $\alpha_0(*) = 0$, $\beta_0(*) = 1$, $\alpha_1(*) = (0, 0)$, and $\beta_1(*) = (1, 1)$.

---

[9] I borrowed this idea from Octavio Agustín-Aquino.

Given a topological category $\mathbb{K}$, the corresponding internal digraph $\overrightarrow{\mathbb{K}}$ of $\mathbb{K}$ in **Cat(Top)** is the tuple $(\mathbb{K}^{\mathbb{I}}, \mathbb{K}, e_0, e_1)$, where $\mathbb{K}^{\mathbb{I}}$ is the category of all topological functors from $\mathbb{I}$ to $\mathbb{K}$ with its set of objects P_0 (that is, of topological functors) topologized as a subspace of $C_1^{E_1} \times C_0^I$ and its set of morphisms P_1 (that is, of natural transformations) topologized as a subspace of $P_0 \times P_0 \times C_1^{E_0}$, and

$$\mathbb{K}^{\mathbb{I}} \xrightarrow{\ e_i\ } \mathbb{K}$$

$$\begin{array}{ccc} F & \longmapsto & F(i) \\ {\scriptstyle \tau}\downarrow & & \downarrow{\scriptstyle \tau_i} \\ G & \longmapsto & G(i), \end{array}$$

for $i = 0, 1$. This internal digraph $\overrightarrow{\mathbb{K}}$ corresponds to the functor $S_{\mathbb{K}}$ defined in Subsect. 2.3, so since **Cat(Top)** is small complete by Theorem 2, for each digraph Γ, we have the topological category of gestures $\Gamma@\mathbb{K}$ with skeleton Γ and body in $\mathbb{K}$. This definition is essentially the same given in [11], where applications of gestures on topological categories in mathematical music theory are discussed.

Example 5. Let Γ be a loop digraph as in the picture

$$a\ \overset{\curvearrowright}{\mathsf{C}}\bullet x\ .$$

Let us make an explicit computation of the topological category $\Gamma@\mathbb{K}$ for any topological category $\mathbb{K} = (C_1, C_0, e, d, c, m)$. First, note that according to the definition of $\Gamma@\mathbb{K}$, it is the equalizer of the diagram

$$\mathbb{K}^{\mathbb{I}} \underset{e_0}{\overset{e_1}{\rightrightarrows}} \mathbb{K}\ .$$

Thus, $\Gamma@\mathbb{K}$ can be described as follows:

(i) Its objects are topological functors $F : \mathbb{I} \longrightarrow \mathbb{K}$, that is, pairs $(F_1, F_0) \in C_1^{E_1} \times C_0^I$ (correspondence on morphisms and on objects) satisfying the functor conditions and $F_0(0) = F_0(1)$. In this way, the set of objects of $\Gamma@\mathbb{K}$ is equipped with the subspace topology of the Tychonoff topology on the product $C_1^{E_1} \times C_0^I$. Here, $C_1^{E_1}$ and C_0^I are function spaces, which are endowed with the compact-open topology.

(ii) A morphism from F to G, where F and G are topological functors as in (i), is a triple (F, G, τ), where $\tau : F \longrightarrow G$ is a natural transformation such that $\tau_0 : F_0(0) \longrightarrow G_0(0)$ and $\tau_1 : F_0(1) \longrightarrow G_0(1)$ are the same morphism. Here we regard τ as a continuous map from I to C_1 satisfying the usual natural transformation conditions. In this way, the set of morphisms of $\Gamma@\mathbb{K}$ is endowed with the subspace topology of the Tychonoff topology on the product
$$C_1^{E_1} \times C_0^I \times C_1^{E_1} \times C_0^I \times C_1^I.\qquad\qquad \square$$

6   Diagrams: Gestures on Categories

Let **Cat** be the category of all small categories, which coincides with **Cat(Set)**. Consider the functor $T : G_1 \longrightarrow \mathbf{Cat}$ identified with the pair of functors F_0, F_1 from the terminal category **1** (just an object and an arrow) to the category of the poset $\{0 < 1\}$, where $F_0(*) = 0$ and $F_1(*) = 1$ (see Fig. 1). Since **Cat** is small cocomplete (Exercise 5 in [7, p. 112]), we know that the realization $\lfloor\_\rfloor_T$ exists according to Subsect. 2.1, but we require a more explicit presentation. Recall (Subsect. 2.1) that $\lfloor\_\rfloor_T$ is left adjoint to the functor $\mathbf{Cat}(T, \_) : \mathbf{Cat} \longrightarrow$ *Digraph* which is essentially the forgetful functor! But we know that it has a left adjoint (unique up to isomorphism), namely the free category functor *Path* (see [7, II.7]), so we can assume that $\lfloor\_\rfloor_T = Path$.

Now **Cat** is cartesian closed by Theorem 1, the categories of functors being the exponentials, so given a category $\mathscr{C}$, we have the internal digraph $S_\mathscr{C}$ from Subsect. 2.3. In this way, we have the category $\Gamma@\mathscr{C}$ of gestures with skeleton Γ and body in $\mathscr{C}$. The interesting fact here is that the reduction to exponentials always holds, that is, we have an isomorphism of categories

$$\Gamma@\mathscr{C} \cong \mathscr{C}^{\lfloor\Gamma\rfloor_T} = \mathscr{C}^{Path(\Gamma)}$$

for any digraph Γ. Therefore, the category $\Gamma@\mathscr{C}$ of gestures with skeleton Γ and body in $\mathscr{C}$ can be identified with the category of all functors from the free category $Path(\Gamma)$ to $\mathscr{C}$. So diagrams are gestures!

Example 6. Let Γ be the digraph $\bullet x \xrightarrow{a} \bullet y$. Its realization in **Cat** is its free category, which is the category with just an arrow plus identities and can be depicted as

$$id_x \,\overset{\curvearrowright}{\subset}\, x \xrightarrow{\;\;a\;\;} y \,\overset{\curvearrowleft}{\supset}\, id_y \;.$$

Note that this category is isomorphic to the category of the poset $\{0 < 1\}$. Moreover, given a small category $\mathscr{C}$, the category $(\bullet x \xrightarrow{a} \bullet y)@\mathscr{C}$ is precisely the category of functors from the category of the poset $\{0 < 1\}$ to $\mathscr{C}$. Thus, the objects of $(\bullet x \xrightarrow{a} \bullet y)@\mathscr{C}$ are essentially morphisms of $\mathscr{C}$ and a morphisms of $(\bullet x \xrightarrow{a} \bullet y)@\mathscr{C}$ from $f : A \longrightarrow B$ to $g : C \longrightarrow D$ is just a pair of morphisms $(h : A \longrightarrow C, k : B \longrightarrow D)$ such that $kf = gh$. That is, our category of gestures is the *category of morphisms of $\mathscr{C}$*. Note that this also exemplifies the exponential reduction.   □

7   Gestures on Linear Categories

Let R be a commutative ring. We define an *R-linear category* to be a category $\mathscr{M}$ with small hom-sets such that for each pair A, B of objects of $\mathscr{M}$ the set of morphisms $\mathscr{M}(A, B)$ is an R-module and such that for each triple A, B, C of objects of $\mathscr{M}$ the composition $\circ : \mathscr{M}(B, C) \times \mathscr{M}(A, B) \longrightarrow \mathscr{M}(A, C)$ is R-bilinear. Given two R-linear categories $\mathscr{M}, \mathscr{N}$, an R-linear functor from $\mathscr{M}$ to $\mathscr{N}$ is a functor $F : \mathscr{M} \longrightarrow \mathscr{N}$ such that for each pair A, B of objects of $\mathscr{M}$ the function $F : \mathcal{M}(A,B) \longrightarrow \mathcal{N}(F(A), F(B))$ is an R-homomorphism of modules. In this way, we have *the category* $\mathbf{Cat}_R$ *of all small R-linear categories and R-linear functors between them.* On the other hand, an ideal $\mathscr{I}$ of an R-linear category consists of a family of subgroups $\mathscr{I}(A,B) \leqslant \mathcal{M}(A,B)$ indexed by all pairs of objects of $\mathcal{M}$ such that $f \in \mathcal{M}(A,B)$ implies $gfe \in \mathcal{M}(D,C)$ for all $e \in \mathcal{M}(D,A)$ and $g \in \mathcal{M}(B,C)$.

Now let k be a commutative field. We say that a small k-linear category $\mathcal{M}$ is a *spectroid* if the non-invertible morphisms of $\mathcal{M}$ form an ideal $Rad(\mathcal{M})$ of $\mathcal{M}$ and if distinct objects of $\mathcal{M}$ are not isomorphic. It can be shown that the first requirement is equivalent to saying that the k-algebras $\mathcal{M}(A,A)$ are local[10] for all $A \in Ob(\mathcal{M})$.

A construction of free R-linear categories is possible in much the same way that in the case of free modules in $\mathbf{Mod}_R$. In fact, there is a functor $R(\_) :$ $\mathbf{Cat} \longrightarrow \mathbf{Cat}_R$ which is left adjoint to the forgetful functor from $\mathbf{Cat}_R$ to $\mathbf{Cat}$. Given a small category $\mathscr{C}$, the R-linear category $R\mathscr{C}$ has as objects the objects of $\mathscr{C}$, for each pair of objects A, B the set $R\mathscr{C}(A,B)$ is defined to be the free module $R^{\mathscr{C}(A,B)}$ on $\mathscr{C}(A,B)$, and the composition is the linear extension of the composition in $\mathscr{C}$. We thus have the functor

$$RT : G_1 \xrightarrow{T} \mathbf{Cat} \xrightarrow{R(\_)} \mathbf{Cat}_R,$$

where T is the functor in Sect. 6; see Fig. 1 for a picture. Further, the realization $\lfloor\_\rfloor_{RT}$ coincides with $R(\_) \circ Path$ since $\lfloor\_\rfloor_T = Path$ and $R(\_)$, as a left adjoint, preserves colimits.

Given an R-linear category $\mathcal{M}$, so as to construct gestures, we consider the external digraph of $\mathcal{M}$ (contravariant functor, Subsect. 2.4)

$$s_{\mathcal{M}} : G_1 \xrightarrow{RT} \mathbf{Cat}_R \xrightarrow{\mathbf{Cat}_R(\_, \mathcal{M})} \mathbf{Set},$$

rather than an internal digraph in $\mathbf{Cat}_R$. So since the functor $\mathbf{Cat}_R(\_, \mathcal{M})$ transforms colimits to limits and the functor $R(\_) \circ Path$ is left adjoint to the forgetful functor $U : \mathbf{Cat}_R \longrightarrow \mathbf{Cat} \longrightarrow Digraph$, we have the bijections

$$\Gamma@s_{\mathcal{M}} \cong \mathbf{Cat}_R(\lfloor\Gamma\rfloor_{RT}, \mathcal{M}) = \mathbf{Cat}_R(RPath(\Gamma), \mathcal{M}) \cong Digraph(\Gamma, U(\mathcal{M})).$$

Note that since right adjoint are unique up to natural isomorphism, the set

$$Digraph(\Gamma, U(\mathcal{M}))$$

is essentially the set of gestures defined in Subsect. 2.4. Moreover, this set of gestures is strongly related to formulas. The difference is that formulas are defined for spectroids $\mathcal{M}$ and that the arrows of the codomain of a formula are only allowed to be non-invertible morphisms of $\mathcal{M}$. A similar result should express formulas as gestures. The better situation would be when the functor Rad (see [10, p. 39]) from spectroids to digraphs has a left adjoint[11]; in such a case, using

---

[10] That is, local rings: all non-invertible elements form a two-sided ideal.

[11] The author ignores whether or not such a left adjoint exists.

the same reasoning from above, this left adjoint could be regarded as a realization such that the associated set of gestures with skeleton Γ and body in a spectroid $\mathscr{M}$ would be isomorphic (as in the above isomorphism) to

$$Digraph(\Gamma, Rad(\mathscr{M})),$$

that is, to the set of formulas! Now the functor $R(\_) \circ Path$ is a naive candidate for such adjoint, but the images of the functor $R(\_) \circ Path$ are not spectroids in general as discussed in the following example and hence we discard it.

Example 7. If Γ is a loop (see the Example 5), then the realization $RPath(\Gamma)$ is isomorphic to the polynomial ring $R[x]$ which is never local since $1 - x$ and x are non-invertible with $1 = 1 - x + x$ invertible. This shows that $RPath(\Gamma)$ is not a spectroid.

However, if R is a field k, the quotient algebra $k[x]/\langle x^2 \rangle$, which can be identified with the algebra of dual numbers, is local with ideal of non-invertible elements generated by the equivalence class of x. Thus, $k[x]/\langle x^2 \rangle$, regarded as the set of morphisms of a category with just an object, is an spectroid.

In this way, a gesture with skeleton a loop and body in the linear category $k[x]/\langle x^2 \rangle$ is just the choice of an equivalence class $[a+bx]$ in $k[x]/\langle x^2 \rangle$. In contrast, a formula in the spectroid $k[x]/\langle x^2 \rangle$ is the choice of a class of the form $[bx]$. For instance, the element $[x]$ is a formula, which can be interpreted as the element x subject to the condition $x^2 = 0$; hence the relation with the intuitive idea of a formula. Finally, note that the class of the unit of k is a gesture that is not a formula.   □

8   Final Comments

Further generalization

The formal definition of gestures in Subsect. 2.2 was deliberately chosen in this form to illustrate the several possibilities of generalizing it. The category G_1 can be replaced by the semi-simplicial category so that we can define gestures whose skeleta are semi-simplicial sets Γ respect to semi-simplicial objects. In that case we can regard digraphs as particular examples of semi-simplicial sets and hence the resultant theory generalizes that for digraphs. This is not only a mathematical fantasy; these generalization have relevant consequences in the theory of gestures for digraphs. For example, in the case of topological spaces, the space of hypergestures $\Gamma'@\Gamma@X$, where Γ', Γ are locally finite digraphs and X is a space, satisfies

$$\Gamma'@\Gamma@X \cong X^{|\Gamma'| \times |\Gamma|} \cong X^{|\Gamma' \times_g \Gamma|},$$

where $\Gamma' \times_g \Gamma$ is the *geometric product* of the digraphs Γ' and Γ, which is usually a semi-simplicial set rather than a digraph. This fact also exemplifies the *combinatorial nature of topological hypergestures with locally finite skeleta*: they basically depend on the digraphs, not on the particular space! Furthermore, the above formula is also valid for gestures on locales.

Gestures and Kan Extensions

The formulas defining objects of gestures and realizations show that the realization functor $|\_|_T$ and the gesture functor $\_@S$ are left and right Kan extensions respectively. In fact, note first that the category of elements $\int \Gamma$ is isomorphic to the comma category $Y \downarrow \Gamma$ and that $(\int \Gamma)^{op}$ is isomorphic to $\Gamma \downarrow Y^{op}$, where $Y : G_1 \longrightarrow \mathbf{Set}^{G_1^{op}}$ is the Yoneda embedding. Thus, from the definitions of realization and gestures in Subsects. 2.1 and 2.2, we obtain the formulas

$$|\Gamma|_T = Colim\ (Y \downarrow \Gamma \xrightarrow{P} G_1 \xrightarrow{T} \mathscr{C}) = Lan_Y(T)(\Gamma)$$

and

$$\Gamma@S = Lim\ (\Gamma \downarrow Y^{op} \xrightarrow{Q} G_1^{op} \xrightarrow{S} \mathscr{C}) = Ran_{Y^{op}}(S)(\Gamma).$$

This means, according to Theorem 1 in [7, X.3], that

> the realization functor $|\_|_T$ is the left Kan extension of T along the Yoneda embedding and, dually, the contravariant gesture functor $\_@S$ is the right Kan extension of S along the opposite of the Yoneda embedding.

This fact helps to locate gesture theory as a particular case of the theory of Kan extensions. Then we have a notion of preservation of gestural structures as shown, for example, by the formula

$$pt(\Gamma@L) \cong \Gamma@pt(L),$$

which says that the space of points of the locale of gestures with skeleton Γ and body in a locale L is homeomorphic to the space of gestures with skeleton Γ and body in the space of points $pt(L)$. Moreover, this viewpoint helps to give a definition of gestures, based exclusively on Kan extensions, that need not deal with limits, that is, there may be objects of gestures that are not pointwise Kan extensions[12].

From the diamond to a category

It is important to make clear that we are not claiming a solution for the so-called 'diamond conjecture', instead we consider that it has not been formulated in a correct way yet. In this way, we hope that the piece of theory presented in this article is useful for giving a more theoretical shape to the diamond diagram [10, p. 43]. In the initial diamond[13], the two vertices related to the category of gestures and the category of formulas should correspond to two particular *realizations* of the category of digraphs (or semi-simplicial objects, if we are more risky).

---

[12] Though probably the more interesting objects to study are the pointwise Kan extensions and hence the realizations and gesture objects defined by means of (co)limits as above.

[13] Which was not precisely a diamond since it is noticed there that there is a possible framework for formulas for each field k.

For gestures it is done, but not for formulas though we are close. Moreover the particular notions of gestures can be compared since we have a notion of preservation of gestural structures, taken from the theory of Kan extensions. Thus we have a category of gestural structures which could be useful to find a precise adjunction between gestures and formulas, allowing us to recover the gestures behind formulas and the formulas behind gestures.

References

1. Arias, J.S.: Gestures On locales and localic topoi. In: Proceedings of the ICMM (2014, to appear)
2. Arias, J.S. : Gesture Theory: topos-theoretic perspectives and philosophical framework. Doctorado Thesis. Universidad Nacional de Colombia (to appear)
3. Escardó, M.H., Heckmann, R.: Topologies on spaces of continuous functions. Topology Proc. **26**(2), 545–564 (2001–2002)
4. Gabriel, P., Roiter, A.V., Keller, B.: Representations of Finite-Dimensional Algebras. Encyclopaedia of Mathematical Sciences, vol. 73. Springer, Heidelberg (1992)
5. Johnstone, P.T.: Sketches of an Elephant: A Topos Theory Compendium, vol. 2. Oxford University Press, Oxford (2002)
6. Johnstone, P.T.: Stone Spaces. Cambridge Studies in Advanced Mathematics, vol. 3. Cambridge University Press, Cambridge (1982)
7. Mac Lane, S.: Categories for the Working Mathematician, 2nd edn. Springer, New York (1998)
8. Mac Lane, S., Moerdijk, I.: Sheaves in Geometry and Logic. Springer, New York (1992)
9. Mazzola, G., et al.: The Topos of Music: Geometric Logic of Concepts, Theory, and Performance. Birkhäuser, Basel (2002)
10. Mazzola, G., Andreatta, M.: Diagrams, gestures and formulae in music. J. Math. Music **1**(1), 23–46 (2007)
11. Mazzola, G.: Categorical gestures, the diamond conjecture, Lewin's question, and the Hammerklavier Sonata. J. Math. Music **3**(1), 31–58 (2009)
12. Migne, J.-P.: Hugonis De S. Victore ... Opera Omnia. Patrologia Latina Tomus CLXXVI, Paris, vol. 2, pp. 925–952 (1854)
13. Pedicchio, M.C., Tholen, W. (eds.): Categorical Foundations: Special Topics in Order, Topology, Algebra, and Sheaf Theory. Encyclopedia of Mathematics and its Applications. Cambridge University Press, Cambridge (2004)

Mathematical Music Theory and the Musical Math Game—Two Creative Ontological Switches

Guerino Mazzola[⊠]

School of Music, University of Minnesota, Minneapolis, USA
mazzola@umn.edu

Abstract. Mathematical Music Theory (MaMuTh) can be understood as a creative support of the musical ontology, a toolset for composition, or a model for theoretical approaches. Several MaMuTh scholars who are also musicians have asked about the opposed possibility, a Musical Math Game (MuMaGm), namely the creative musical support of the mathematical ontology, setting up conjectures, mathematical theories and eventually helping solve mathematical problems. We discuss this idea and our related proposal of music and mathematics being adjoint functors between (the categories of) formulas and gestures. We illustrate this bidirectional ontological shift of creativity between music and mathematics through the history of counterpoint.

Keywords: Creativity · Mathematical Music Theory
Musical mathematics game · Ontology · Counterpoint

1 Introduction

The recent success of Mathematical Music Theory (MaMuTh) has always been relativized by the failure of an "adjoint" movement that one could call Musical Math Game (MuMaGm), a musical *movens* behind mathematical theory. This caveat has been forwarded (orally) in different ways also by my fellow scholars, namely (among others) Moreno Andreatta, Emilio Lluis Puebla ("Mathematics and music are both fine arts."), and Octavio Alberto Agustín-Aquino. Historically, this imbalance of MaMuTh versus MuMaGm is also traced in Leibniz' statement that "Musica est exercitium arithmeticae occultum nescientis se numerare animi." Accordingly, music is only a superficial activity that is driven and caused by a hidden mathematical machine or program. In this perspective, MaMuTh makes perfectly sense and MuMaGm doesn't.

The MuMaGm perspective is however not inexistent, at least as a philosophical approach. The most important philosopher of German romanticism, Georg Philipp Friedrich von Hardenberg, aka Novalis, in his philosophical writings (Fragmente - Kapitel 10) says: "Aller Genuss ist musikalisch, mithin mathematisch." (All enjoyment is musical, consequently mathematical.) He explicitly invokes a "musical mathematics." The English poet and mathematician James Joseph Sylvester in his *Philosophical Transactions* called mathematics the

© Springer International Publishing AG 2017
O. A. Agustín-Aquino et al. (Eds.): MCM 2017, LNAI 10527, pp. 201–212, 2017.
https://doi.org/10.1007/978-3-319-71827-9_15

"music of reason". Leibniz's famous words could be counterbalanced by "Mathematica est ludus musicalis animi se nescientis" (Mathematics is a musical game of the unconscious soul.).

In a more mathematical perspective, Guerino Mazzola has argued [10] that music and mathematics could be mutually adjoint functors between gestures and formulas:

$$\text{formulas} \underset{\text{mathematics}}{\overset{\text{music}}{\rightleftarrows}} \text{gestures}$$

In this wording, mathematics would no longer be the in-depth structure whose surface produces music. Both fields would be different, but balanced and interdependent movements of human expressivity. Of course, the above setup is not strictly mathematical since one would have to specify the categories of formulas and gestures. We will make this topic more precise in Sect. 3.

Although MuMaGm seems to share some reality, one major reason for its historiographic absence is that—as opposed to theories—games are rarely documented if their rules are difficult to explain. The only explicit textual reference to such a game is, ironically, Hermann Hesse's novel *Das Glasperlenspiel*, published in 1943, and describing a fictitious game of the future, where mathematics and music would interact as in supreme human intellectuality. We shall discuss a historical example of MuMaGm when tracing the development of counterpoint in Sect. 4.

In this paper we want to investigate MaMuTh and its adjoint MuMaGm from a particularly important point of view, namely as creative processes that are characterized by a *switch of ontologies* in the following sense. Creativity in MaMuTh starts with a situation of the musical ontology, be it a type of musical structures, such as chords, intervals, motives, a task of musical composition, or a question regarding sound colors, etc. One then transfers such ontological instances to the ontology of mathematics, thereby generating calculations, formulas, theorems, or mathematical models, which then, when transfered back to the corresponding musical entities, generate a creative musical output. The essential point of this type of creativity resides in the switch of ontologies. This MaMuTh creativity restates musical instances in a powerful mathematical ontology and provides *via mathematical procedures* a background for the musical output. In Sect. 2, we want to discuss such creative actions, one example from musical composition, and two examples from the prehistory and history of counterpoint.

In the concluding Sect. 5, we want to present the next big step for a future counterpoint, a step that is enabled by theories and software for contrapuntal composition based upon different interval concepts and also extensions to microtonal pitch systems.

Despite the philosophical flavor of our paper we propose a number of operational and experimental initiatives in favor of a deeper understanding of the creative MuMaGm switch.

2   Three Examples of MaMuTh Creativiy

It is evident that our discourse will not focus on the role of music as a tool for daydreams, a role which is acceptable and was important for Albert Einstein's creative work. For him, music was a useful tool for inspiration and intuition, only, not more than an ornament of psychological relaxation. Our discussion of creativity with music aims at an understanding of its balanced interplay with mathematics. Before we embark in the discussion of creativity we should recall the definition of a creative process that was developed in [11] and has been applied, among others, in the field of computational creativity research [4].

In [11], creativity is defined as a process that comprises seven successive steps:

1. Exhibit an open question
2. Define its (semiotic) context
3. Find a core concept
4. Describe its 'walls'
5. Soften the walls
6. Extend them
7. Evaluate the extended concept

This in particular describes a process of a semiotic nature. The context of the open question (not "problem", but "question", which is less restrictive) is a semiotic system (step 2). At the end of the creative process we have produced new signs that extend the system. Creativity is a strict extension of a semiotic system via new signs that result from extended conceptual walls (step 6). This is the reason why computational creativity has to solve the hard problem of computational semiotics as a preliminary *conditio sine qua non*.

The important concept here is "wall". Let us explain this. A concept has its determining attributes. They delimitate the concept from others. For example, in pre-Einsteinian physics, the Newtonian concept of time was a real number that was essentially the same for all inertial system, it was "God's one and only time." The critical wall of this concept was that it is a *singular* noun. Time was not understood as something that could have a plural. Einstein's creative extension was to admit that time could have a *plural* case, to admit a plurality of times, one for each inertial system, and to describe the transformation between such time instances by the Lorentz transformation. The point here is to understand that this creative Einsteinian switch from singular to plural was difficult because the wall was thought to be an essential attribute of the very concept of time.

For musical practice, the creative process looks like a loop that successively improves the relevant concepts, as shown in Fig. 1. We come back to this loop in Sect. 5.

A last preliminary example of creativity is the following theorem from algebraic topology, which we shall use in Sect. 3:

Theorem 1. *Every group is isomorphic to a fundamental group $\pi_1(X)$ of a topological space X.*

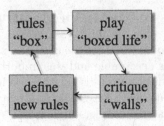

Fig. 1. The creative loop of musical performance.

The creative point here is that the abstract concept of a group is covered by the concept of the homotopy class of a closed curve in X, which comprises an "elastic collection" of curves. It would seem that algebraic abstraction is opposed to the unprecise object of a homotopy class of curves. This theorem proves that this is wrong. So the wall "abstraction" must be opened to include fuzzy objects.

2.1 MaMuTh for Composition: Beethoven's "Hammerklavier Sonate" Revisited

This example refers to Mazzola's sonata op. 3 (l'essence du bleu) [6], an Allegro movement that was composed following a detailed analysis of the Allegro movement of Beethoven's op. 106 (Hammerklavier). The wall in the musical ontology was the impossibility to be more creative than Beethoven in his sonata that is accepted as a most difficult and musically also unsurpassable creation. Mazzola's challenge was to break down this wall. This was done by an ontological switch from music to mathematics. The mathematical analysis of the sonata's harmonic and motivic architecture revealed a mathematical group, namely the symmetry group $Sym_\mathbb{Z}(C\#^{-7})$ of the diminished sevenths chord $C\#^{-7} = \{C\#, E, G, A\#\}$, in the role of defining all possible tonal modulations as well as the motivic kernels, see [7, Ch. 28.2] for details.

In the environment of the mathematical theory of groups, it was quite straightforward to envisage another group, namely the symmetry group $Sym_\mathbb{Z}(C\#^+)$ of the augmented triad $C\#^+ = \{C\#, F, A\}$. These two symmetry groups of chords are fundamentally related to the Sylow groups $\mathbb{Z}_4$ and $\mathbb{Z}_3$ of the pitch class group $\mathbb{Z}_{12} \xrightarrow{\sim} \mathbb{Z}_4 \times \mathbb{Z}_3$. It was therefore mathematically stringent to replace the group of op. 106 by the group $Sym_\mathbb{Z}(C\#^+)$ with the aim of building a modulatory and motivic architecture in analogy to Beethoven's architecture that is derived from $Sym_\mathbb{Z}(C\#^{-7})$. Therefore the ontological switch (as shown in Fig. 2) to mathematics enabled a musical creativity in the composition of a new sonata Allegro movement, Mazzola's op. 3. This composition would have been psychologically and creatively impossible if starting directly from op. 106, this Mount Everest of sonata compositions.

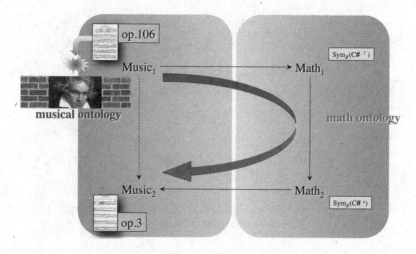

Fig. 2. The creative ontology switch for the composition op. 3 (l'essence du bleu) as derived from Beethoven's op. 106 (Hammerklavier).

2.2   The Pythagorean Prehistory of Counterpoint

This example deals with the prehistory of counterpoint, namely the development of the concept of consonant or dissonant intervals. We want to include the time interval from Pythagoras (around 570 B.C.) to the initiation of polyphony around 900 A.D. We have to start at the initial musical setup which is the monochord, on which Pythagoras and his students were "listening to the first principles of the universe." We have to be aware that they were hindered by a major wall, namely the wall of a total absence of acoustics. No Fourier decomposition, overtones and the like were available 570 B.C. The ontological switch from music to mathematics performed by the Pythagoreans was the mathematical interpretation of musical intervals as found on the monochord's string: dividing the string by two produced an agreeable octave, taking two thirds of the string produced an agreeable fifth, and taking three quarters of the string length produced an equally agreeable fourth. The musical impression was therefore transfered to a mathematical structure, the tetractys. The tetractys was the world formula of ancient Greece. It was the background of all phenomena. This 'formula' consists of ten points (a holy number in ancient Greece) which are arranged in a triangle of one on top, then two, three, and four points stapled on four rows. The ratios 2/1, 3/2, 4/3 of successive row point numbers were interpreted as the rational background for monochord sound intervals (Fig. 3).

This initial switch from musical to mathematical ontology was the basis of the entire concept architecture of interval constructions. The creativity for musical intervals was directed by their mathematical representation. The Pythagorean

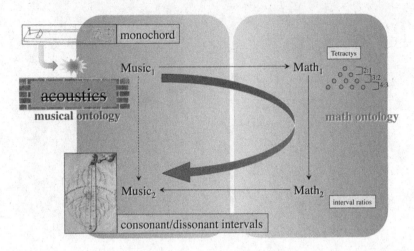

Fig. 3. The creative ontology switch for the development of interval categories following the Pythagorean prehistory.

tuning is built upon the interval types that one may deduce from the prime numbers 2 and 3, which appear in the tetractys.[1]

2.3   From Palestrina-Fux to New Counterpoint Worlds

A third example of MaMuTh creativity starts with the established dichotomy of consonant versus dissonant intervals by Palestrina in the 16th century and then canonized by Johann Joseph Fux in his famous catechism-styled *Gradus ad parnassum* [5]. The consonant intervals modulo octave are prime, minor and major third, fifth, minor and major sixths, denote their set by K. The other six intervals are dissonant, their set is denoted by D. In particular the fourth, which was consonant for Pythagoreans, turned out to be dissonant. We come back to the time interval from 900 A.D. to Palestrina's time in Sect. 4.

Starting from the Palestrina-Fux dichotomy K/D of consonances and dissonances, we are given a musical situation; this dichotomy is the basis of the contrapuntal theory with its five species of increasing complexity. This theory is the endpoint of the contrapuntal development since Fux. In fact, students of music still learn this model in their counterpoint courses. The wall here is the musicological termination and the related educational dead end. And it is a real *dead* end in the sense that the Fux model is not even critically questioned. Fourths are dissonant, opposed to the Pythagorean and also the physical approaches, a fact that has also been discussed as an unsolved problem of music theory by Carl Dahlhaus [3]. Parallels of fifths are forbidden, because they are

---

[1] It was only in the renaissance that the Pythagorean tetractys was extended to what may be called Zarlino's "pentactys", and which added the prime number 5 as a fifth row.

boring. Parallels of thirds are less boring? Here psychology is infused into music theory. But the system of education simply fixes these axioms without any critical analysis. The wall is this teaching of interval processes as a (historically?) sanctioned catechism. And, correlated to this rationale, an extension or variation of the given approach is nearly impossible, except, perhaps, Charles Louis Seeger's idea of a dissonant counterpoint [14], which exchanges K and D. But this is not a creative extension at all (Fig. 4).

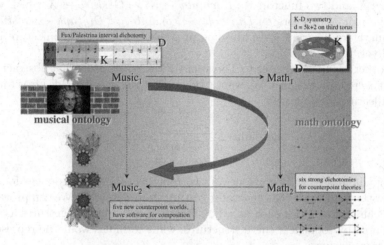

Fig. 4. The creative ontology switch for the Fux-Palestrina classification of intervals to the present five-world theory of also microtonal intervals.

In view of this wall it was reasonable to repeat the Pythagorean MaMuTh and transfer the contrapuntal kernel K/D to the mathematical ontology. This has been realized by Mazzola's mathematical theory of counterpoint, which now is described together with its generalizations in collaboration with Agustín-Aquino and Julien Junod in [1]. This theory exhibits a unique *autocomplementarity* symmetry $A(k) = 5k+2$ on the pitch class group $\mathbb{Z}_{12}$, exchanging the two components K and D. This symmetry is used to model interval successions and in particular *implies* the rule of forbidden parallels of fifth. The dichotomy K/D is also recognized as a geometrically distinguished dichotomy on the toroidal interpretation $\mathbb{Z}_{12} \xrightarrow{\sim} \mathbb{Z}_4 \times \mathbb{Z}_3$ of $\mathbb{Z}_{12}$. Therefore the dissonant fourth is a *consequence* of this geometric fact.

This mathematical theory exhibits five new dichotomies for new worlds of contrapuntal composition, including the corresponding rules of allowed interval successions, and is extended to arbitrary microtonal pitch class groups $\mathbb{Z}_{2n}$ for $n > 2$. The theory has also been implemented in Mazzola's rubato composer software by Junod and is now ready for practical compositions. We come back to this perspective in Sect. 5.

3   Can We Define Math and Music as Adjoint Functors?

The above examples all pertain to the MaMuTh switch. Before we discuss a very important MuMaGm switch, we should present some mathematical arguments for a more balanced switch dynamics between mathematics and music.

A first argument relates to Mazzola's *diamond conjecture* [8] which involves two types of categories, the category *Formula* of formulae and the category *Gesture* of gestures. The diamond conjecture argues that there should be a big category X and two functors $\phi : Formula \to X, \gamma : Gesture \to X$ which would close the pair of functors $r : Digraph \to Formula, g : Digraph \to Gesture$ to a commutative diagram. Despite some progress (see [9]), this conjecture is not proved yet.[2]

Another argument relates to Theorem 1 which guarantees that abstract algebra is 'covered' by gestures. We want to show that this result yields hints towards the conjectured adjointness

$$\text{formulas} \underset{\text{mathematics}}{\overset{\text{music}}{\rightleftarrows}} \text{gestures}$$

Our adjointness conjecture needs two functors, one that produces formulas from gestures, and one that generates gestures from formulas. We can prove that there is an argument for such an adjointness when thinking of gestures as being represented by toplogical curve structures, while formulas would be represented by abstract groups.

Let us discuss in more detail how Theorem 1 is technically demonstrated. We shall see that from the demonstration it also follows that the functor *formulas* $\to$ *gestures* is also musically meaningful. This had already been observed in [8, Section 6.1]. The proof of Theorem 1 for the group $\mathbb{Z}$ is that $\mathbb{Z} \xrightarrow{\sim} \pi_1(1, S^1)$, the fundamental group of the circle S^1. In this case the elements $\sum_n \gamma_n[n]$ of the abstract group algebra $\mathbb{C}\mathbb{Z}$ correspond to elements $\sum_n \gamma_n e^{2\pi n t}$ of the group algebra $\mathbb{C}\pi_1(1, S^1)$, which are precisely the Fourier expressions of a wave of frequency one. The gestures, i.e., loops in $\pi_1(1, S^1)$, are interpreted as partials of the Fourier representation, a thoroughly musical perspective. We want to show now that *this musical interpretation also holds for general groups*.

The theorem we can prove is a weaker statement than adjointness, namely a natural transformation

$$Mu2Ma : HTop(H, F(G)) \to Grp(\pi_1(H), G)$$

---

[2] The well-known adjointness

$$Hom(SZ, Y) \xrightarrow{\sim} Hom(Z, \Omega Y)$$

of the suspension functor S that generates the Homotopy cogroup SZ from a topological space Z, and the loop space functor Ω that generates the Homotopy group ΩY from topogical space Y could be thought as an additional argument, but we refrain from this argument here.

with a functor $F : Grp \rightarrow HTop$ from the category Grp of groups to the category $HTop$ of homotopy classes of pathwise connected topological spaces. The fundamental group functor $\pi_1 : HTop \rightarrow Grp$ is viewed as acting on pointed topological space classes, but as we suppose that such spaces are pathwise connected, the fundamental group is unique up to isomorphism.

Proof. The proof of the existence of the natural transformation $Mu2Ma$ resides on the functor F. If we can show that $\pi_1 \circ F \xrightarrow{\sim} Id_{Grp}$, the natural transformation can be defined by the fundamental group functor, i.e., $Mu2Ma(f : H \rightarrow F(G)) := \pi_1(f) : \pi_1(H) \rightarrow \pi_1(F(G)) \xrightarrow{\sim} G$.

The functor F is derived from the classical construction of a topological space from a given group G. This one goes as follows, refer to [13, Chapter 3, Sect. 8] for a thorough presentation. We first need the construction for free groups. To this end, we first use the evident functor $Free : Grp \rightarrow Grp$ that sends a group G to the free group $Free(G)$ generated by the elements of G. The group G is then recovered by the kernel diagram $Ker(p) \rightarrowtail Free(G) \xrightarrow{p} G$ induced by the identity on G. Given a free group $Free(G)$ that is generated by the set G, we first define the wedge space $Wedge(G) := \coprod_{g \in G}^1 S_g^1$, which is the coproduct of G copies S_g^1 of the unit circle S^1, glued together at point 1. It is then straightforward that $\pi_1(Wedge(G)) \xrightarrow{\sim} Free(G)$. The elements of $\pi_1(1, Wedge(G))$ are the loops at 1, an evident generalization of the fundamental group construction for the free group $\mathbb{Z} = Free(1)$. The less trivial part of Theorem 1 is the management of the kernel $Ker(p)$.

To this end, one uses the method of adjoining 2-cells. One first defines a continuous map $a_x : S^1 \rightarrow Wedge(G)$ that maps the unit circle to the loop which is defined by the element x of $Wedge(G)$. One then embeds S^1 in the closed unit disk E^2, whose boundary is S^1. Finally, one takes the topological quotient space deduced from $Wedge(G) \sqcup \coprod_{x \in Wedge(G)} E_x^2$, one copy E_x^2 of $\dot{E}^2$ for every element $x \in Wedge(G)$, by the identifications of boundary elements of copies E_x^2 with their images via a_x, and taking the coherent topology for the maps a_x. This topology conserves the topology of the interiors $E_x^2 \setminus S_x^1$. Geometrically speaking, we glue the copies E_x^2 to the loops x of $Wedge(G)$ along their boundaries S_x^1. Call this space $F(G)$. The crucial step in this proof is to show that the space $F(G)$ has in fact the fundamental group $\pi_1(F(G)) \xrightarrow{\sim} G$. More precisely, the canonical continuous map $F_G : Wegde(G) \rightarrow F(G)$, when given the fundamental group evaluation $\pi_1(F_G)$, yields a group homomorphism diagram

$$Ker(\pi_1(F_G)) \rightarrowtail \pi_1(Wedge(G)) \xrightarrow{\pi_1(F_G)} \pi_1(F(G))$$

whose kernel is $Ker(p)$, the normal subgroup of $\pi_1(Wedge(G)) \xrightarrow{\sim} Free(G)$ defined above. This implies $\pi_1(F(G)) \xrightarrow{\sim} G$, and we are done. It is straightforward that the space $F(G)$ is functorial in G, being mediated by the functor $Free$ of free groups. QED.

The *musical interpretation* of this construction is now immediate: Coming back to the idea of a group algebra $\mathbb{C}G$, we may interpret the circles S_g^1 of

$Wedge(G)$ as fundamentals of different frequencies f_g, one for each g. The monomials in $\mathbb{C}G$ are related to products of n_ith powers of these fundamentals, i.e., $\prod_i e^{2\pi n_i f_{g_i} t}$. However, these Fourier products are commutative, while this should be avoided for general groups G. We may therefore step over to non-commutative compositions of functions $e^{2\pi n_i f_{g_i} t}$ by juxtaposing these functions in time, unfolding their values as spirals along a time line. The relations given by the 2-cells would then generate a deformation of the cells' boundaries to singular points, as shown in Fig. 5.

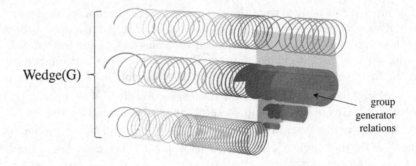

Fig. 5. The spiral representation of fundamental group algebra elements.

The elements $Q(t) = \sum_j c_j F_j(t)$ of the group algebra $\mathbb{C}G$ would represent the time deployment of chord type events, which are defined by the sound events of different summands $F_j(t)$ of $Q(t)$ at time t.

4   The MuMaGm of Medieval Counterpoint

In view of the previous arguments for adjunction between mathematics and music, we take a look at the development of counterpoint from the beginning with Gregorian choral around 900 A.D. to the final stage developed in the 16th century by Palestrina. The central result of this development is the establishment of a stable concept of consonances and dissonances, the Fuxian dichotomy K/D. Initially, the concept of consonances and dissonances was not the final one, fourths were consonant (recall the fourth and fifth *organum*: parallels of a very early type). The process of stabilization of basic contrapuntal concepts took more than 600 years. The mathematical wall was that consonant intervals were seen as individual entities instead of members of a set of a particular quality. The history of this huge field of theoretical and practical experimentation is in part traced in books, e.g., Ernst Apfel's *Diskant und Kontrapunkt des 12. bis 15 Jahrhunderts* [2] and Klaus-Jürgen Sachs: *Der Contrapunctus im 14. und 15. Jahrhundert* [12]. The historiography shows an extremely complex meandering movement, where very differing theoretical approaches, e.g. sixths being dissonant, were tested by a plethora of composers, refused, replaced by other approaches, tested, and so forth. The final result, especially the confirmation of K/D and the forbidden parallels of fifths, was reached without an evident logic. But it is important to understand that the mathematically distinguished structure K/D was reached through a complex *musical* process, where *individual* interval qualities were replaced by a quality of a *collaborative set* of intervals. We may view this type of musical creativity as an excellent example of MuMaGm, a musical game of 600 years duration (!) that eventually provided us with the mathematically excellent solution. *The investigation of whatever logic was responsible for this MuMaGm is an important open research field that could be supported by the experts in MaMuTh in collaboration with musicologists and perhaps also AI simulation technology* (Fig. 6).

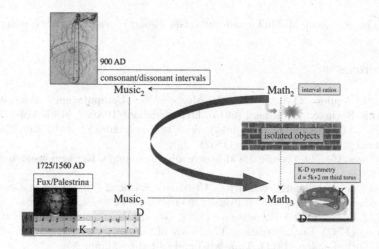

Fig. 6. The MuMaGm from 900 A.D. to Palestrina.

5   The MuMaGm of Future Counterpoint Worlds

The present state of the art of counterpoint seems to envisage a second MuMaGm epoch, where the 'universe' of microtonally extended contrapuntal worlds as described in [1] would be combined and tested by new compositions and eventually lead to new mathematical insights that transcend our present understanding of the contrapuntal universe. See Fig. 7 for the entire processual display of past to future counterpoint.

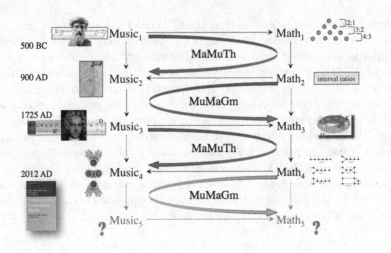

Fig. 7. The processual MaMuTh and MuMaGm display of past to future counterpoint.

References

1. Agustín-Aquino, O.A., Junod, J., Mazzola, G.: Computational Counterpoint Worlds. Springer, Heidelberg (2015). https://doi.org/10.1007/978-3-319-11236-7
2. Apfel, E.: Diskant und Kontrapunkt in der Musiktheorie des 12. bis 15. Jahrhunderts. Heinrichshofen, Wilhelmshafen (1982)
3. Dahlhaus, C.: Zur Theorie des klassischen Kontrapunkts. Kirchenmusikalisches Jb 45 (1961)
4. From Computational Creativity to Creativity Science.: Conference at Center for Interdisciplinary Research, University Bielefeld, 19–22 September 2016
5. Fux, J.J.: Gradus ad Parnassum (1725). Dt. und kommentiert von Mitzler, L., Leipzig (1742). English edition: The Study of Counterpoint. Norton & Company, New York, London (1971). Translated and edited by Mann, A
6. Mazzola, G.: L'Essence du Bleu (sonate pour piano). Acanthus, Rüttenen (2002)
7. Mazzola, G., et al.: The Topos of Music-Geometric Logic of Concepts, Theory, and Performance. Birkhäuser, Basel (2002)
8. Mazzola, G., Andreatta, M.: Diagrams, gestures, and formulas in music. J. Math. Music **1**(1), 23–46 (2007)
9. Mazzola, G.: Categorical gestures, the diamond conjecture, Lewin's question, and the Hammerklavier Sonata. J. Math. Music. **3**(1), 31–58 (2009)
10. Mazzola, G.: Musica e Matematica: due movement aggiunti tra formula e gesti. In: Bertocci, C., Odifreddi, P (eds.) La Matematica: Suoni, Forme, pp. 159–198. Parole, Einaudi (2011)
11. Mazzola, G., Park, J., Thalmann, F.: Musical Creativity. Springer, Heidelberg (2011). https://doi.org/10.1007/978-3-642-24517-6
12. Sachs, K.-J.: Der Contrapunctus im 14. und 15. Jahrhundert. AMW, Franz Steiner, Wiesbaden (1974)
13. Spanier, E.H.: Algebraic Topology. McGraw-Hill, New York (1966)
14. Spilker, J.D.: The origins of "Dissonant Counterpoint": Henry Cowell's unpublished notebook. J. Soc. Am. Music **5**(4), 481–533 (2011)

Graph Theory and Combinatorics

Hamiltonian Graphs as Harmonic Tools

Giovanni Albini[1]([✉]) and Marco Paolo Bernardi[2]([✉])

[1] Department of Music Theory, Conservatorio "F.A. Bonporti", Trento, Italy
mail@giovannialbini.it
[2] Department of Mathematics, University of Pavia, Pavia, Italy
marco.bernardi@unipv.it

Abstract. This article introduces a method for building and studying various harmonic structures in the actual conceptual framework of graph theory. Tone-networks and chord-networks are therefore introduced in a generalized form, focusing on Hamiltonian graphs, iterated line graphs and triangles graphs and on their musical meaning. Reference examples as well as notable music-related Hamiltonian graphs are then presented underlining their relevance for composers.

Keywords: Tone-networks · Note-based graphs · Chord-based graphs
Chord-networks · Lattices · Generalized interval system · Tonnetz
Graph theory · Hamiltonian cycles · Graph dual · Triangles graph
Line graph

1 Introduction

Hamiltonian graphs, also known as Hamilton graphs are graphs possessing a **Hamiltonian cycle**, i.e. a circuit through a graph that visits each node exactly once. The concept is strongly related to those of **Eulerian paths** and **cycles**, i.e. paths and cycles which visit every edge exactly once. Hamiltonicity has been widely investigated and exploited in several branches of applied mathematics and computer science. Given that in the general case testing whether a graph is Hamiltonian is an NP-complete problem [14], the issue of describing efficient procedures for finding such graphs under specific conditions still arouses interest. Additionally, also the pursuit of new necessary and sufficient conditions for a graph to be Hamiltonian is of interest today: developments on the topic are widely summarized in [8–10].

Furthermore, it is well-known that over the last few decades a geometrical approach to music theory has led to several noteworthy examples of music object models availing of the theoretical framework of graph theory or implicitly leading to its representations, as for example in the works by Albini, Baroin, Bigo, Brower, Callender, Cohn, Douthett, Giavitto, Gollin, Lewin, O'Connell, Quinn, Spicher, Steinbach and Tymoczko.

Therefore, this article has two correlated aims. On the one hand, it intends to define a specific and detailed paradigm within the distinctive framework of

O. A. Agustín-Aquino et al. (Eds.): MCM 2017, LNAI 10527, pp. 215–226, 2017.
https://doi.org/10.1007/978-3-319-71827-9_16

graph theory in order to represent music objects related to harmony. On the other hand, it makes use of Hamiltonicity to model, build, study and eventually enumerate certain music structures from the point of view of the paradigm just defined. According to the Authors' view, the overall outcome may be of interest not only in the context of abstract Music Theory, but also that of Composition and Music Analysis.

Tone-networks and chord-networks will be therefore introduced, deepened and classified referring to Lewin's Generalized Interval System (GIS). Their Hamiltonicity will be then studied focusing on two characteristic classes of chord-networks obtained from given tone-networks: iterated line graphs and the triangles graph, that will be defined and examined. Their musical properties will be then shown with the aim of generalizing some concepts introduced in [1]. In conclusion considerations will be made on the advantages given by approaching music-theoretical graphs directly from the point of view of graph theory.

2   Tone-Networks

Let a **tone-network** $T(Q, H)$ be a simple vertex labeled graph whose vertices represent and are labeled as notes (pitches or pitch-classes) and whose edges correspond to intervals (or interval-classes). Q is the set of the vertices, H the set of the edges: the former, in order to not build an empty graph, must contain at least one element, while the connections are arbitrary.[1]

A more formal definition of a tone-network can be formulated by considering a Generalized Interval System and generalizing the definition presented in [1].

Let us recall the definition of a **Generalized Interval System (GIS)**. A GIS is an ordered triple (P, I, φ), where P is a set of pitches (or pitch classes), the **pitch set**, I is an abelian group, the **group of intervals**, and φ is an **action** of I on P which is free and transitive.[2]

Let Q be a subset of P, $Q \subseteq P$, and H a subset of I, $H \subseteq I$: a **tone-network** $T(Q, H)$ is a simple vertex labeled graph which has precisely one vertex for each of the elements of Q; moreover, an edge between two different vertices is present in a tone-network if, and only if, an element in H which maps one of them into the other, exists.[3]

---

[1] The well-known term *lattice* has been deliberately avoided in favor of the more abstract term *tone-network* to define our general note-based graphs. In fact, all lattices are tone-networks as we defined them (in particular, the vertex-transitive ones) but not all tone-networks are, or can be seen as, lattices. The term *chord-network* followed accordingly.

[2] A definition of the Generalized Interval System equivalent to the one given in [15].

[3] Although a GIS in its original formulation admits more general musical elements in its set, a tone-network defined as such admits only Generalized Interval Systems so that P is a set of pitches, pitch classes or similar one-note musical elements (such as for example scale degrees). This allows us to build a framework in which certain graphs obtained from tone-networks always represent chords or general n-note musical elements (such as for example a collection of scale degrees).

Since a tone-network is a simple graph, the identity element has no importance in defining it. Furthermore, the inverse of an element of I always connects the same pair of vertices. In order to avoid confusion, we will suppose that the identity element is thereby always in H and, if an element of I is in H, then its inverse is always in it too. This definition allows us to relate tone-networks to a mathematically clear and consistent model of pitches and intervals, showing some underlying properties of the graph itself. For this purpose, let us now classify tone-networks over their GIS.

Given a GIS (P, I, φ), we shall say that a tone-network $T(Q, H)$ is:

- (P, I, φ)-**complete** if $Q = P$ and $H = I$;
- (P, I, φ)-**proper** if H is a set of generators of I and $Q = P$;
- (P, I, φ)-**unproper** if H is a proper, finite and not empty subset of I but not a set of its generators and $Q = P$;
- (P, I, φ)-**arbitrary** if Q is a proper, finite and not empty subset of P ($Q \subset P$) and H is a not empty subset of I ($H \subseteq I$).

Let us show some results, achieved by relating tone-networks to their GIS, that exhibit tone-network graphical characterizations and that will be important in order to study their Hamiltonicity. The following Proposition 1 is well-known, but for a more self-contained exposition we prove it in details.

Proposition 1. *If a tone-network is (P, I, φ)-complete, its graph is complete and its order is such as $|T(P, I)| = |P| = |I|$.*

Proof. To show that for any GIS (P, I, φ) it is always true that $|P| = |I|$, let's consider $p_0 \in P$ and define a map $f : I \to P$ such as for any $g \in I$ $f(g) = g(p_0)$. φ is transitive then f is surjective; since φ is free then f is injective: therefore $|P| = |I|$ for any GIS. Since the action of φ is free and transitive, given two elements of P there always is a unique element of I which maps the first in the latter. Thus every vertex of $T(P, I)$ is connected with all the other ones and the graph is complete. Hence $|T(P, I)| = |P|$. •

It is worth noting that, from a musical point of view, a (P, I, φ)-complete tone-network is the graphical representation of the GIS (P, I, φ) itself. Moreover, if a tone-network is a complete graph, then a GIS (P, I, φ) for which that tone-network is (P, I, φ)-complete, exists.

Proposition 2. *If a tone-network $T(Q, H)$ is (P, I, φ)-arbitrary, its graph is k-regular with $k = |H \setminus \{e\}| = |\{h \in H \mid h \neq e\}|$ if and only if, for any $q \in Q$ and for any $h \in H$, $h(q) \in Q$.*

Proof. First let's prove that if, for any $q \in Q$ and for any $h \in H$, $h(q) \in Q$, then the graph of a (P, I, φ)-arbitrary tone-network $T(Q, H)$ is k-regular. Let's consider $q_0 \in Q$ and define a map whose domain is $H \setminus \{e\}$ and whose codomain is the set of edges of $T(Q, H)$ to which q_0 belongs. For any $h \in H \setminus \{e\}$, it is then true that $h(q_0) \neq q_0$, so q_0 and $h(q_0)$ are connected; the edge between them is written as $\{q_0, h(q_0)\}$. Let's now define $f(h) = \{q_0, h(q_0)\}$. Since φ is free, f is injective; f is also surjective because of the properties of H: in fact any edge q_0 belongs to is of the type $\{q_0, h(q_0)\}$, or, if $q_0 = h(q_1)$, it is of the type $\{q_1, q_0\}$, and the latter is such that $\{q_1, q_0\} = \{q_0, h^{-1}(q_0)\}$. So domain and codomain have the same cardinality which is the one of $H \setminus \{e\}$, and what has been proved is true for any $q_0 \in Q$: hence the graph is k-regular and $k = |H \setminus \{e\}|$.

Let's now prove by contradiction that if $T(Q, H)$ is k-regular with $k = |H \setminus \{e\}|$ then, for any $q \in Q$ and for any $h \in H$, $h(q) \in Q$. Suppose there are $q_0 \in Q$ and $h_0 \in H \setminus \{e\}$ such as $h_0(q_0) \notin Q$. How many edges q_0 belongs to? All of them must be of the type $\{q_0, h(q_0)\}$ with $h \in H \setminus \{e\}$. So they are no more than $|H \setminus \{e\}| - 1$, since $h_0(q_0) \notin Q$. This contradicts the supposition that the graph is regular with valency $|H \setminus \{e\}|$. •

Note that the only hypothesis that a tone-network $T(Q, H)$ is regular does not imply that, for any $q \in Q$ and for any $h \in H$, $h(q) \in Q$ as well. In effect, a counterexample can be the following: let us consider a $T(Q, H)$ regular tone-network built over the GIS of the twelve equally tempered pitch classes $(\{C, C\#, ..., B\}, I \cong \mathbb{Z}/12\mathbb{Z}, \varphi)$ such that $Q = \{C, D, E, F, G, A, B\}$ and $H = \{0, 1, 2, 10, 11\}$. Although its graph is 2-regular, it is evident that for example $1(C) = C\# \notin Q$.

Proposition 3. *If a tone-network $T(Q, H)$ is (P, I, φ)-proper or is (P, I, φ)-unproper, its graph is k-regular with $k = |H| - 1$.*

Proof. Since for (P, I, φ)-proper and for (P, I, φ)-unproper tone-networks $Q = P$, it is always true that for any $q \in Q$ and for any $h \in H$, $h(q) \in Q$. •

In addition, being that a **Cayley graph**[4] $X(G, S)$ is the graph with vertex set G and edge set $\{gh : hg^{-1} \in S\}$, where G is a group and S a subset of G that is closed under taking inverses and does not contain the identity, we can thus show the following.

Proposition 4. *All the (P, I, φ)-unproper, (P, I, φ)-proper and (P, I, φ)-complete tone-networks are isomorphic to Cayley graphs.*

Proof. Given a GIS (P, I, φ), $H \subseteq I$ and $H^* = H \setminus \{e\}$, we will show that the tone-network $T(P, H)$ is isomorphic to the Cayley graph $X(I, H^*)$. Since all (P, I, φ)-unproper, (P, I, φ)-proper and (P, I, φ)-complete tone-networks are different instances of $T(P, H)$, this will prove the Proposition.

Let's consider $v \in P$ and define a map $f : I \to P$ such as for any $g \in I$ $f(g) = g(v)$. Since the action φ is transitive, f is surjective; since φ is free, f is injective: therefore f is a bijection from the vertex set of the Cayley graph $X(I, H^*)$ to the one of the tone-network $T(P, H)$.

Let's now show that f preserves adjacency. Let gh be an edge in $X(I, H^*)$: we need to show that $g(v)$ and $h(v)$ are connected in $T(P, H)$. In fact $g(v) \neq h(v)$, because the identity is not in H^*, and $(hg^{-1})(g(v)) = h(v)$, hence there is $l \in H$

---

[4] Proper directed Cayley graphs are also known as Cayley graphs. The definition we present of undirected Cayley graphs is the one offered in [7].

such as $l(g(v)) = h(v)$. Vice versa let's suppose there is an edge in $T(P, H)$ connecting $g(v)$ and $h(v)$. We know that $g(v) \neq h(v)$ and also that there is $k \in H$ such as $k(g(v)) = h(v)$; k cannot be the identity element, so $k \in H^*$. Furthermore $hg^{-1}(g(v)) = h(v)$, and, since φ is free, $k = hg^{-1}$. Hence $k \in H^*$ and gh is an edge of the Cayley graph $X(I, H^*)$. •

A Cayley graph $X(G, S)$ is connected if, and only if, S is a set of generators of G, cf. [7]. Furthermore it is well-known that the complement of a disconnected graph is connected. Thus, the following corollaries derive directly from the definitions of (P, I, φ)-complete, (P, I, φ)-proper and (P, I, φ)-unproper tone-networks.

Corollary 5. *All the (P, I, φ)-proper and (P, I, φ)-complete tone-networks $T(Q, H)$ are connected.*

Corollary 6. *All the (P, I, φ)-unproper tone-networks $T(Q, H)$ are disconnected and their complements are (P, I, φ)-proper.*[5]

3   Chord-Networks

Likewise we define a **chord-network** $T(C, R)$ as a simple vertex labeled graph whose vertices represent and are labeled as chords (ordered or unordered set of pitches or pitch-classes) and whose edges correspond to chord transformations.

Note that a chord-network, as it has just been defined, can potentially include sets of notes of different cardinalities, representing and mapping transformations between chords of different sizes. Chord-networks can then be built from scratch or, more interestingly, they can be derived from tone-networks under some certain kinds of graph duality or any type of construction. In this paper, two building methods of chord-networks will be introduced: chord-networks as iterated line graphs and as triangles graphs of tone-networks.

Given a graph W, its **line graph** $L(W)$ is a new simple undirected graph such that each vertex of $L(W)$ is an edge of W, and where two vertices of $L(W)$ are adjacent if, and only if, the corresponding edges are incident in W. Thus $L(T(Q, H))$, the line graph of a tone-network, is a labeled chord-graph whose vertices are all the unordered bichords comprising connected notes in $T(Q, H)$ so that two bichords are adjacent in $L(T(Q, H))$ if, and only if, they share an element of Q. Line graphs present some peculiar properties that make them easy to handle. First of all, it is always easy to determine the number of vertices $v_{L(W)}$ and edges $e_{L(W)}$ of a line graph $L(W)$ where W consists of v_W vertices, that have valency d_i, and e_W edges. Indeed, $v_{L(W)} = e_W$ and $e_{L(W)} = -e_W + \frac{1}{2} \sum d_i^2$. Determining them is easier if W is a complete graph K_n, because $L(K_n) = J(n, 2)$ where $J(n, k)$ is the Johnson graph, such that $v_{J(n,2)} = \binom{n}{2}$ and $e_{J(n,2)} = (n - 2)\binom{n}{2}$, cf. [11]. Furthermore, if a graph W is k-regular, then its line graph $L(W)$ is regular as well with valency $2k - 2$, cf. [7]. This leads us to the following proposition, without the need of a proof.

---

[5] Note that the complement of a (P, I, φ)-proper tone-network is not always (P, I, φ)-unproper!.

Proposition 7. *If a tone-network $T(Q,H)$ is (P,I,φ)-complete, (P,I,φ)-proper or (P,I,φ)-unproper, its line graph is regular with valency $2k - 2$, where k is the valency of $T(Q,H)$, i.e. the number of intervals in H which are different from the identity.*

The process of building line graphs can indeed be iterated[6]. $L^2(T(Q,H)) = L(L(T(Q,H)))$ is then the line graph of the line graph of the tone-network $T(Q,H)$, the **2-iterated line graph**. Its vertices are all the trichords (labeled with one of the three inversions of the trichord) which can be seen as paths of length 2 on $T(Q,H)$ that are connected on $L^2(T(Q,H))$ if, and only if, they share an edge. Musically speaking, connections on $L^2(T(Q,H))$ represent relations between trichords with two or all three notes in common. This graph is quite large and detailed if compared for example to a topological dual under some immersion: every possible trichord that can be built in $T(Q,H)$ is presented in all the eventual inversions[7], implying a much increased set of different chord transformations. Enumerating its vertices and edges is again practical, iterating the calculation made for $L(T(Q,H))$. Moreover, Proposition 7 can be generalized with the following one.

Proposition 8. *If a tone-network $T(Q,H)$ is (P,I,φ)-complete, (P,I,φ)-proper or (P,I,φ)-unproper, then for any $n > 0$ its n-iterated line graph is regular with valency $2^n k - 2(2^n - 1)$, where k is the valency of $T(Q,H)$.*

Proof. Let v_n be the valency of the n-iterated line graph. The statement holds for $n = 1$, in fact: $v_1 = 2^1 k - 2(2^1 - 1) = 2k - 2$. Let's now show that if v_n holds, then also v_{n+1} holds. In fact, $v_{n+1} = 2v_n - 2 = 2(2^n k - 2(2^n - 1)) - 2 = 2(2^n k - 2^{n+1} + 2 - 1) = 2^{n+1} k - 2(2^{n+1} - 1)$.   ●

Finally, the **triangles graph** $\Delta(G)$ of a given graph G is defined as the simple undirected graph having as vertices all the 3-cycles of G and so an edge between two vertices exists in $\Delta(G)$ if, and only if, the corresponding 3-cycles in G share an edge. Thus, the triangles graph of a tone-network, $\Delta(T(Q,H))$, is such that its vertices are all the trichords (labeled with an unordered set of three notes) that can be formed and closed[8] in $T(Q,H)$ and such that two vertices are connected by an edge if, and only if, their trichords share two notes out of three. Clearly, the topological dual of a tone-network $T(Q,H)$ under a triangular immersion in a two-dimensional surface is a subgraph of $\Delta(T(Q,H))$.

---

[6] This paper is limited to n-iterated line graphs with $n < 3$, but it is possible to extend the results and study the cases for $n \geq 3$.

[7] A chord can be redundantly presented in case it is a limited transposition one.

[8] $L^2(T(Q,H))$ represents paths of length two on $T(Q,H)$, while $\Delta(T(Q,H))$ represents 3-cycles on $T(Q,H)$. It means that the former admits all the trichords in all the possible inversions that can be built on $T(Q,H)$. In fact, it could be that some inversions are not possible because of missing edges/intervals. The latter consider only the trichords that admit all the inversions. As a matter of fact a vertex in $\Delta(T(Q,H))$ represents all of them.

4   Hamiltonicity of Tone-Networks and Chord-Networks

Tone-network and chord-network Hamiltonicity shows interesting properties. First of all, Hamiltonian cycles in a tone-network represent a complete and cyclic sequence of notes that recalls and extends the concept of a twelve-tone row. In fact, the Hamiltonian cycles of the complete tone-network $T(\{C, C\#, ..., B\}, I \cong \mathbb{Z}/12\mathbb{Z})$, built over the GIS of the twelve equally tempered pitch classes $(\{C, C\#, ..., B\}, I \cong \mathbb{Z}/12\mathbb{Z}, \varphi)$, are all the twelve-tone rows. Building and enumerating twelve-tone rows - or similar sequences of notes - with specific needs is then simple to do just by shaping the starting tone-network. The issue itself of knowing if a row exists under certain conditions brings us to the problem of checking the existence of a Hamiltonian cycle.

On the other hand, Hamiltonian cycles of chord-networks represent complete sequences through all the admitted chords which only consider certain kinds of transformations. For example, in [1], Hamiltonian cycles in the Chicken-Wire Torus[9] have been deeply studied from a theoretical point of view as well as from that of composition and music analysis. They represent complete sequences through all twenty-four major and minor triads employing neo-Riemannian PLR-transformations and in which each major and minor triad is used once and once only. "These cycles are exclusively triadic and overall completely chromatic, since every pitch class is used exactly six times. As stated in [1]: the succession can also be more or less diatonic, depending on the patterns of the transformations that are employed. So these classes of cycles could be a useful compositional device to define harmonic structures that are triadic (and in some cases locally diatonic) but without any real tonal center."[10]

From a mathematical point of view, checking if a tone-network or a chord-network are Hamiltonian can be quite easy in some specific cases. In fact, despite testing whether a graph is Hamiltonian is an NP-complete problem, cf. [14], and although it is difficult to decide whether a graph is Hamiltonian or not in the general case, the pursuit of necessary and sufficient conditions for a graph to be Hamiltonian continues to provide new results year after year, see [8–10]. Let us present some of them that are useful for our purposes.

Well-known historical results of wide application are, in cronological order, Dirac's [5] and Ore's [16] theorems.[11]

Theorem 9 (Dirac, 1952). *A graph with $n \geq 3$ vertices is Hamiltonian if each of its vertices has degree greater than or equal to $n/2$.*

Theorem 10 (Ore, 1960). *A graph with $n \geq 3$ vertices is Hamiltonian if for every pair of non-adjacent vertices u and v it is true that $d_u + d_v \geq n$.*

---

[9] A representation introduced by Douthett and Steinbach in [6].

[10] Cf. [1].

[11] Note that Dirac's theorem is a corollary of Ore's. In 1976 Bondy and Chvátal's proved a more general result of which Dirac's and Ore's theorem are both corollaries.

A number of authors have independently proved the following result regarding Hamiltonicity of Cayley graphs on abelian[12] groups [18].

Proposition 11. *Every connected Cayley graph on an abelian group is Hamiltonian.*

This along with Proposition 4 and its corollaries leads to the following fundamental corollary.

Corollary 12. (P, I, φ)-*complete and* (P, I, φ)-*proper tone-networks are Hamiltonian.*

In Sect. 6, we will need to study a Cayley graph built on the dihedral group D_{24}. Since D_{24} is not abelian, Proposition 11 cannot be applied. However, the following, shown in [12], is likewise true.

Proposition 13. *Every connected 3-regular Cayley graph on a dihedral group is Hamiltonian.*

In general, determining if a line graph $L(W)$ is Hamiltonian can be much simplified by knowing if W is Eulerian or Hamiltonian itself. In fact, as shown in [11], if W has an Euler cycle (i.e. if it is connected and has an even number of edges at each vertex) then its line graph $L(W)$ is Hamiltonian. Moreover, the line graph of a Hamiltonian one is itself Hamiltonian. The following propositions are then always true.

Proposition 14. *If a tone-network* $T(Q, H)$ *is Hamiltonian, then* $L(T(Q, H))$ *is Hamiltonian as well.*

Proposition 15. *If a tone-network* $T(Q, H)$ *is connected and has an even number of edges at each vertex, then* $L(T(Q, H))$ *is Hamiltonian.*

Finally, the following result, shown in [7], can be useful to prove the Hamiltonicity of certain chord-networks.

Proposition 16. *If a group acts in a freely and transitive way on the vertices of a graph, then that graph is a Cayley one.*

5   Twelve-Tone Rows as Hamiltonian Cycles

Several twelve-tone rows used by composers in notable scores are limited to certain intervals and are characterized by a unique sonority which sometimes reveals a mixture of serialism and tonality. Alban Berg's *Violin Concerto* (1935) and its system of triads or Arvo Pärt's *Symphony No.2* (1966) are just two of the many scores with these features.

---

[12] Lewin did not require the group of a GIS to be abelian [15], but we think that commutativity is a strong requirement for the intuitiveness and consistency of a group of interval. Nevertheless, the result of Proposition 10 seems to apply also to the general case: excluding K_2 all the known non-Hamiltonian vertex transitive graphs are not Cayley graphs and this leads to the conjecture that all Cayley graphs are Hamiltonian [12].

3 4 4 3   3 4 4 3 2 2 2 2 (2)   3 10 1 2 3 10 1 2 3 10 1 (2)

Fig. 1. Twelve-tone rows from Alban Berg's *Violin Concerto* (left) and from Arvo Pärt's *Symphony No.2* (right).

As shown in Fig. 1, they are built on a limited set of intervals - the former: 2, 3, 4 and implicitly their inverses 10, 9 and 8; the latter: 1, 2, 3 and their inverses 11, 10 and 9 - and they link the last tone of the row with the first one again with one interval of the limited set. Thus, both can be represented as Hamiltonian cycles of two proper tone-networks built over the GIS of the twelve equally tempered pitch classes ($\{C, C\#, ..., B\}, I \cong \mathbb{Z}/12\mathbb{Z}, \varphi$). Clearly, both the rows feature also other interesting structural qualities: that of Berg's presents four overlapping triads built on the four open strings of the violin, while Pärt's is made up of one single tetrachord which has been then transposed twice.

Composers, who want to know if such kinds of rows exist[13] under specific conditions and who eventually wish to enumerate and study them, could find a powerful tool in our theoretical framework. For example, thanks to Corollary 6, they can be sure that no rows can be built on unproper tone-networks. Indeed, both Berg's and Pärt's examples consider a tone-network $T(Q, H)$ where H is a set of generators for the group of intervals of the underlying GIS. Moreover, thanks to Theorem 9, they can always begin from a complete tone-network and make it into an arbitrary one, cutting edges under their needs while keeping their valency major or equal to half the total number of vertices. In this way they are sure they can find Hamiltonian cycles, thus rows, on it. The practical computation of the rows - seen as Hamiltonian cycles in suitable tone-networks - can be then done by making use of certain graph theory computer programs[14]. In fact, if answering the question 'does a twelve-tone row under certain conditions exist' is in some interesting cases easy, the same cannot be said for the issue of counting them. The NP-completeness of the general problem imposes the use of a computer to accomplish the task. At least though, the graph-theoretical framework reformulates the problem in a way that makes it easy and quick to represent it in order to implement a solution.

6   The Tonnetz, Its Chord-Networks and Their Hamiltonicity

As reference examples we present the **Tonnetz**, perhaps one of the older and most iconic music-theoretical graph - that in our theoretical framework is the

---

[13] This can be obviously done also in an arbitrary n-tone system.

[14] The software employed by the Authors is *Groups and Graph* version 3.6 by William Kocay.

tone-network $Ton = T(\{C, C\#, ..., B\}, \{0, 3, 9, 4, 8, 5, 7\})$ built over the GIS of the twelve equally tempered pitch classes $(\{C, C\#, ..., B\}, I \cong \mathbb{Z}/12\mathbb{Z}, \varphi)$ - and some derived chord-networks: $L(Ton)$, $L^2(Ton)$, $\Delta(Ton)$ and its topological dual under its immersion in the torus, $D(Ton)$, the Chicken-Wire Torus by Douthett and Steinbach, cf. [6].

Since $\{0, 3, 9, 4, 8, 5, 7\}$ is a set of generators of $\mathbb{Z}/12\mathbb{Z}$ and its vertices are all the elements of the underlying GIS, Ton is a $(\{C, C\#, ..., B\}, I \cong \mathbb{Z}/12\mathbb{Z}, \varphi)$-proper tone-network, a 6-regular graph (Proposition 3) with 12 vertices and 36 edges and it is isomorphic to a connected Cayley graph (Proposition 4 and Corollary 5). If we had not already known Ton, we could have found all these properties and we could have imagined its shape without drawing a single vertex. This provides substantive information for testing its Hamiltonicity and the Hamiltonicity of some of the chord-networks we can build from it.

The same applies to its n-iterated line graphs. $L(Ton)$ is a 10-regular graph with 36 vertices and 180 edges. It means that there are 36 different bichords in the Tonnetz and 10 different transformations joining them preserving one pitch-class on the two. Consequently $L^2(Ton)$ is a quite big 18-regular graph: it has 180 vertices that are all the possible trichords that can be built on the Tonnetz in all their possible inversions and it considers 1620 edges totally.

Being a $(\{C, C\#, ..., B\}, I \cong \mathbb{Z}/12\mathbb{Z}, \varphi)$-proper tone-network Ton is Hamiltonian (Proposition 11), as are $L(Ton)$ and $L^2(Ton)$ accordingly (Corollary 12).

$D(Ton)$, whose Hamiltonicity has been already studied in [1], is a 3-regular connected graph with 24 vertices - all the major and minor triads - and 36 edges - which consider just three parsimonious voice leading transformations: the neo-Riemannian P (*Parallel*), R (*Relative*) and L (*Leading-Tone Exchange*). In [1] the Hamiltonicity of $D(Ton)$ was established by an explicit computation of all its Hamiltonian cycles, but it is also possible to prove it from a mathematical point of view. In fact, the dihedral group D_{24} is the group of automorphisms of $D(Ton)$ which acts freely and in a transitive way on it and P, L and R constitute a set of its generators. Thus, thanks to Proposition 16 and to Proposition 13, we can prove that $D(Ton)$ is a Cayley graph, thus Hamiltonian.

Finally, $\Delta(Ton)$ (Fig. 2) is a graph with 28 vertices, the 24 major and minor triads plus the 4 augmented ones, and 60 edges. It considers nine types of parsimonious voice leading transformations, the neo-Riemannian P, L and R plus six that map an augmented triad in a major or in a minor one preserving two notes. $D(Ton)$ - the 3-regular graph that considers only major and minor triads and PLR-transformations - and the well-known Douthett and Steinbach's Cube Dance, [6], are subgraphs of $\Delta(Ton)$. The former can be obtained deleting the four vertices labeled with the four augmented triads, the latter cutting the edges that represent R transformations. The Hamiltonicity of $\Delta(Ton)$ can be easily derived from the one of its subgraphs $D(Ton)$ and Cube Dance. In fact $\Delta(Ton)$ has four more vertices than $D(Ton)$, the four augmented triads, and since they are connected to major and minor triads that are connected to each other by R transformations and since the cycle that alternates L and R - cycle #41 in [1] - is

Hamiltonian in $D(Ton)$, then the four augmented triads lie in a Hamiltonian cycle in $\Delta(Ton)$. In an even simpler way, the Hamiltonicity of $\Delta(Ton)$ can be deduced from the one of Cube Dance, which in turn is easy to check. In fact, in a cube any pair of opposite vertices is connected by a (not unique) Hamiltonian path. Four such paths form a Hamiltonian cycle in the Cube Dance graph, and therefore in $\Delta(Ton)$ too.

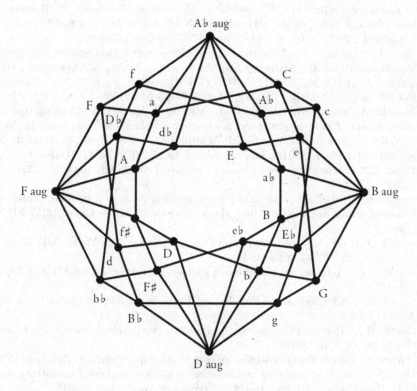

Fig. 2. $\Delta(Ton)$.

7   Conclusions

Tone-networks and chord-networks have been introduced with the aim to offer a paradigm in the distinctive framework of graph theory in order to represent music objects related to harmony. Two new music graphs have been introduced: the iterated line graph and the triangles graph of a tone-network. Tone-network and chord-network Hamiltonicity has then been studied in the general case and two distinctive examples have been analyzed: twelve-tone rows and the Tonnetz and some of its tone-networks. Hamiltonicity has also served to show the potential of the paradigm introduced as a tool not only for music theorists but also for composers. Future developments might try to extend the results to other well-known music-theoretical graphs and to other graphical properties - distinct from Hamiltonicity - with some musical meaning.

226 G. Albini and M. P. Bernardi

References

1. Albini, G., Antonini, S.: Hamiltonian cycles in the topological dual of the Tonnetz. In: Chew, E., Childs, A., Chuan, C.H. (eds.) MCM 2009. Communications in Computer and Information Science, vol. 38, pp. 1–10. Springer, Heidelberg (2009). https://doi.org/10.1007/978-3-642-02394-1_1
2. Baroin, G.: The planet-4D model: an original hypersymmetric music space based on graph theory. In: Agon, C., Andreatta, M., Assayag, G., Amiot, E., Bresson, J., Mandereau, J. (eds.) MCM 2011. LNCS (LNAI), vol. 6726, pp. 326–329. Springer, Heidelberg (2011). https://doi.org/10.1007/978-3-642-21590-2_25
3. Bigo, L., Giavitto, J.-L., Spicher, A.: Building topological spaces for musical objects. In: Agon, C., Andreatta, M., Assayag, G., Amiot, E., Bresson, J., Mandereau, J. (eds.) MCM 2011. LNCS (LNAI), vol. 6726, pp. 13–28. Springer, Heidelberg (2011). https://doi.org/10.1007/978-3-642-21590-2_2
4. Bigo, L., Andreatta, M., Giavitto, J.-L., Michel, O., Spicher, A.: Computation and visualization of musical structures in chord-based simplicial complexes. In: Yust, J., Wild, J., Burgoyne, J.A. (eds.) MCM 2013. LNCS (LNAI), vol. 7937, pp. 38–51. Springer, Heidelberg (2013). https://doi.org/10.1007/978-3-642-39357-0_3
5. Dirac, G.A.: Some theorems on abstract graphs. Proc. Lond. Math. Soc. 2, 69–81 (1952)
6. Douthett, J., Steinbach, P.: Parsimonious graphs: a study in parsimony, contextual transformations, and modes of limited transposition. J. Music Theor. 42(2), 241–63 (1998)
7. Godsil, C., Royle, G.: Algebraic Graph Theory. Springer, New York (2001). https://doi.org/10.1007/978-1-4613-0163-9
8. Gould, R.J.: Updating the Hamiltonian problem - a survey. J. Graph Theor. 15(2), 121–157 (1991)
9. Gould, R.J.: Advances on the Hamiltonian Problem - A Survey. Emory University, Atlanta (2002)
10. Gould, R.J.: Recent advances on the Hamiltonian problem: survey III. Graphs Comb. 30(1), 1–46 (2014)
11. Harary, F.: Graph Theory. Addison-Wesley Publishing Company, Reading (1969)
12. Heus, A.: A study of necessary and sufficient conditions for vertex transitive graphs to be Hamiltonian. Master's thesis, University of Amsterdam (2008)
13. Hiller, P.: Arvo Pärt. Clarendon Press, Oxford (1997)
14. Karp, R.M.: Reducibility among combinatorial problems. In: Miller, R.E., Thatcher, J.W. (eds.) Complexity of Computer Computations, pp. 85–103. Plenum, New York (1972)
15. Lewin, D.: Generalized Musical Intervals and Transformations. Yale University Press, New Haven (1987)
16. Ore, O.: A note on Hamiltonian circuits. Am. Math. Mon. 67, 55 (1960)
17. Tymoczko, D.: A Geometry of Music: Harmony and Counterpoint in the Extended Common Practice. Oxford University Press, Oxford (2011)
18. Witte, D.S., Gallian, J.A.: A survey: Hamiltonian cycles in Cayley graphs. Discrete Math. 51, 293–304 (1984)

New Investigations on Rhythmic Oddity

Franck Jedrzejewski$^{(\boxtimes)}$

CEA-INSTN, Universit Paris Saclay, Paris, France
franckjed@gmail.com

Abstract. The "rhythmic oddity property" (rop) was introduced by ethnomusicologist Simha Aron in the 1990s. The set of *rop words* is the set of words over the alphabet $\{2, 3\}$ satisfying the rhythmic oddity property. It is not a subset of the set of Lyndon words, but is very closed. We show that there is a bijection between some necklaces and rop words. This leads to a formula for counting the rop words of a given length. We also propose a generalization of rop words over a finite alphabet $\mathcal{A} \subset \{1, 2, \ldots, s\}$ for some integer $s \geq 2$. The enumeration of these generalized rop words is still open.

Keywords: Combinatoric on words · Lyndon words · Rhythmic oddity Music formalization

The *rhythmic oddity property* was discovered by ethnomusicologist Simha Aron [2] in the study of Aka pygmies music. The rhythms satisfying this property are refinement of aksak rhythms described by C. Brăiloiu [5] in 1952, and are also used by Turkish and Bulgarian music. They have been studied by G.T. Toussaint in [11] , M. Chemillier and C. Truchet in [7] and André Bouchet in [4] who gave some important characterizations. Related problems of asymmetric rhythms have been studied by Rachael Hall and Paul Klingsberg [8,9]. In this paper, we carry on these studies by showing a one to one correspondence between some necklaces and words satisfying the rhythmic oddity property. In the last section, a generalization of rop words is proposed.

1 The Rhythmic Oddity Property

Patterns with rhythmic oddity property are combinations of durations equal to 2 or 3 units, such as the famous Aka pygmies rhythm 32222322222, and such that when placing the sequence on a circle, "one cannot break the circle into two parts of equal length whatever the chosen breaking point." In other words, no two of the onsets of the rhythm are located diametrically opposite to each other on the circle. In the language of combinatorics of words, this property in terms of words ω over the alphabet $A = \{2, 3\}$ is defined as follows. The height $h(\omega)$ of a word $\omega = \omega_0 \omega_1 \ldots \omega_{n-1}$ of length n is by definition the sum of its letters $h(\omega) = \sum_{j=0,\ldots,n-1} \omega_j$. A word ω satisfies the rhythmic oddity property

© Springer International Publishing AG 2017
O. A. Agustín-Aquino et al. (Eds.): MCM 2017, LNAI 10527, pp. 227–237, 2017.
https://doi.org/10.1007/978-3-319-71827-9_17

(rop) if $h(\omega)$ is even and no cyclic shift of ω can be factorized into two words uv such that $h(u) = h(v)$. For short, we call *rop* word a word over the alphabet $\{2,3\}$ satisfying the rhythmic oddity property. For instance, the word 32322 of height 12 is a rop word, as well as all words of the form $32^n 32^{n+1}$ for all non-negative integers n (the notation 2^n means the letter 2 is repeated n times). The properties of rop words have been outline in [7]. The set of rop words is not a subset of the set of Lyndon words: the words 222 and 233233233 are rop words, but not Lyndon words. A *Lyndon word* (see for example textbooks [10] or [3]) is a string that is strictly smaller in lexicographic order than all of its rotations. Conversely, the set of Lyndon words is not included in the set of rop words, since 2233 is a Lyndon word, but not a rop word (the words 23 and 32 have the same height and 2332 is a rotation of 2233). A *Lyndon rop word* is a word of the monoid $\{2,3\}^*$ that is both a Lyndon word and a rop word. The aim of this paper is to count the number of Lyndon rop words and the number of rop words of length n.

2   Properties of rop Words

Let $\mathcal{A}$ denote a finite set of symbols. The elements of $\mathcal{A}$ are called letters and the set $\mathcal{A}$ is called an alphabet. A word over an alphabet $\mathcal{A}$ is an element of the free monoid $\mathcal{A}^*$ generated by $\mathcal{A}$. The identity element ε of $\mathcal{A}^*$ is called the *empty word*. A word $\omega \in \mathcal{A}^*$ is written uniquely by $\omega = a_0 a_1 ... a_{r-1}$ with letters $a_j \in A$ for $j = 0, 1, ..., r-1$. The length of ω is r and denoted by $|\omega|$. If $\omega \in \mathcal{A}^*$ and $a \in \mathcal{A}$ is a letter, $|\omega|_a$ denotes the number of occurrences of the letter a in the word ω.

$$|\omega| = \sum_{a \in \mathcal{A}} |\omega|_a \tag{1}$$

If $\omega \in \mathcal{A}^*$, and if the alphabet $\mathcal{A}$ is a set of integers, the height $h(\omega)$ of $\omega = a_0 a_1 ... a_{r-1}$ is the sum of its letters

$$h(\omega) = \sum_{j=0}^{r-1} a_j = a_0 + a_1 + \cdots + a_{r-1} \tag{2}$$

Until the last section, the alphabet will always be the set $\mathcal{A} = \{2,3\}$ of two letters. The cycle δ of ω is defined by $\delta(\varepsilon) = \varepsilon$ and $\delta(a\omega) = \omega a$, for $a \in \{2,3\}$. The rotations of ω are the words $\delta^i(\omega)$, for all positive integer $i > 0$.

Definition 1. *A word $\omega \in \{2,3\}^*$ satisfies the rhythmic oddity property (rop) if*

(i) $h(\omega)$ is even
(ii) no cyclic shift of ω can be factorized into two words uv such that $h(u) = h(v)$.

This definition excludes the trivial case of a rop word ω with odd height. If the height of ω is odd, the second condition is automatically verified. In his article [4] of 2010, André Bouchet gave the following characterization of a rop word.

Theorem 1. *Let $\omega = \omega_0\omega_1 \ldots \omega_{n-1}$ be a word over the alphabet $\mathcal{A} = \{2,3\}$ of height $2h$. The word ω is a rop word if and only if the two conditions are satisfied:*

(i) The length of ω is odd, say $2\ell + 1$.

(ii) The height of the prefixes of length ℓ of the rotations $\delta^i(\omega)$ of ω is equal to $h - 2$ or $h - 1$.

The proof of this theorem is based on the two following properties.

Proposition 1. *Let ω be a rop word of length n and let i and λ be two integers such that $0 \leq \lambda < n$.*

(1) If the height of the prefix of length λ of the word $\delta^i(\omega)$ is $h - 1$ or $h - 2$, then the height of the prefix of $\delta^{i+1}(\omega)$ is $h - 1$ or $h - 2$.

(2) The word ω has a prefix of height $h - 1$ or $h - 2$.

Summarising, the properties of rop words are the following.

Proposition 2. *If $\omega \in \{2,3\}^*$ of height $2h$ satisfies the rhythmic oddity property then*

(i) the number $|\omega|_2$ of symbol 2 in ω is odd

(ii) the number $|\omega|_3$ of symbol 3 in ω is even

(iii) the length of ω is odd

(iv) $\delta^i(\omega)$ is a rop word for any i

(v) ω and $\delta^i(\omega)$ have a prefix of height $h - 1$ or $h - 2$.

Some of these properties are consequences of the previous results. Since the height of ω satisfies $2|\omega|_2 + 3|\omega|_3 = 2h$, the number of occurrences of letter 3 is even. Thus, since the length of ω is odd by the previous proposition, the number of occurrences of letter 2 is even.

Furthermore, André Bouchet defines d-pairing. Let ω be a word of length n and d be an integer such that $0 < d \leq n/2$. A d-pairing of ω is a partition of the subset of indices $\{i : 0 \leq i < n, \omega_i = 3\}$ into pairs of indices $\{j, j+d\}$. Arithmetic operations on indices are to be understood mod n. Let $\omega = \omega_0\omega_1 \ldots \omega_{n-1}$ be a word of $\{2,3\}^*$ and d a positive integer coprime with n. Denote by $\omega^{(d)} = x_0x_1 \ldots x_{n-1}$ the word obtained by reading all letters of ω by step d, starting from ω_0. Namely, each letter of $\omega^{(d)}$ is $x_j = \omega_k$ with $k = jd \bmod d$, $0 \leq j < d$. For instance, the word $\omega = 2233233$ depicted on Fig. 1 below with $n = 7$ and $d = 3$ admits a 3-pairing and $\omega^{(3)} = 2333322$. As a corollary of the previous result, A. Bouchet shows the following result, which is the theoretical meaning of the *Hop-and-jump* algorithm given by Godfried T. Toussaint in [11].

Theorem 2. *Let ω be a word of even height. ω is a rop word if and only if the two conditions are satisfied:*

(i) The length of ω is odd, say $2\ell + 1$.

(ii) ω admits a ℓ-pairing.

Some years before, Marc Chemillier and Charlotte Truchet [7] gave another characterization of rop words by introducing asymmetric pairs.

Definition 2. *The words* (u, v) *form an asymmetric pair if no pair of prefixes* (u', v') *of u and v respectively exist such that* $h(u') = h(v') + 1$.

For instance, $(2233, 233)$ is an asymmetric pair, but $(2232, 333)$ is not since the pair of prefixes $(22, 3)$ verifies $h(22) = h(3) + 1$.

Theorem 3. *A word* ω *satisfies the rhythmic oddity property if and only if there exists an asymmetric pair* (u, v) *such that* $\omega = uv$ *or* $\omega = vu$ *with* $h(v) = h(u) + 2$.

A construction of asymmetric pairs, given by Chemillier and Truchet, uses two functions over the monoid $\mathcal{A}^* \times \mathcal{A}^*$ into itself, namely

$$r(u, v) = (3u, 3v), \qquad s(u, v) = (v, 2u) \tag{3}$$

and identifies any word ω of $\mathcal{A}^*$ with a word α over $\{r, s\}^*$ by the map $\omega \rightarrow f(\omega)$ such that $f(\omega) = \alpha(\varepsilon, \varepsilon)$ where ε is the empty word. For instance, the Cuban *tresillo* rhythm represented by the word 332 is associated with the word sr, since $(3, 32) = s(3, 3) = sr(\varepsilon, \varepsilon)$. The characterization of rop words ω is then moved to a property of the associated word α.

Theorem 4. *A word* ω *satisfies the rhythmic oddity property if and only if there exists a word* $\alpha \in \{r, s\}^*$ *with* $|\alpha|_s$ *being odd, such that* $\delta^n(\omega) = uv$ *or* $\delta^n(\omega) = vu$ *for some n with* $(u, v) = \alpha(\varepsilon, \varepsilon)$.

3   A Bijection Between Some Necklaces and rop Words

Let n_2 and n_3 be the number of symbols 2 and 3 in ω and $n = n_2 + n_3$ be the length of ω. For a given n_2, we will use the d-pairing to show that there is a one to one correspondence between aperiodic necklaces of length n with n_2 black beads (represented by letter 2) and $(n - n_2)$ white beads (represented by letter 3) and Lyndon rop words of length $n' = 2n - n_2$ with $n'_2 = n_2$ letters 2 and $n'_3 = 2(n - n_2)$ letters 3. And also a one-to-one correspondence between necklaces (eventually periodic) of length n with n_2 black beads (represented by letter 2) and $(n - n_2)$ white beads (represented by letter 3) and rop words. The correspondence is obtained by adding or removing the letters 3 coming from the pairing. Let us examine an example (See Fig. 1).

Fix n_2, for instance $n_2 = 3$, and let n be $n = 5$. The word 2233233 is a (Lyndon) rop word with odd length 7 ($n'_2 = 3$, $n'_3 = 4$) since it has a 3-pairing. Put the word on a circle, starting from the bottom and turn counterclockwise as shown on the Fig. 1. Now discard the second 3 of each pair $(3, 3)$ turning counterclockwise. Reading the remaining word clockwise starting from the bottom gives the word 22332, one of the two necklaces of length 5 with 3 letters 2. Conversely, starting from the word 22233, it is easy to add a 3-pairing by doubling each letter 3, with respect to the counterclockwise tour.

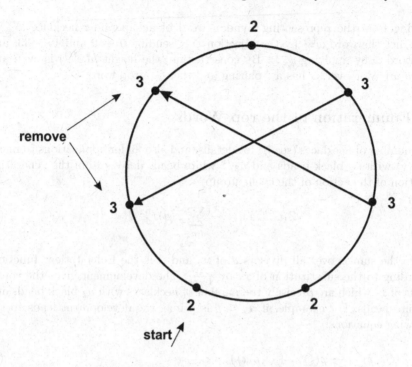

Fig. 1. Cyclic representation of rop words

Thus, we can always transform a (resp. Lyndon) rop word ω of length $2\ell + 1$ by a one-to-one map ϕ such that the letters 3 in $\omega^{(\ell)}$ are always coupled by subwords 33. The bijection ψ sending $2 \to 0$ and $33 \to 1$ maps $\omega^{(\ell)}$ to a word $\omega' \in \{0,1\}^*$ corresponding to a (resp. aperiodic) necklace.

$$\omega \xrightarrow{\phi} \omega^{(\ell)} \xrightarrow{\psi} \omega'$$

The Table 1 shows the first Lyndon rop words for $n_2 = 3$ and the corresponding aperiodic necklaces. The first Lyndon rop word of Table 1 is a rhythm 22323, sometimes called *fume-fume* and used by Ewe people of Ghana. Conversely,

Table 1. Correspondence for $n_2 = 3$

Aperiodic necklaces	n	Lyndon rop words	n'
0001	4	22323	5
00011	5	2233233	7
00101	5	2323233	7
000111	6	223332333	9
001011	6	232332333	9
001101	6	232333233	9

starting from the representing Lyndon word of an aperiodic necklace w', we construct the word $w^{(\ell)}$ by the bijection ψ^{-1} sending $0 \to 2$ and $1 \to 33$, and the word w by applying ϕ^{-1}. By construction, the height $h(w^{(\ell)})$ is even and also $h(w)$. Moreover, w has a ℓ-pairing and then is a rop word.

4   Enumeration of the rop Words

The number of necklaces (see [1] for details, and also [6] for applications to music theory) with n_2 black beads and $n_3/2$ white beads derives from the generating function of the action of the cyclic group

$$Z(C_{n_2}, x) = \frac{1}{n_2} \sum_{d|n_2} \varphi(d) x_d^{n_2/d} \tag{4}$$

where the sum is over all divisors d of n_2 and φ is the Euler totient function, according to the substitution of x_j by $\frac{1}{1-x^j}$. The development gives the coefficients of x_j which are precisely the number of necklaces with n_2 black beads and j white beads. For example, if $n_2 = p$ is prime, the development leads to the following equations:

$$Z(C_p, x) = \frac{1}{p}\left(\varphi(1)x_1^p + \varphi(p)x_p\right) \tag{5}$$

$$= \frac{1}{p}\frac{1}{(1-x)^p} + \frac{p-1}{p}\frac{1}{1-x^p}$$

$$= \frac{1}{p}\left(1 + \sum_{n=1}^{\infty}\frac{p(p+1)\dots(p+n-1)}{n!}x^n\right) + \frac{p-1}{p}\left(\sum_{n=0}^{\infty}x^{np}\right)$$

$$= 1 + \frac{1}{p}\sum_{n=1}^{\infty}\binom{p+n-1}{n}x^n + \frac{p-1}{p}\left(x^p + x^{2p} + x^{3p} + \dots\right)$$

$$= 1 + \sum_{n=1}^{\infty}a_n x^n$$

with

$$a_n = \begin{cases} \binom{p+n-1}{n} & if\ n \not\equiv 0\ mod\ p \\ \binom{p+n-1}{n} + p - 1 & if\ n \equiv 0\ mod\ p \end{cases} \tag{6}$$

The Table 2 with n_2 on the horizontal axis and n_3 on the vertical axis shows the number of rop words for n_2 and n_3 fixed. The number of rop words of length n is given by summing along the diagonal $n_2 + n_3 = n$. Each column of the table is obtained from the development of the generating function $Z(C_{n_2}, x)$. For n_2 prime, the coefficients agree with the formula of a_n given above. In each column, we recover the number of binary necklaces with length n_2+q and density $q = n_3/2$ given by the right hand side of the next formula. From the bijection of the previous section, it follows that the number $R(n_2, n_3)$ of rop words with n_2

Table 2. Number of rop words (n_2, n_3)

	1	3	5	7	9	11	13	15	17
2	1	1	1	1	1	1	1	1	1
4	1	2	3	4	5	6	7	8	9
6	1	*4*	7	12	*19*	26	35	*46*	57
8	1	5	14	30	55	91	140	204	285
10	1	7	*26*	66	143	273	476	*776*	1197
12	1	*10*	42	132	*335*	728	1428	*2586*	4389

symbols 2 and n_3 symbols 3 is the number of binary necklaces of length $n_2 + n_3/2$ with fixed density $n_3/2$,

$$R(n_2, 2q) = \frac{1}{n_2 + q} \sum_{d \mid \gcd(n_2+q, q)} \varphi(d) \binom{(n_2 + q)/d}{q/d}, \qquad q = 1, 2, 3, \ldots \quad (7)$$

By computing Lyndon words on alphabet $\{2, 3\}$ and deleting those which are not rop words, we get the Table 3 of the number of Lyndon rop words for n_2 and n_3 fixed, with n_2 on the horizontal axis and n_3 on the vertical axis. The total number of Lyndon rop words of length n is obtained by summing along diagonal $n_2 + n_3 = n$. The differences between the Tables 2 and 3 are in italics. In each column of Table 3, we recover the number of fixed density Lyndon words given by the following formula, with $n_3 = 2q$. It follows from the previous section, that the number $L(n_2, n_3)$ of Lyndon rop words with n_2 symbols 2 and n_3 symbols 3 is

$$L(n_2, 2q) = \frac{1}{n_2 + q} \sum_{d \mid \gcd(n_2+q, q)} \mu(d) \binom{(n_2 + q)/d}{q/d}, \qquad q = 1, 2, 3, \ldots \quad (8)$$

where μ is the Möbius function.

Table 3. Number of Lyndon rop words (n_2, n_3)

	1	3	5	7	9	11	13	15	17
2	1	1	1	1	1	1	1	1	1
4	1	2	3	4	5	6	7	8	9
6	1	*3*	7	12	18	26	35	*45*	57
8	1	5	14	30	55	91	140	204	285
10	1	7	*25*	66	143	273	476	*775*	1197
12	1	*9*	42	132	333	728	1428	*2584*	4389

By summing these formulas along a diagonal $n = n_2 + n_3$, we get the number $\mathcal{L}_n$ of Lyndon rop words of length n and the number $\mathcal{R}_n$ of rop words of length n:

$$\mathcal{L}_n = \sum_{n_2+n_3=n} L(n_2, n_3) = \sum_{p=0}^{(n-3)/2} L(2p+1, n-2p-1) \tag{9}$$

and

$$\mathcal{R}_n = \sum_{n_2+n_3=n} R(n_2, n_3) = 1 + \sum_{p=0}^{(n-3)/2} R(2p+1, n-2p-1) \tag{10}$$

These numbers are tabulated as follows: If n is prime, the difference between the cardinal of the two sets is 1, since the word 2^n (where the letter 2 is repeated n times) is a rop word but not a Lyndon word. If n is a product or a power of primes, some periodic words appear that are rop words but not Lyndon words. This explains the differences between the set of rop words and the set of *Lyndon* rop words. For instance, if $n = 9$, $(233)^3$ is a rop word but not a Lyndon word. The same is true for the words $(22323)^3$, $(233)^5$ and $(23333)^3$ of length 15. For $n_2 = 9$ and $n_3 = 12$, there are 333 Lyndon rop words and 335 rop words. The two non Lyndon rop words are: $(2233233)^3$ and $(2323233)^3$.

Table 4. Numbers of Lyndon rop words and rop words of length n

n	3	5	7	9	11	13	15	17	19	21	23	25	27
$\mathcal{R}_n$	2	3	5	10	19	41	94	211	493	1170	2787	6713	16274
$\mathcal{L}_n$	1	2	4	8	18	40	90	210	492	1164	2786	6710	16264

5   Generalized rop Words

Let s be an integer ≥ 2. The rhythmic oddity property could be generalized over any alphabet of positive integers as follows.

Definition 3. *Let $\mathcal{A}$ be an alphabet $\mathcal{A} \subset \{1, 2, \ldots, s\}$. A word $\omega \in \mathcal{A}^*$ is a generalized rop word of parameters (p, q) or a (p, q)-grop word for short, if*

(i) $h(\omega) \equiv 0 \mod p$
(ii) *No cyclic shift of ω can be factorized into q words $u_1, u_2, \ldots, u_q$ such that*

$$h(u_1) = h(u_2) = \ldots = h(u_q)$$

A Lyndon grop word is both a grop word and a Lyndon word. For instance, if $\mathcal{A} = \{2, 3\}$ the first grop words of parameters $(2,3)$ are: 2223, (3333), 22233, 22323, (33333), 222333, 2222223, 2232333, 2233233, 2233323, 2323233, (3333333). Non Lyndon words are given in parenthesis. A computation of the number of the first $(2, 3)$-grop words of length n is shown in Table 5. The first $(2, 4)$-rop words are: 22233, 22323*, (222222), 223333, 232333, (233233), 2222233,

2222323, 2223223*, 2333333*. Non Lyndon words are in parenthesis. The star indicates $(2,2)$-rop words. A computation of the number of the first $(2,4)$-grop words of length n is shown in Table 6. Over the alphabet $\{1,2,3\}$, Olivier Messiaen uses in *Cinq Rechants* the indian rhythm *simhavikrama* 2221323. It is a $(3,2)$-grop word, but not a $(2,2)$-grop word. And for instance, the words 111 and 333 are not $(2,3)$-grop words, but 123 and 132 are. The word 11133 is not a $(2,3)$-grop word since the subwords $u_1 = 111$, $u_2 = 3$ and $u_3 = 3$ have the same height, but the word 11313 is a grop word. By definition, a *Lyndon grop word* is both a grop word and a Lyndon word. For instance, 11133, 11313, 11322 are Lyndon $(2,3)$-grop words.

Table 5. Number of $(2,3)$-rop words over $\{2,3\}$

n	4	5	6	7	8	9	10	11	12	13	14	15	16	17
$\mathcal{R}_n^{(3)}$	2	3	1	6	11	6	25	46	41	117	232	278	631	1237
$\mathcal{L}_n^{(3)}$	1	2	1	5	9	6	22	45	40	116	226	278	620	1236

Table 6. Number of $(2,4)$-rop words over $\{2,3\}$

n	5	6	7	8	9	10	11	12	13	14	15	16	17
$\mathcal{R}_n^{(4)}$	2	4	4	5	13	27	47	50	131	284	479	685	1450
$\mathcal{L}_n^{(4)}$	2	2	4	5	12	24	47	50	131	279	473	683	1440

In the following, we consider grop words of parameters $(2,2)$ over the alphabet $\{a,b\}$ with a and b positive integers in $\{1,2,...,9\}$ and coprime. Most of the previous results can not be extended. For example, if we consider words over alphabet $\{1,4\}$ instead of $\{2,3\}$, the number of rop words of a given length is the same: there are 3 rop words of length 5 (22222, 22323, 23333) over $\{2,3\}$ and 3 rop words of length 5 (11444, 14144, 44444) over the alphabet $\{1,4\}$. However, there is no trivial bijection between the two sets. The criteria of ℓ-pairing does not work and similarly for asymmetric pairs. Nevertheless, we have the following result.

Proposition 3. *If ω is a $(2,2)$-grop word over the alphabet $\{a,b\}$ with a and b positive coprime integers, $a < b$, then there exists a unique pair (u,v) with*

$$h(v) = h(u) + 2m \tag{11}$$

such that $\omega = uv$ or $\omega = vu$, for some $m \in \{1,2,\ldots,\lfloor b/2 \rfloor\}$.

Proof. The uniqueness of the factorization is trivial. To prove the existence of such a factorization, consider the first longest prefix of $\omega = uv$ such that $h(u) > h(v)$. Denote x the last symbol of u, $\omega = u'xv$. Since u is maximal, $x + h(v) \geq$

$h(v')$ and $h(u) < h(v') + x$. The rhythmic oddity property implies $x + h(v) > h(v')$. Thus,

$$-x < h(u) - h(v') < x$$

Let t be an integer such that $h(u) - h(v') = t$. The value of this integer depends on x and varies between 0 and $\pm(b-1)$. The height of v is then $h(v) = t + h(u') = h(u) - x + t$. Since the height $h(\omega) = h(u) + h(v) = 2h(u) + t - x$ is even because ω is a rop word, $x - t$ is even, and equal to $2m$ for some m positive integer less that $\lfloor b/2 \rfloor$.

For example, consider the alphabet $\{2, 7\}$. The rop word 222 has a unique pair $(2, 22)$ such that $h(22) = h(2) + 2$. The rop word $\omega = 22277$ has a unique pair $v = 2227, u = 7$ such that $\omega = vu$ and $h(v) = h(u) + 6$.

Proposition 4. *Let $\omega = uv$ be a factorization of a word such that $h(v) = h(u) + 2m$, for some m. There exists a pair of prefixes (u', v') of (u, v), namely $u = u'u''$ and $v = v'v''$ such that $h(v') = h(u) + m$ if and only if there exists a cyclic shift $v''u'u''v'$ such that $h(v''u') = h(u''v')$.*

Proof. The proof is just a formal computation.
$$h(v') = h(u') + m \Leftrightarrow 2h(v') = 2h(u') + 2m \Leftrightarrow h(u') - h(v') + 2m = h(v') - h(u')$$
$$\Leftrightarrow h(v'') - h(u'') = h(v') - h(u') \Leftrightarrow h(v''u') = h(u''v').$$

We end this section by a computation with Maple Software of the number of rop words of parameters (2, 2) over the alphabet $\{a, b\}$. If $(a + b)$ is odd (resp. even) the length of rop words is odd (resp. even). By comparing Table 4 and Table 7, the computation shows that the number of rop words over $\{2, 3\}$ is the same as the number of rop words over $\{1, 4\}$. In Table 7, we compute the number of (2, 2)-grop words of length n over $\{a, b\}$ for n and $(a + b)$ odd.

Table 7. Number of (2, 2)-grop words of odd length n over $\{a, b\}$

$\{a,b\}\backslash n$	3	5	7	9	11	13	15	17	19	21
$\{1,4\}$	2	3	5	10	19	41	94	211	493	1170
$\{1,6\}$	2	4	9	23	59	162	459	1308	3802	11179
$\{1,8\}$	2	4	10	29	85	262	823	2596	8290	26684
$\{2,9\}$	2	4	10	30	93	305	1019	3416	11554	39281
$\{4,9\}$	2	4	10	30	94	315	1083	3752	13135	46235
$\{7,8\}$	2	4	10	30	94	316	1095	3841	13663	48990
$\{8,9\}$	2	4	10	30	94	316	1096	3855	13781	49770

And in Table 8, the same is done for n and $(a + b)$ even. None of these sequences are referenced in the database of Neil J. Sloane.

Perspectives. The next step of this study is to show the following statement. The number of (2, 2)-grop word of length n over the alphabet $\{a, b\}$ with a, b

Table 8. Number of (2, 2)-grop words of even length n over $\{a, b\}$

$\{a, b\}\backslash n$	2	4	6	8	10	12	14	16	18	20
$\{1, 5\}$	1	2	5	10	25	62	157	410	1097	2954
$\{1, 7\}$	1	2	6	15	44	128	378	1138	3478	10712
$\{1, 9\}$	1	2	6	16	51	162	521	1698	5586	18464
$\{5, 7\}$	1	2	6	16	52	171	574	1958	6742	23309
$\{5, 9\}$	1	2	6	16	52	172	585	2034	7167	25418
$\{7, 9\}$	1	2	6	16	52	172	586	2047	7270	26064

positive coprime integers is equal to the number of (2, 2)-grop word of length n over $\{a', b'\}$ for all n if and only if

$$a + b = a' + b' \tag{12}$$

For instance, the computation shows that the numbers of (2, 2)-grop words of a given length over alphabet $\{1, 6\}$, $\{2, 5\}$ or $\{3, 4\}$ are equal. The problem is still open.

Acknowledgements. The author thanks Marc Chemillier and André Bouchet for stimulating discussions and Harald Fripertinger for valuable comments and remarks.

References

1. Aigner, M.: A Course in Enumeration. Springer, Heidelberg (2007). https://doi.org/10.1007/978-3-540-39035-0
2. Arom, S.: African Polyphony and Polyrhythm. Cambridge University Press, Cambridge (1991)
3. Berstel, J., Lauve, A., Reutenenauer, C., Saliola, F.: Combinatorics on Words: Christoffel Words and Repetitions in Words. CRM Monograph Series, vol. 27. American Mathematical Society, Providence (2008)
4. Bouchet, A.: Imparité rythmique. ENS, Culturemath, Paris (2010)
5. Brăiloiu, C.: Le rythme aksak. Revue de musicologie **33**, 71–108 (1952)
6. Case, J.: Necklaces and bracelets: enumeration, algebraic properties and their relationship to music theory. Master Thesis, University of Maine, Markowsky, G. (advisor) (2013)
7. Chemillier, M., Truchet, C.: Computation of words satisfying the "rhythmic oddity property". Inf. Process. Lett. **86**(5), 255–261 (2003)
8. Hall, R.W., Klingsberg, P.: Asymmetric rhythms, tiling canons, and Burnside's lemma. In: Sarhangi, R., Sequin, C. (eds.) Bridges: Mathematical Connections in Art, Music, and Science, pp. 189–194. Winfield, Kansas (2004)
9. Hall, R.W., Klingsberg, P.: Asymetric rhythms and tiling canons. Amer. Math. Monthly **113**(10), 887–896 (2006)
10. Lothaire, M.: Combinatorics on Words. Addison-Wesley, Reading (1983)
11. Toussaint, G.T.: The Geometry of Musical Rhtythm. Chapman & Hall, CRC Press, Boca Raton (2013)

Polytopic Graph of Latent Relations
A Multiscale Structure Model for Music Segments

Corentin Louboutin[✉] and Frédéric Bimbot

IRISA (CNRS - UMR 6074 & Université Rennes 1), Campus de Beaulieu,
263 avenue du Général Leclerc, 35042 Rennes cedex, France
{corentin.louboutin,frederic.bimbot}@irisa.fr

Abstract. Musical relations and dependencies between events within a musical passage may be better explained as a graph rather than in a sequential framework. This article develops a multiscale structure model for music segments, called Polytopic Graph of Latent Relations (PGLR) as a way to describe nested systems of latent dependencies within the musical flow. The approach is presented conceptually and algorithmically, together with an extensive evaluation on a large set of chord sequences from a corpus of pop songs. Our results illustrate the efficiency of the proposed model in capturing structural information within such data.

1 Presentation

It is quite common sense that listeners do not perceive music only as a mere sequence of sounds, nor composers conceive their works as such. Music is essentially the result of patterns whose inner organization and mutual relationships participate to the overall structure of the musical content, at different time-scales simultaneously.

What is exactly music structure remains an open scientific question. This article is a contribution towards one particular aspect of music structure: it proposes and investigates a multiscale model of the inner organization of musical segments, which we call Polytopic Graph of Latent Relations (PGLR).

The musical content observed at a given instant t within a music segment obviously tends to share privileged relationships with its immediate past, hence the sequential perception of the music flow. But music content at instant t also relates with distant events which have occurred in the longer term past, especially at instants which are metrically homologous to t, in previous bars, motifs, phrases, etc. This is particularly evident in strongly "patterned" music, such as pop music, where recurrence and regularity play a central role in the design of cyclic musical repetitions, anticipations and surprises. But it is also discernable in a number of other music genres, which rely abundantly on all sorts of multiscale similarities, progressions, expectations and denials.

To overcome the limitations of purely sequential models in music content descriptions, hierarchical models are often resorted to, in order to provide a

O. A. Agustín-Aquino et al. (Eds.): MCM 2017, LNAI 10527, pp. 238–249, 2017.
https://doi.org/10.1007/978-3-319-71827-9_18

representation framework for the grouping structure of a musical passage. The most famous hierarchical approach is undoubtedly the Generative Theory of Tonal Music (GTTM) by Lerdahl and Jackendoff [8], which has been for many years a source of inspiration for a wide variety of work in music structure modeling. However, hierarchical approaches such as GTTM rely axiomatically on an adjacency hypothesis, under which the grouping of elements into a higher level object is strictly limited to neighbouring units.

In this work, we develop a different view as regards the structural association of elements forming music segments. We describe the "web" of musical elements as a *Polytopic Graph of Latent Relations* (PGLR) which models relationships developing predominantly between homologous elements within the metrical grid.

For most segments of 2^n events, the PGLR lives on an n-dimensional cube (square, cube, tesseract, etc.), n being the number of scales considered simultaneously in the multiscale model. By extension, the PGLR can be generalized to a more or less regular n-polytope.

Each vertex in the polytope corresponds to a low-scale musical element, each edge represents a relationship between two vertices and each face forms an elementary system of relationships. In addition, one variant of the proposed model views the last vertex in each elementary system as the denied realization of a (virtual) expected element, itself resulting from the implication triggered by the combination of former relationships within the system.

The estimation of the PGLR structure of a musical segment can be obtained computationally as the joint estimation of:

1. the description of the polytope (as a more or less regular n-polytope),
2. the nesting configuration of the graph over the polytope, reflecting the flow of dependencies and interactions between the elementary implication systems within the musical segment (this flow being assumed to be causal),
3. the set of relations between the nodes of the graph, with potentially multiple possibilities which need to disambiguated (hence the "latent" nature of the relations, as they are not actually observed).

In this paper, the shape of the polytope is assumed to be a tesseract (4-cube) and we focus our study on the modeling of meter-synchronous chord sequences of 16 chords. However, the general framework encompassed by PGLR is potentially applicable to many other musical dimensions (rhythm, melody, etc.) as soon as relevant latent relations can be defined.

In Sect. 2, we introduce the main concepts and formalism related to the model. Section 3 covers computational aspects of the approach, including optimality criteria and algorithmic design. In Sect. 4, we present a series of experimental results which assess the advantages of the PGLR model. We conclude with perspectives outlined by the proposed approached.

2   Concepts and Formalism

2.1   Chord Representation and Relations

Strictly speaking, a chord, in music, is any harmonic set of notes (or "pitches") that are heard as if sounding simultaneously. However, in tonal western music, chords are more generally conceived as sets of *pitch classes* supporting the local harmonic groundplan of the music. In particular, chords play a strong role in the accompaniment of the melody in pop songs. The most frequently encountered chords are triads (i.e. sets of three pitch classes), with a predominance of major and minor triads. More sophisticated chords contain combinations of 4 pitch classes or even more.

Chords can be represented in various ways. In this article, we consider two types of representations: (*i*) the complete set of pitch classes forming the chord (PC description) and (*ii*) the tabular notation of the major or minor triadic reduction of the chord (TR description). Assuming 4 or 5 pitch classes per chord, this leads to potentially several hundreds of different PC descriptions (much less in practice), but only 24 distinct TRs.

A number of formalisms exists to describe chord relations, either in classical musicology (through chromatic relations or via the circle of fifths) or in the framework of more recent theories, in particular Wietzmann regions [21] or neo-Riemannian theory [3]. Tymockzo [18,19] also proposes a model based on combinations of chromatic and scalar transpositions.

Depending on the formalism under consideration, the property of uniqueness of the relation between two chords may or may not be satisfied.

Triad Circles. We call *triad circle* any circular arrangement of triads aimed at reflecting some proximity relationship between triads along its circumference. The circle of thirds is formed by alternating major and minor triads with neighbouring triads sharing two common pitch classes. The circle is shown on Fig. 1. This representation provides a way to express the relationship between two TRs – in a unique way – as the angular displacement around the circle. Alternatively, the chromatic circle is arranged according to a chromatic progression (not represented on Fig. 1).

Optimal Transport. If two chords X and Y are represented as a set of pitch classes x_i and y_j, the set of *transports* between X and Y can be defined as:

$$T = \{t_k = (x_{i_k}, y_{j_k}) \mid x_{i_k} \in X, \ y_{j_k} \in Y\} \tag{1}$$

that is, pairs of notes across the two chords indexed by an integer k which represents a virtual mapping between their respective pitch classes. This is a simplified model that can be used to represent "voices" in chord sequences. We consider complete transports, i.e. each note is associated to at least one voice. Examples of transports are given on Fig. 2.

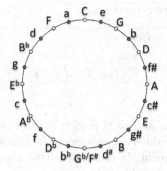

Fig. 1. Triads: circle of thirds.

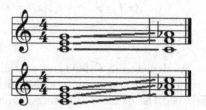

Fig. 2. Two possible transports between C and Fm.

The *cost* of a transport is defined as the sum of the costs associated with each pair of notes in the transport: $|T| = \sum_{(x,y)\in T} |d(x,y)|$. In this paper we use two types of distances:

- the *chromatic distance* (or smoothness) [3,9,16], which is the shortest displacement in semitones from pitch class x to pitch class y. In Fig. 2 the first transport is minimal for the chromatic distance (cost equal to 2).
- the *harmonic distance*, where the displacement is considered on the circle of fifths instead of the chromatic scale. In Fig. 2, the second transport is minimal for the harmonic distance (cost equal to 6).

2.2 Systemic Organization

Based on the hypothesis that the relations between musical elements in a segment are not necessarily sequential, the System & Contrast (S&C) model has been recently formalized [1] as a generalization and an extension of Narmour's Implication-Realization model [14]. Its applicability to various music genres for multidimensional and multiscale music analysis has been explored in [4] and algorithmically implemented in an early version as "Minimal Transport Graphs" [10].

The S&C model primarily assumes that relations between 4 elements in a musical segment x_0 x_1 x_2 x_3 can be viewed as relying on a matrix-based *system* of relations in reference to the first element x_0 (the *primer*), which thus plays the role of a common *antecedent* to all other elements in the system. This is the basic principle that enables the joint modeling of two timescales simultaneously.

Moreover, in the S&C approach, it is further assumed that latent relationships $x_1 = f(x_0)$ and $x_2 = g(x_0)$ trigger a process of implication:

$$x_0 \; f(x_0) \; g(x_0) \overset{implies}{\Longrightarrow} g(f(x_0)) = \hat{x}_3$$

Virtual element $\hat{x}_3$ may be more or less strongly denied by a *contrast*: $x_3 \neq \hat{x}_3$, which creates a potential closure to the segment.

Table 1. Antecedent function for the various models.

Sequential	Systemic	System & Contrast
$\phi_{Seq}(x_i) = x_{i-1}$	$\phi_{Sys}(x_i) = x_0$	$\phi_{S\&C}(x_i) = \begin{cases} x_0 & \text{if } i = 1, 2 \\ g(f(x_0)) & \text{if } i = 3 \end{cases}$

As depicted in Table 1, sequential, systemic and S&C models studied in this article are all first-order models which assume different antecedent functions, Φ, between the elements forming a musical segment. It is worth noting that the antecedent function summarizes the entire history of x_i into a single element.

2.3 Polytopic Representation and Nested Configurations

Polytopic Representation. Elementary systems of 4 elements, as described in the previous section, can further be used to describe longer sequences of musical events. In particular, sequences of 2^n elements can be arranged as an n-dimensional cube, within which each face potentially forms a S&C at time instants that share specific relationships in the metrical grid.

For instance, a sequence of 16 chords can be divided into four sequences of four successive chords, each of them being described as separate systems. Then, these four S&Cs, taken as elementary objects, can be related by forming an upper-scale S&C, linking the four primers of the 4 lower-scale S&Cs. Figure 3 represents such a description projected on a tesseract, in the case of the chord sequence from the chorus section of *Master Blaster* by Stevie Wonder:

$$Cm \; Cm \; Cm \; Bb \quad Ab \; Ab \; Ab \; Gm \quad F \; F \; F \; F \quad Cm \; Cm \; Bb \; Bb$$

System Nesting. However, depending on the sequence, other arrangements of the systems may prevail. If we now consider the following example:

$$Bm \; Bm \; A \; A \quad G \; Em \; Bm \; Bm \quad Bm \; Bm \; A \; A \quad G \; Em \; Bm \; Bm$$

a different configuration appears to be more efficient to explain the sequence with a multiscale model. In fact, grouping chords [0, 1, 8, 9], [2, 3, 10, 11], [4, 5, 12, 13] and [6, 7, 14, 15], and then relating these four faces of the polytope by an upper-scale system [0, 2, 4, 6] leads to a less complex (and therefore more economical) description of the relations between the data within the systems. This nesting configuration is called P^* in the rest of the paper and is distinct from the configuration considered in the first example, P_0, where the upper-scale system [0, 4, 8, 12] links four lower-scale nested systems $[4k + j]_{0 \leq j < 4}$ for $0 \leq k < 4$. Figure 5 illustrates these two configurations.

Therefore, multiscale polytopic descriptions involve different possible flows of dependencies and interactions between systems, which correspond to distinct *nesting configurations*. A nesting configuration is characterized by its corresponding antecedent function (as defined in Sect. 2.2). We furthermore assume that

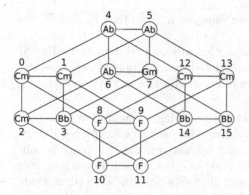

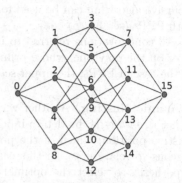

Fig. 3. Polytopic representation of the chord sequence taken from *Master Blaster* by Stevie Wonder.

Fig. 4. Tesseract where elements of the same depth are aligned vertically.

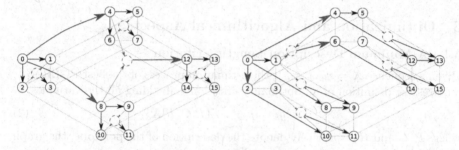

Fig. 5. Representations of the relations used by a multiscale analysis of a sequence of 16 events projected on a tesseract: P_0 (left), P^* (right).

nesting configurations must respect a causality principle: that is, the antecedent of any element in a system must have been observed before that element. This leads to a partial order between elements in the tesseract, as depicted on Fig. 4.

Static Configurations. Among all possible ways to construct nested configurations, an interesting subset consists in nesting faces of the polytope such that all vertices are used once and only once. In that case, valid nesting configurations consist in specific permutations of the initial index sequence. As, for each cube in the tesseract, there are three possible pairs of square systems corresponding to parallel faces of the cube, there is a total of $4 * 3 * 3 = 36$ possible permutations such that each lower scale system contains only causal flows.

Among these 36 possibilities, 6 are dual solutions. For 6 others, which we call Primer Preserving Permutations (PPPs), the system formed by the primers of each lower-scale system is itself a face in the polytope. PPPs preserve the role of elements with index 2^p as being primers of one of the system in the configuration. Whereas the list of PPPs is easy to tabulate for a tesseract, a recursive algorithm can be used for larger values of n. Note that P_0 and P^* are both PPPs (see Fig. 5).

All configurations referred to in this section are made of four non-adjacent faces on the polytope, whose primers are related by a fifth upper-scale system. In the case of PPPs, the upper-scale system is itself a face in the polytope.

Dynamic Nesting. Another way to define a nesting configuration is to construct it *on-the-fly*, by determining successively for each element placed in contrastive position, which of the possible implication systems it is more advantageous to relate it to. In this case, the cost function is used for each system hypothesis, to select the optimal one and disambiguate the antecedent function when several options are possible. Looking at Fig. 4, it appears that nodes $7, 11, 13, 14$ are contrastive in three different implication systems and 15 in 6 implication systems. Therefore, there exists $3^4 * 6 = 486$ distinct dynamic nesting configurations.

3   Optimization and Algorithmical Aspects

3.1   A Minimum Description Length Criterion

Given a sequence $X = x_0 \ldots x_{l-1}$, the estimation process of the best PGLR, S^X, requires the definition of an optimality criterion embedding all the variables:

$$S^X = argmin_{P,G,R} \; \mathcal{F}(P, G, R|X)) \tag{2}$$

where P, G and R respectively denote the description of the polytope, the graph and the latent relations for sequence X.

Assuming that $\mathcal{F}$ is measuring the complexity of the sequence structure, S^X can be defined as the shortest description of the sequence. Therefore, searching for S^X can be seen as a Minimum Description Length (MDL) problem [20] and $\mathcal{F}$ can be understood as a function that evaluates the size of the "shortest" program needed to reconstruct the data [6]. This is strongly related to the concept of Kolmogorov complexity, which has received increasing interest in the music computing community over the past years [11–13,17].

The exact computation of S^X cannot be achieved and it is approximated in the following way:

– the description cost of P can be estimated as a function of the regularity of the polytope. In this work it is discarded because all polytopes are tesseracts.
– the description cost of G can be assumed to be constant for all configurations within a model class. It is related to the number of distinct possible graphs (DPG) in the PGLR.
– the cost ($\mathcal{F}_R$) of the relations associated with a given nested configuration:

$$R^X = argmin_R \; \{\mathcal{F}_R(R|G, X)\} \text{ with } \mathcal{F}_R(R|G, X) = \sum_{i=1}^{l-1} |r(\Phi_G(x_i), x_i)| \tag{3}$$

where Φ_G is the antecedent function associated to G and $|r(x, y)|$ is the cost of the relation between x and y.

3.2   Optimization Process

Given that the cost of P and G are assumed to be constant, the aim of the optimisation process is to estimate the set of latent relations.

In the case of TRs, the process is rather straightforward: a relation between two chords in a triad circle is unique.

Conversely, optimal transport provides multiple possibilities of connecting chords together. The exhaustive optimization over the whole sequence would require to consider all combinations of transports. However, to make the computation tractable, the process is divided in several simpler sub-problems as follows.

For the sequential model, the chord sequence is processed as groups of 4-chord progressions (fusing beforehand identical neighboring chords, for which the transport is trivially determined). Then the last chord of each group is related to the first chord of the next group by minimal transport.

For the static systemic models, each elementary problem corresponds to a square system to optimize. Upper-scale systems are optimized first and then each lower-scale system is estimated independently. This process is repeated for each possible configuration. Details can be found in [10], with two adjustments which do not significantly impact the performance but save a lot of computation load: (i) for square systems, the contrast relation is optimized aside from the other systemic relations, (ii) the set of static configurations is restricted to PPPs.

For the dynamic nested S&C model, each chord is considered successively in a chronological order. Those which are directly related to the primer (nodes 1, 2, 4 and 8) enable the estimation of the corresponding latent relation. Those who are in a single contrastive position (nodes $3, 5, 6, 9, 10, 12$) are used to complete the estimation of the corresponding systems. Some chords in contrastive position belong to several systems (nodes $7, 11, 13, 14$ to 3 systems and node 15 to 6 systems): in these cases, the system with minimal cost is chosen. The whole process therefore results in a graph which has been built dynamically by successive optimisations of square systems.

4   Experimental Validation

4.1   Methodology

Experimental Setups. To assess the ability of the PGLR model to capture structural information in chord sequences, we have carried out a set of experiments on a corpus of 727×16 beat-synchronous chord sequences from the RWC POP dataset [5].

These experiments aim at evaluating the relevance of the PGLR model and at comparing different chord representations, types of models and optimization schemes.

The two types of chord representations presented in Sect. 2.1 (PCs and TRs) are considered in conjunction with optimal transport (for PCs and TRs) and triad circle relations (for TRs only). We compare the sequential bi-gram model

(*Seq*) – a very common approach in MIR [15] – with different types of systemic models (*Sys* and *S&C*) as defined in terms of their antecedent functions in Table 1, as well as the dynamic approach (*Dyn*).

For the systemic models, three types of system optimization are considered:

- S_0 which corresponds to the static configuration P_0 (see Fig. 5, left);
- S^* which corresponds to the globally optimal PPP over the whole corpus which happens to be P^* (see Fig. 5, right);
- S^X: in this case, the optimal PPP is chosen a posteriori as the one that optimizes the description of X, which varies across all Xs.

Perplexity. As there exists no ground truth as of the actual structure of a chord sequence, we compare the different models as regards their prediction ability. This is done by calculating for each model the *perplexity* [2], B, derived from the *negative log likelihood* (NLL), H. The aim is to measure how well an unseen sequence, $X = x_0 \ldots x_{l-1}$, can be predicted by the model:

$$H(X) = -\frac{1}{l} \sum_{i=0}^{l-1} \log P(x_i | \Phi(x_i)) \tag{4}$$

with the convention $\phi(x_0) = x_0$ and $P(x_0|x_0) = P(x_0)$.

For the triad circle relations, $P(y|x)$ is estimated as the relative frequency of $r(x, y)$ (and $P(x_0)$ is set to $1/24$). Similarly, for a pitch class distance d, $P(y|x)$ is also estimated as the frequency of $d(x, y)$ (and here, $P(x_0) = 1/12$). The learning phase for r and d is done using a 2-fold cross-validation strategy: probabilities are estimated on one half of the corpus (even numbered songs) and used on the other half (odd numbered songs) to compute H and vice-versa.

For optimal transport, X is viewed as a set of simultaneous "voices", X^k, and we compute H as the average voice NLL:

$$H(X) = \frac{1}{k} \sum_k H(X^k) \tag{5}$$

where each term $H(X^k)$ can be computed horizontally, using Eq. 4.

Ultimately, the performance is reported in terms of *perplexity*, B, which can be understood as an estimation of the average branching factor in predicting the sequence knowing its PGLR structure:

$$B(X) = 2^{H(X)} \tag{6}$$

Note that, whereas PGLR is fundamentally optimized on the basis of a complexity criterion, its impact is evaluated in a probabilistic framework, so as to measure its capacity to compress the data information in a meaningful way.

4.2   Results

Table 2 summarizes the perplexity figures obtained for a variety of experimental setups, from which a number of observations can be made.

Table 2. Average perplexity obtained with 2-fold cross-validation for the different models on RWC POP. *DPG* stands for Distinct Possible Graphs.

	Triad circle rotation on TR	Optimal transport				DPG
		Chromatic on PC	Chromatic on TR	Harmonic on PC	Harmonic on TR	
Seq	8.00	3.32	3.58	4.11	4.50	1
Sys_0	8.88	3.43	3.68	4.32	4.72	1
Sys^*	7.62	3.12	3.11	3.86	4.23	1
Sys^X	5.78	2.66	2.73	3.18	3.41	6
$S\&C_0$	6.68	2.97	3.16	3.92	4.06	1
$S\&C^*$	5.35	2.60	2.71	3.39	3.56	1
$S\&C^X$	**4.63**	**2.39**	**2.48**	**2.99**	**3.12**	6
Dyn^X	4.82	2.55	**2.44**	4.29	4.32	486

Benefit of Systemic Organizations. Systemic models globally outperform the sequential one[1]: all perplexity values are lower, except for the basic Sys_0 configuration. In particular, the $S\&C^X$ model provides the most spectacular perplexity improvement for all types of chord representations and relations (at the expense of a very limited number of DPGs). Note that the P^* configuration provides a noticeable advantage over Seq and P_0 configurations. The last row of the table also shows that the dynamic nesting approach is an interesting alternative as it provides perplexity scores almost as favorable as $S\&C^X$.

Predictive Support of the Virtual Element. The effectiveness of the virtual element in the S&C scheme is underlined by the systematic improvement observed when shifting from Sys to $S\&C$ results. The virtual element, $\hat{x_3}$, in $S\&C$ appears globally as a better antecedent for x_3 than does the primer, x_0, in Sys. However, for about one third of test sequences Sys^X outperforms $S\&C^X$ (figure not reported in Table 2), in particular for *aaba* structures.

Triad Circle Relations vs. Optimal Transport. The performance of triadic circle relations (TCRs) is based on a global sequence entropy while the optimal transport (OT) approach is evaluated in terms of average "per voice" entropy. In particular, the maximal branching factor of TCRs is 24 instead of 12 for OT. Therefore, the two perplexity scores cannot be compared. However, both approaches show similar trends w.r.t. the relative model performance. This supports the hypothesis of a general benefit of the multiscale approach rather independently from the way the chord information and relations are encoded.

---

[1] This confirms preliminary results formerly obtained on a much smaller corpus of 45 chord sequences [10].

In Table 2, results are also provided for optimal transport on triadic reductions (TRs) treated as PC description. Here too, the relative performance levels across models show the same trends. Note that the perplexity on TRs is slightly higher because the average pitch class distance between triads tends to be larger than that between chords with 4 notes or more.

Harmonic vs. Chromatic Transport. In chromatic optimal transport, the distance is computed from the set of note displacements measured on a semitone scale. We also tested a harmonic distance by considering displacements on the circle of fifths. Results in Table 2 show that this globally degrades the performance. Conversely, there is no need to consider triad rotations on a chromatic circle, as this is formally equivalent to modeling systems on the circle of thirds.

5   Conclusions

Both from the conceptual and experimental viewpoints, the PGLR approach appears as an efficient way to model multiscale relations in music segments. It is expected to provide a useful framework for a number of tasks in automatic music processing, as well as offering an interesting tool for music analysis.

Given that its core principles are not specific to a particular type of musical information, the application of PGLR to other types of musical objects, such as melodic motives and rhythmic patterns is a rather natural extension, currently under investigation. Ongoing work also includes the extension of the PGLR model to a larger range of timescales (n-cubes) and to chord patterns of other lengths (using irregular polytopes, by truncating or duplicating vertices, edges or faces, as in [7]).

References

1. Bimbot, F., Deruty, E., Sargent, G., Vincent, E.: System & Contrast: a polymorphous model of the inner organization of structural segments within music pieces. Music Percept. **33**, 631–661 (2016). (former version published in 2012 as Research Report IRISA PI-1999. hal-01188244)
2. Brown, P.F., Pietra, V.J.D., Mercer, R.L., Pietra, S.A.D., Lai, J.C.: An estimate of an upper bound for the entropy of English. Comput. Linguist. **18**(1), 31–40 (1992)
3. Cohn, R.: Audacious Euphony: Chromatic Harmony and the Triad's Second Nature. Oxford University Press, Oxford (2011)
4. Deruty, E., Bimbot, F., Van Wymeersch, B.: Methodological and musicological investigation of the System & Contrast model for musical form description. Research Report RR-8510, INRIA (2013). hal-00965914
5. Goto, M., Hashiguchi, H., Nishimura, T., Oka, R.: RWC music database: popular, classical and jazz music databases. In: ISMIR 2002, pp. 287–288 (2002)
6. Grunwald, P., Vitányi, P.: Shannon information and Kolmogorov complexity. arXiv preprint arXiv:cs/0410002 (2004)

7. Guichaoua, C.: Modèles de compression et critères de complexité pour la description et l'inférence de structure musicale. Ph.D. thesis, Rennes 1, 19 September 2017. Directed by F. Bimbot
8. Lerdahl, F., Jackendoff, R.: An overview of hierarchical structure in music. Music Percept. Interdisc. J. **1**(2), 229–252 (1983)
9. Lewin, D.: Some ideas about voice-leading between PCSets. J. Music Theory **42**(1), 15–72 (1998)
10. Louboutin, C., Bimbot, F.: Description of chord progressions by minimal transport graphs using the System & Contrast model. In: ICMC, Utrecht, NL (2016)
11. Louboutin, C., Meredith, D.: Using general-purpose compression algorithms for music analysis. J. New Music Res. **45**, 1–16 (2016)
12. Mavromatis, P.: Minimum description length modelling of musical structure. J. Math. Music **3**(3), 117–136 (2009)
13. Meredith, D.: Music analysis and Kolmogorov complexity. In: Proceedings of the 19th Colloquio d'Informatica Musicale (XIX CIM) (2012)
14. Narmour, E.: The Analysis and Cognition of Melodic Complexity: The Implication-Realization Model. University of Chicago Press, Chicago (1992)
15. Pearce, M.T.: The construction and evaluation of statistical models of melodic structure in music perception and composition. City University London (2005)
16. Straus, J.N.: Uniformity, balance, and smoothness in atonal voice leading. Music Theory Spectrum **25**(2), 305–352 (2003)
17. Temperley, D.: Probabilistic models of melodic interval. Music Percept. **32**(1), 85–99 (2014)
18. Tymoczko, D.: The geometry of musical chords. Science **313**(5783), 72–74 (2006)
19. Tymoczko, D.: Scale theory, serial theory and voice leading. Music Anal. **27**(1), 1–49 (2008)
20. Vitányi, P.M., Li, M.: Minimum description length induction, Bayesianism, and Kolmogorov complexity. IEEE Trans. Inf. Theory **46**(2), 446–464 (2000)
21. Weitzmann, C.F., Saslaw, J.K.: Two monographs by Carl Friedrich Weitzmann: Part I: "The Augmented Triad" (1853). Theory Pract. **29**, 133–228 (2004)

Machine Learning

Dynamic Time Warping for Automatic Musical Form Identification in Symbolical Music Files

Cristian Bañuelos[1]([✉]) and Felipe Orduña[2]

[1] Universidad Nacional Autónoma de México, México City, México
holomorfo@comunidad.unam.mx
[2] Universidad Nacional Autónoma de México,
Centro de Ciencias Aplicadas y Desarrollo Tecnológico, México City, México

Abstract. Music information retrieval techniques are used to automatically extract structural data of a piece, however there have been few attempts to study ways to automatically identify the musical form of digital files. In this work we present an implementation of the dynamic time warping algorithm for the automatic identification of musical form structure by means of a segmentation matrix in which we group elements according to maximal similarity. The system was implemented in symbolic files parsed with the music21 library. We tested it in two pieces: Bagatelle No. 25 in A minor by L.V. Beethoven, and Piano Sonata No. 11 in A major, K331, movement 3 by W.A. Mozart. The system obtained a correct identification of the similar sections, both with a rondo form. We foresee that this algorithm can be extended to measure harmonic similarity and with this be able to analyze more complex forms, like a sonata.

Keywords: Musical form · Dynamic time warping
Music information retrieval

1 Introduction

Music Information Retrieval uses different algorithms and techniques to extract structural content from musical files. There are mainly two approaches to this, on one hand, we can study an audio signal, like in [9,10], on the other, we can take a symbolical musical file, like MIDI or MusicXML, and apply pattern recognition techniques to extract harmonic or melodic content, this approach is taken in [3,5].

Many of the Music Information Retrieval problems addressed revolve around developing genre recommendation systems [7], style imitation with artificial intelligence and machine learning [4], chord recognition and harmonic extraction [12], instrument recognition from an audio file [14]. However there have been few

C. Bañuelos—This work was made possible by the doctoral financial support of the Consejo Nacional de Ciencia y Tecnología (CONACYT), México. Grant number: CVU 662627 no: 583329.

© Springer International Publishing AG 2017
O. A. Agustín-Aquino et al. (Eds.): MCM 2017, LNAI 10527, pp. 253–258, 2017.
https://doi.org/10.1007/978-3-319-71827-9_19

attempts to analyze and extract musical form, among these [8], in which they find internal similarity within an audio file to find sections that relate among themselves and can be used as transitions between them. In [1], they analyze internal similarity of an audio file to create a general thumbnail, that is the minimal amount of music that represents the whole piece, they use this in order to simplify the search in extensive databases.

In symbolical music files analysis, there is a need to apply these kind of similarity analysis to create a system that classifies internal similarity of a musical piece in order to find its musical form. We argue that a system that can extract and identify the formal structure of symbolical music files can be useful to simplify other tasks, like automatic harmonic and chord labeling, removing redundancies in the calculation of repeating or similar sections. It can also be a useful pedagogical tool that can be used in music education contexts to help the student to better understand the concept of form.

In this work, we propose to apply the time series technique Dynamic Time Warping to a symbolical musical file. This algorithm is normally used in audio signals and it's useful to calculate a measure of how different two signals are by finding the cost of transforming one into another by means of time stretching. The novelty of this work is that instead of working with audio files, we will use the much smaller symbolical representation of music. In order to classify the musical form, we run through all possible sub segments of the piece and compare them in a similarity matrix, similar to the one used in [1], we then group segments with maximal similarity and label them as a new section. The process is repeated until all the maximal similarities have been found. In the present work we apply the algorithm to find the repeating section of a rondo form.

2   Dynamic Time Warping

The DTW algorithm uses dynamic programming to find the optimum alignment between two time series. It does this by calculating the cost to align each point of the first to the second. Afterwards, it takes the minimal path of change needed to transform each point to the other. It's very useful in cases where we need to compare sequences that are time stretched or transposed. Because of this flexibility we consider that its application to a musical context would bring optimal results.

The following is a brief review of the DTW as presented by Müller in [11]. If we have sequences $X := (x_1, x_2, ..., x_N)$ and $Y := (y_1, y_2, ..., y_N)$, the warping path $p = (p_1, ..., p_L)$ is defined as the assignment of x_{nl} to y_{ml}. The path must satisfy the boundary condition that it always aligns the first elements, and the last elements of the sequences, respectively; it must be monotonic; and only move by unitary steps; also all elements from X and Y must be paired and there can be no repetitions. Müller also defines the total cost as

$$c_p(X, Y) := \sum_{l=1}^{L} c(x_{nl}, y_{ml}). \tag{1}$$

where c is the local cost, in this case we will use the euclidean distance as a measure of difference. To find the optimal cost we search for the path with the minimum value.

$$DTW(X,Y) = \min\{c_p(X,Y) \,|\, p \text{ is a transformation path}\} \qquad (2)$$

With this we find the minimum cost of the different possible paths of transforming X into Y. This measure will be useful to compare similarity segments within the piece. The system developed in this work uses the Fast DTW [13], which is based on dynamic programming concepts. The DTW measure can be used for query systems by comparing one small segment to different windows of a larger segment, this methodology is used in [6]. In order to apply this to our symbolic musical file, we shall use the sequence of notes given by a MusicXML file and treat them as a time series.

3   Segmentation Matrix

The proposed system was implemented in Python with the library Music21 [2], which is optimal for the parsing and processing of symbolical musical files like MIDI and MusicXML. In order to apply the DTW algorithm we need to prepare the data so it takes the form of a time series. We use the *flat* functionality in the music21 library to convert the xml file to a linear representation of notes and time offsets. Next, we translate the musical information into a time series of all the notes, thus having a sequence of order pairs giving the time stamp and the pitch of a particular event.

$$P = \{(time_i, pitch_i)\} \qquad (3)$$

Where $0 \leq i \leq N$, and N is the total number of notes in the piece to analyze; we will call this: the events list.

As a next step we will create a list of all the possible sub segmentations of the events list, that is all the possible subsets of P in which all consecutive elements from i to j are present.

$$Segs = \{U(i,j) \subset P| \text{ if } i \leq k \leq j, \implies (t_k, p_k) \in U(i,j)\} \qquad (4)$$

We will use the notation $Segs_{i,j}$ to indicate the segment time consecutive notes from element i to j in the events list. We have N^2 different segments, and we need to group them according to their similarity. In order to do this, we will create a similarity matrix in which each entry (p,q) will have the similarity measure assigned between $Segment_p$ and $Segment_q$, that is

$$(a_{p,q}) \in \mathbb{R}^{N \times N} \qquad (5)$$

$$a_{p,q} = DTW(Seg_p, Seg_q) \qquad (6)$$

In this matrix we calculate the DTW similarity measure between all segments and thus can be used to obtain a measure to classify and group similar sections of the piece. The method for computing the DTW similarity value is exactly the same used for acoustic time series as shown by Müller [11], which uses dynamic programming to find the cost of transforming one series to the other, as was explained in the previous section.

4   Musical Tests

For the purpose of testing the system we choose two pieces that have a rondo form, in order to see if the algorithm correctly predicts the repeating pattern of the analysis.

In the Bagatelle in A minor, Fur Elise, from Beethoven we have a rondo structure in which there is a section A at the beginning, and the musical form as a whole is A - B - A - C - A. When we apply the DTW measures to this piece we obtain a clear indication of a repeating section. In this work we show an example of use of the algorithm to find the repeating section of the rondo piece, that is, the A part. In Fig. 1a, we have a plot of time offset to the cost of the current test window. The graph takes an initial window of size 30, and runs through the rest of the piece in steps of 10 notes, obtaining the cost of comparing the test window to each of the other segments, the initial window size and step length were chosen empirically. The program then increases the window size to check if a larger frame would give a better matching result for some sections. The windows change in size from 30 to 100 in steps of 10, Fig. 1 show the costs of different windows sizes. Given the local minimums that appear around offsets 20, 55 and 130 we see that there appears a clear similarity between these sections and the test window. Given that the test window was the initial part of the piece and repeats several times, we will call it section A, that in fact corresponds to the repeating section of the rondo. To analyze if the parts that differ from A are in fact sections B and C, we would need to apply the algorithm starting in the first offset not classified as section A, and search for similarities, this will be addressed in future work, but currently the algorithm shows a good result in classifying one section at a time.

In Mozart's Piano Sonata No. 11 in A major, K331, movement 3, also known as Rondo Alla Turca, we also have a similar form like in Beethoven, and if we apply the DTW algorithm to the piece we get the cost graph shown in Fig. 1b. We can see that the beginning section of the piece repeats itself at time offsets around 40, 140, and 160. In 140 we have an almost identical similarity regardless of window size, while in the other cases we have some minor variation in the structure, as shown by the higher cost obtained. We will call this repeating section A. Again, to find the rest of the sections we would need to apply the algorithm to the rest of the piece that was not labeled as A, this analysis will be applied in future work.

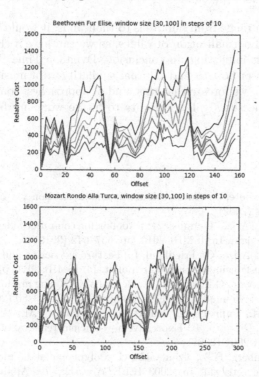

Fig. 1. Cost analysis of an initial test window in: (a) Fur Elise and, (b) Rondo Alla Turca. The lines show the costs of different window sizes. The bottom line has a window size 30, while the top is 100. The middle lines increase in steps of 10 from the initial window size.

5   Conclusions

The use of DTW to compare internal sections of a musical piece gives us enough flexibility to compare sections of a work by allowing us to compare all the possible combinations of windows sizes, we use this to find section repetitions, variations and changes. The current work presents a way to identify a repeating section, and it will be further developed to find and classify all repeating sections. One of the main challenges to solve is that we don't have yet a clear maximum value for the window size, so we have to decide what is the cost tolerance of a section and assign a degree of belonging of each part of the score, to the test window. In future work, this will lead to establish some fuzzy belonging functions to help us decide how we should segment the similarity sections given by the DTW algorithm. The current implementation only takes into account the height and time stamp of the note, however, in future work we could implement other kind of metrics to include more parameters, like harmonic distances, which can help to further classify similar parts and sections of a symbolical music file using features others than just the pitch of the notes. Another thing to consider is the range of the costs values, in this version, the cost is given directly by the DTW, but we can see that it can range from hundreds to thousands, it would be a real benefit to find an optimal normalization of values, so we can have a clearer estimation of what the cost is indicating. In conclusion, Dynamic Time Warping gives us a good estimation of how to find internal similarity on a musical work to find its structural form, we foresee that this kind of approaches, complemented with other similarity measures will contribute to future work in Music Information Retrieval research.

References

1. Cooper, M.L., Foote, J.: Automatic music summarization via similarity analysis. In: ISMIR 2002 (2002)
2. Cuthbert, M.S., Ariza, C.: music21: a toolkit for computer-aided musicology and symbolic music data. In: ISMIR 2010, pp. 637–642 (2010)
3. Cuthbert, M.S., Ariza, C., Friedland, L.: Feature extraction and machine learning on symbolic music using the music21 toolkit. In: ISMIR 2011, pp. 387–392 (2011)
4. Dubnov, S., Assayag, G., Lartillot, O., Bejerano, G.: Using machine-learning methods for musical style modeling. IEEE Comput. **36**(10), 73–80 (2003)
5. Droettboom, M., Fujinaga, I., MacMillan, K., Patton, M., Warner, J., Sayeed Choudhury, G., DiLauro, T.: Expressive and efficient retrieval of symbolic musical data. In: ISMIR 2001 (2001)
6. Hu, N., Dannenberg, R.B., Tzanetakis, G.: Polyphonic audio matching and alignment for music retrieval. In: 2003 IEEE Workshop on Applications of Signal Processing to Audio and Acoustics, pp. 185–188. IEEE (2003)
7. Kaminskas, M., Ricci, F.: Contextual music information retrieval and recommendation: state of the art and challenges. Comput. Sci. Rev. **6**(2), 89–119 (2012)
8. Lamere, P.: The Infinite Jukebox (2012). www.infinitejuke.com. Accessed 24 May 2016
9. Lee, K., Slaney, M.: Automatic chord recognition from audio using a HMM with supervised learning. In: ISMIR 2006, pp. 133–137 (2006)
10. McVicar, M., Santos-Rodríguez, R., Ni, Y., De Bie, T.: Automatic chord estimation from audio: a review of the state of the art. IEEE/ACM Trans. Audio Speech Lang. Process. **22**(2), 556–575 (2014)
11. Müller, M.: Dynamic time warping. In: Information Retrieval for Music and Motion, pp. 69–84. Springer, Heidelberg (2007)
12. Pardo, B., Birmingham, W.P.: Algorithms for chordal analysis. Comput. Music J. **26**(2), 27–49 (2002)
13. Salvador, S., Chan, P.: FastDTW: toward accurate dynamic time warping in linear time and space. In: 3rd Workshop on Mining Temporal and Sequential Data, ACM KDD 2004, Seattle, Washington, 22–25 August 2004 (2004)
14. Park, T., Lee, T.: Musical instrument sound classification with deep convolutional neural network using feature fusion approach. arXiv preprint arXiv:1512.07370 (2015)

Identification and Evolution of Musical Style I: Hierarchical Transition Networks and Their Modular Structure

Pablo Padilla[1,2(✉)], Francis Knights[1], Adrián Tonatiuh Ruiz[3], and Dan Tidhar[1]

[1] University of Cambridge, Cambridge, UK
[2] Institute for Applied Mathematics and Systems, IIMAS,
Universidad Nacional Autónoma de México, Mexico City, Mexico
pablo@mym.iimas.unam.mx
[3] Institute of Neural Information Processing, Ulm University, Ulm, Germany

Abstract. The problem of identifying musical styles using mathematical tools is central not only in musicology and the mathematical theory of music, but also in applications to music pattern recognition and automated music generation in a particular idiom. In this paper we propose a methodology related to the transition network approach developed by D. Cope in his Experiments on Musical Intelligence, EMI. This extension allows for the possibility of defining stylistic cells at different scales as motifs and moduli of networks at the corresponding scale. It can be applied to study recursivity aspects of music. We also outline how this methodology can be used to systematically study stylistic changes in different contexts by incorporating probabilistic and statistical tools and connections with other approaches.

> Misura ciò che è misurabile, e rendi misurabile ciò che non lo è.
> (Measure what is measurable, and make measurable what is not).
> Attributed to Galileo Galilei.

1 Introduction

One of the most interesting problems in musicology and the history of music is the identification of a particular musical style. Developing a systematic methodology to identify and classify stylistic trends has deep theoretical implications not only in analysis, composition and musicology, but is also relevant in specific applications such as automated music generation and authorship validation. The problem is not new and has been addressed in different ways (see Sect. 3.4.2 on stylistic classification in (Nierhaus 2009) for a historical perspective and (Hardooon et al. 2014) for a recent approach using machine learning and information theoretical tools). In the context of automated music generation in a particular idiom much has been done, particularly using Markov chains. We refer to (Collins and Laney 2017) and the bibliography therein for an up-to-date account of the problem, in particular for new approaches dealing with the non-Markovian nature of music. In particular David Cope used

© Springer International Publishing AG 2017
O. A. Agustín-Aquino et al. (Eds.): MCM 2017, LNAI 10527, pp. 259–278, 2017.
https://doi.org/10.1007/978-3-319-71827-9_20

transition networks in his Experiments in Musical Intelligence, EMI, to generate music in the style of a particular composer (e.g. *The Well Programmed Clavier* or *Virtual Bach*). In this methodology a formal grammar with the incorporation of probabilistic elements is applied in order to compose stylistically admissible pieces. As Cope points out, recombination plays an important role in the emulation and composition of music in a particular style (Cope 2004). In EMI, Cope proposes a way in which basic elements can be recombined by means of transition networks and a careful selection of the units and patterns to be used. Similar approaches have been implemented incorporating also probabilistic elements (see also (Loy 2006)). The basic idea is that transitions between different elements, for instance notes, chords or rhythmic values are studied and their probabilities computed based on a music fragment or piece(s). In a more technical language, a transition matrix is thus obtained and a corresponding Markov chain associated. The states of this chain correspond to the elements being considered. In the case of chords a matrix element would provide the probability of a specific chord, say the dominant, being followed by another, the tonic. In this way concatenations of melodic and rhythmic motifs or harmonic progressions can be compared in terms of their likelihood. This already provides a first element to understand some stylistic features. For instance, in baroque or classical music melodic transitions corresponding to stepwise motion would be more likely than others, whereas in serial music transitions among wider intervals will have a not negligible probability (see the example in the next section). In (Farber 2001) a very interesting study of Palestrina's counterpoint is made. The authors incorporate stylistic rules in a probabilistic model for the generation of *cantus firmi* in the style of Palestrina, encoding strict rules as forbidden transitions, i.e. transitions with probability zero. They also mention that the inverse process of inferring the rules from computed probabilities is possible.

In recent years, several classification tools have been applied to music. The use of machine learning techniques is promising and genre classification constitutes a very active area of research (Bassiou et al. 2015). Although a little surprising, similar methods have been less used in musicological analysis.

In the present work we also use transition matrices, and take them as the starting point to construct graphs. Our goal is to show that important structural and stylistic features of a piece of music can be defined and inferred from these graphs.

We outline now how this is done. In a forthcoming paper, this ideas are systematically applied to study the stylistic differences in the keyboard music attributed to Charles and Louis Couperin (Knights et al. 2017a), but for the sake of completeness include an example here. Take for instance, the transition matrix associated to a melodic line; rows and columns will represent notes, so an element in the E row and the G column in this 12 by 12 matrix is simply the probability in the given melody that E will be followed by a G. We can now construct a graph in which the nodes are the notes and the edges are arrows with a weight (the weight being the transition probability just described). There are several advantages in using this graph theoretical approach. Once a graph for the melody has been obtained, the topological and connectivity properties of this graph contain important information on the melodic features and stylistic analysis and comparisons can be systematically made using them. To mention a specific example in the melodic case, moduli or communities in the graph might be associated with different tonal regions or the detection of small repeated patterns in the graph provides also a way of extracting information about the motivic structure of the piece.

An element that can also be incorporated in these models is the hierarchical nature of music. In (Tidhar 2005) a systematic approach using music grammars is presented to study the unmeasured preludes attributed to Louis Couperin (Tidhar 2005). In these works a hierarchical structure is determined by the slurs in the score, but in general, and given a music fragment, it is not straightforward to establish the different levels of organization. This aspect plays an essential role not only in style recognition, but it is fundamental also in the way composers, listeners and performers perceive music. It can be said that one of the more elusive aspects of music is precisely the interplay between several structural levels and its recursive character. Using a graph theoretical approach there is a simple way of identifying hierarchical structural features by studying the topological properties of the constructed graphs at different scales. In other words, we suggest a procedure for identifying the basic elements at different scales. In a more standard musical terminology and again referring to the melodic case, we will detect typical motifs at a first hierarchy, then phrases at a higher level and so on. These elements will be the motifs and modules (in a graph theoretical sense) of a graph associated with a particular fragment of music. The possibility of studying different hierarchical levels in this way could also be compared with other musicological approaches such as the structural Schenkerian analysis.

As we remarked above, there is a natural and standard way to associate either a weighted directed graph or just a directed graph. A weighted directed graph can be obtained by considering the transition matrix as representing the weights of the adjacency matrix of a graph (see the next section for further details). A directed graph (with no weights) can also be constructed by keeping only the edges whose weight is bigger than a certain threshold. We stress again that standard graph theory methods can be employed in order to determine the moduli (or communities) of the graph[1] and other properties. One such modulus constitutes what one might call a structural cell or motif[2]. This might seem technical, but in the next section we illustrate this methodology by means of an example comparing two contrasting musical fragments, one from several *cantus firmi* and another from a serial melody by A. Schoenberg. Before doing so, and summarizing our approach: we encapsulate the style of a piece by extracting information on the probability of the choices that the composer makes, consciously or unconsciously, in different musical aspects (melodic, harmonic, rhythmic, textural or timbrical) at different hierarchical levels.

Although the theoretical aspects are very interesting, it must be said that an essential motivation for us were several concrete musicological questions. These involve authorship attribution and are dealt with in different papers (Knights et al. 2017a, b)

---

[1] Methods based on random walks provide a way of generating music fragments (see for instance (Collins and Laney 2017)).

[2] Not to be confused with the motifs introduced by U. Alon in the context of systems biology, which will also be used. In order to avoid confusion we employ the term structural cells for moduli of graphs.

The rest of the paper is organized as follows. In the next section a simple melodic example is presented in order to introduce the methodology. We do this by comparing the results in two contrasting music fragments. In the final section we describe a few applications to the Allemandes attributed to Louis Couperin. Some musicological and historical implications are discussed and further research is also considered.

2 Hierarchical Moduli of Networks

We now proceed to describe the methodology outlined in the introduction. For simplicity we restrict the detailed analysis to melodic aspects and comment on how the methodology can be applied to take into account rhythmic, harmonic or other structural considerations. Moreover, a clear understanding of how to combine these elements, either in automated composition or analysis is still missing and one of the fundamental questions in the area. The first melodic lines are five *cantus firmi* in Dorian mode (see Fig. 1). In order to have sufficient melodic material we concatenate these fragments and consider them as one, avoiding the repetition of the final and initial D's between segments.

Fig. 1. Cantus firmi in Dorian as in (Jeppesen 1992)

The second melody is taken from *The book of Hanging Gardens,* op. 15 by A. Schoenberg (Fig. 2).

Fig. 2. A. Schoenberg's Book of the Hanging Gardens, measures 8–16. From (Domek 1979)

We notice how some basic statistical quantities already contain useful information about these fragments. In Figs. 3 and 4, a histogram of the pitch classes for each of the fragments is shown. More precisely, the percentage of the pitch classes appearing in the examples.

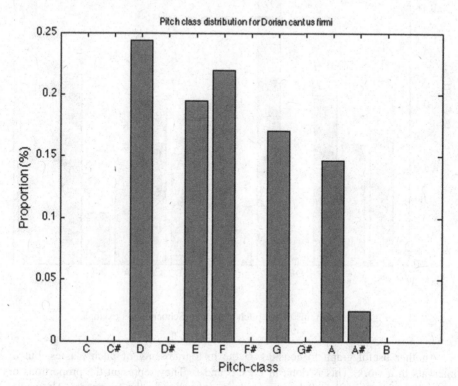

Fig. 3. Histograms of the pitch classes in the Dorian *cantus firmi*

Notice that the distribution of the notes is much more uniform in Schoenberg's example, as it is expected in a serial piece, with no functional tonal center, as opposed to the *cantus*, in which the most frequent pitch is precisely D, the tonal center. In particular, in the Dorian *cantus firmi* no chromatic notes appear in the histogram (the only accidental in the histogram is B flat, or rather its enharmonic A#, which belongs to the mode). It is clear that the histogram provides ús with a basic and simplified summary of the melodic material used. This may already help in establishing some genre or author differences, even in fragments not as contrasting as the ones discussed so far (see the discussion at the end of this paper and (Knights et al. 2017a)).

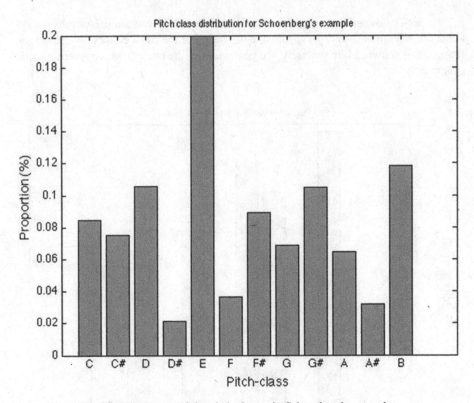

Fig. 4. Histograms of the pitch classes in Schoenberg's example.

Another useful graph to consider is the histograms not of pitch classes, but of intervals in a work. This is done in Figs. 5 and 6. They represent the proportions of intervals, rather than the pitches themselves, appearing in the fragments. These distributions provide systematic information about the melodic contours that would be difficult to assess otherwise.

From these histograms we see that in the Dorian *cantus* the preponderant interval is the descending second (either major or minor) and that intervals of a second account for approximately 80% of the intervallic material used. This is in agreement with the usual prescription for composing a *cantus firmus,* which should consist essentially of stepwise motion. Not only that, but the fact that melodic jumps upwards tend to be compensated by descending stepwise downwards, making the descending seconds the most common intervals in the histogram. This feature clearly distinguishes the fragment from the Schoenberg example in which the interval distribution is also much more uniform. We comment on these features not because they are surprising or new, but to illustrate the fact that even simple quantitative indicators might be helpful in distinguishing stylistic characteristics. In fact, in stylometric measurements in literature, even the frequency of words used by different authors might serve as a basis for classification (Stamatatos 2009).

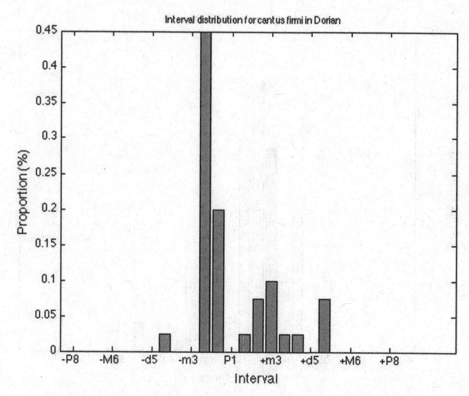

Fig. 5. Interval distribution for the intervals in the Dorian *cantus firmi* example.

Finally, we explain what the transition distribution matrix of pitch classes is, since it is the basic element that will allow us to apply statistical analysis, generate graphs and use standard tools from graph theory to propose what might be called stylistic signatures. We consider any given note in a music fragment, let say E, in the *cantus firmi* example and look for the probability that it goes to a D, where we count the total number of transitions in the piece (intervals). We compute it by counting the number of times E is followed by D and divide by the total number of transitions. A graphic representation of these matrices is presented in Figs. 7 and 8. For the specific example this would be represented by the square on the E row (pitch class 1) and the D column (pitch class 2). It can bee seen at first glance that the one corresponding to the *cantus* is sparser than the other. Also the fact that elements in Fig. 7 are concentrated in the upper and lower diagonals is related to the preponderance of stepwise motion. However, other relevant features such as the existence of a tonal center, D in this case, and its fundamental structural role can also be inferred from the graphical representation of the transitions. Of course such a representation can provide useful visual information,

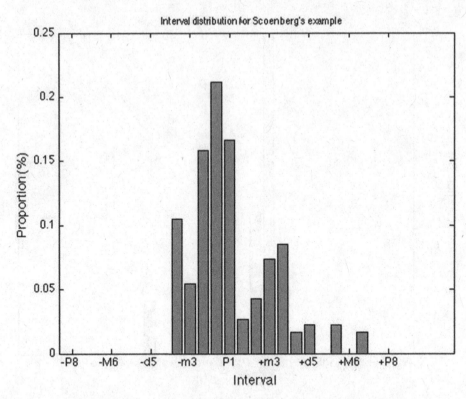

Fig. 6. Interval distribution for the intervals in Schoenberg's example.

but as such is limited (Fig. 9). However, the matrix itself as an array of numbers, provides the raw material for a more quantitative analysis that can include standard techniques[3]. For the sake of completeness, we include the corresponding matrix for the *cantus firmi* example as computed with Miditoolboox[4]. Notice the different row order, which is more conventional for a transition matrix. The empty spaces correspond to 0 and are left blank for clarity. With the matrix for this example, we now proceed to construct the associated graph. We explain in detail the procedure for the *cantus firmi*. The other example is done in a completely analogous way. Each note can be taken as a node and every element of the matrix as an edge joining the two corresponding elements. An empty space or 0 means that there is no connection between the notes. For instance, given that from D to E there is a nonzero element, an edge from the former to the latter note has to be drawn. After doing this for all the notes we obtain the graph

---

[3] For instance, singular value decomposition, spectral analysis, principal component analysis, etc. See (Knights et al. 2017a). For more details on statistical methods.

[4] The previous histograms for pitch clases and intervals were also generated using Miditoolbox (Tolviainen and Eerola 2016).

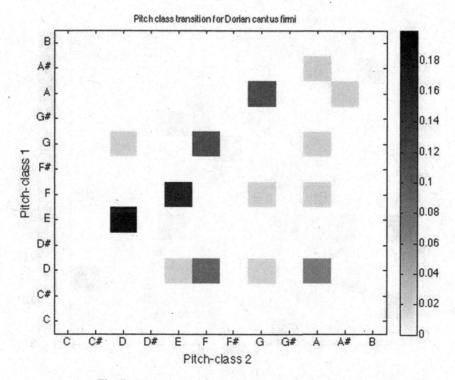

Fig. 7. Interval transition matrix for the Dorian *cantus.*

shown in Fig. 10a. In standard graph theoretical terminology the transition matrix is called the adjacency matrix. Notice that in the figure neither the weight of the connection (the actual numerical value of the corresponding element), nor the direction of the transition is shown. As before, we point out a few interesting facts that can be observed right away. For instance, there are six nodes with no edges leading to or coming out of them (C, C#, D#, F#, G# and B) which except for the C# correspond to notes not belonging to the mode[5]. The nodes with a larger number of edges are D, G and A, which even without any harmonic consideration or reference to tonic, subdominant and dominant can be seen to play an important role. The notions of connected components (disjoint parts in which the graph can be naturally separated, 7 in this case), degree of a node (number of edges associated to a node), the indegree or outdegree (number of edges going into or out of a node respectively) and many other quantities can be computed in order to characterize a musical fragment in this way (Fig. 10b).

---

[5] C# enters the mode as a leading tone, although it does not belong to it in a strict sense.

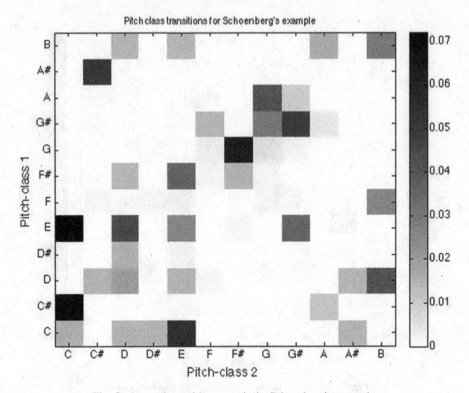

Fig. 8. Interval transitions matrix in Schoenberg's example.

	C	C#	D	D#	E	F	F#	G	G#	A	A#,Bb	B
C												
C#												
D					0.025	0.1		0.025		0.075		
D#												
E			0.2									
F					0.175			0.025		0.025		
F#												
G			0.025			0.125				0.025		
G#												
A								0.125			0.025	
A#,Bb										0.025		
B												

Fig. 9. Interval transition matrix for the Dorian *cantus* (numeric values).

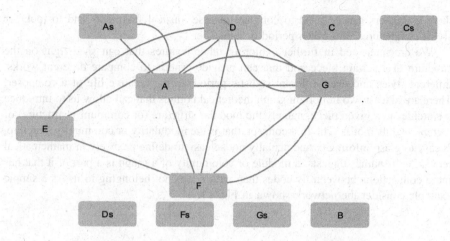

Fig. 10a. Graph associated to the *cantus firmi* example.

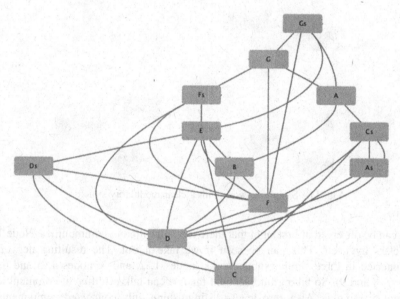

Fig. 10b. Graph associated to the Schoenberg example.

For the corresponding analysis in the Schoenberg example, we see that the nodes are much more interconnected and that the degree of a node does not vary much.

Other less intuitive parameters include the clustering coefficient or the network centralization, which could be obtained using standard network packages[6]. Each one of

---

[6] In particular, we used Cytoscape in order to analyze the networks. This open source software has already a built in analyzer.

these estimators can be used to compare diverse musical fragments and to look for stylistic signatures of a given period or composer[7].

We are interested in further exploring other features that can shed light on the structure of a certain piece and that can provide criteria to compare different works, different styles and even different stylistic periods in the creative life of a composer. There are at least two important graph theoretical notions that can allow us to introduce hierarchies in a given piece, namely the modular structure (or community structure) of a graph and its motifs. The first concept, that is the modularity or community structure, is easy to grasp informally and actually not so easy to define precisely in mathematical terms. As its name suggests, a module or community of a graph is a part of it that has more connections between its nodes than with nodes not belonging to it. As a simple example consider the network shown in Fig. 11.

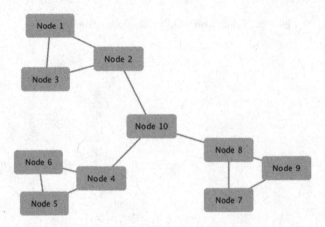

Fig. 11. Simple network with a clear modularity structure.

It can be observed at first sight that there are four different communities. Node 10 is in a class by itself. This can be seen if we take it out. The resulting network is disconnected in three similar subnetworks (nodes 1, 2, and 3, nodes 4, 5 and 6 and nodes 7, 8 and 9). So taking into account this we can talk of different communities at different hierarchies. At a zero level (distinguishing only connected components), a single community, the whole graph is obtained. At hierarchy level 1, a distinguished module (node 10) is detected and the other three already mentioned subnetworks (see Fig. 11).

---

[7] The same consideration can be applied to different performers, focusing on the rhthmic aspects, rather than the melodic ones, which in principle are fixed. Of course ornamentation aspects or improvised music can also be approached using these tools.

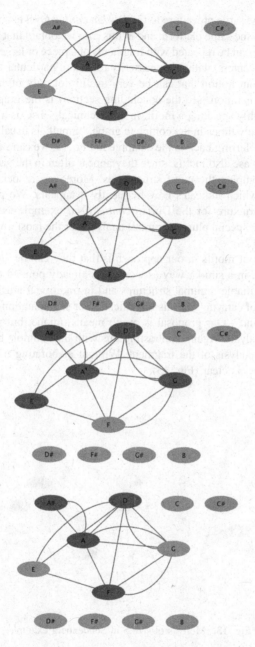

Fig. 12. Four motives of 5 and 4 nodes.

We can apply the same procedure to the *Dorian cantus firmi* example. This analysis was carried out with the same platform and in this case we obtain that D plays the central role. This procedure can be repeated with fragments of a piece or larger pieces in order to establish structural centers, without necessarily refer to a particular tonal idiom.

Another important notion that can be very useful not only in analysis but also in automatic music generation (see the concluding section) is the graph theoretical concept of motif. Roughly speaking a motif, in this technical sense, is a small subgraph or unit that appears many times in the complete graph. "Small" is usually taken to mean 3 or 4. For instance, referring back to the example in Fig. 11, we can say that the modules of size 3 are in this case also motifs, since they appear often in the graph. The larger the graph, the more complex the motif structure is. Moreover, associated to each motif there is a weight, which indicates how frequently it appears. We present in the final figures the motif structure for the Dorian *cantus firmi* example as obtained with the same platform and a special plugin for motif analysis for the most important motifs of 4 and 5 nodes (Fig. 12).

The importance of motifs in our approach is that they can be identified with basic units with meaning, in a similar way as words. As already pointed out, in music there are no clearly identifiable minimal structures and in traditional music analysis there is no systematic way of carrying out this identification or segmentation and in many cases this is done "by hand". Our proposal is by no means unique, but provides a starting point for further analysis. Again, contrasting this with the example by Schoenberg, we provide the motif analysis of the fragment. Without elaborating more, the difference with the *cantus firmi* is clear (Fig. 13).

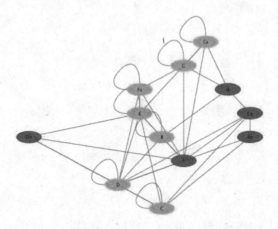

Fig. 13. Motives of size 5 in Schoenberg example.

As a final illustration we present a more realistic example. We consider all the Allemandes attributed to Louis Couperin and using the pitch class distributions for both the bass and the melodic lines we use several classification techniques ranging from principal components analysis to machine learning. We leave the discussion of the musicological implications of this example as well as the historical context and attribution problem that motivated it for another paper (Knights et al. 2017a) and (Wilson). However, we mention that even at this level of generality the classification schemes correctly separate major from minor pieces as well as melodic and bass lines and allows for the identification of outliers (Fig. 15).

The table in Fig. 14 presents all used pieces. The number refers to the catalogue number in A. Curtis edition (Curtis 1970).

Piece	Tonality	Voice	Tag	Cluster	Distance cluster	Distance data	1-class SVM score	Outlier score$\lesssim 0$
CC 02*	A minor	Bass line	CC 02 (b)	3	50.7	51.2	0.581 ± 0.029	
		Melody	CC 02 (m)	1	35.9	37.8	0.015 ± 0.002	*
CC 08	A minor	Bass line	CC 08 (b)	3	17.4	27.8	0.581 ± 0.029	
		Melody	CC 08 (m)	1	33.5	29.1	0.581 ± 0.029	
CC 13	A minor	Bass line	CC 13 (b)	3	29.6	45.5	0.638 ± 0.047	
		Melody	CC 13 (m)	1	12.0	10.7	1.229 ± 0.039	
CC 14	A minor	Bass line	CC 14 (b)	3	47.9	45.5	0.806 ± 0.045	
		Melody	CC 14 (m)	1	18.3	13.7	0.581 ± 0.029	
CC 19	B minor	Bass line	CC 19 (b)	3	39.2	40.5	0.301 ± 0.022	
		Melody	CC 19 (m)	1	16.7	22.9	0.755 ± 0.045	
CC 23*	C mayor	Bass line	CC 23 (B)	4	40.0	30.0	-0.036 ± 0.006	*
		Melody	CC 23 (M)	2	16.0	15.0	1.008 ± 0.045	
CC 34*	C minor	Bass line	CC 34 (b)	–	65.2	493.5	-0.053 ± 0.001	*
		Melody	CC 34 (m)	1	47.8	50.5	0.581 ± 0.029	
CC 40*	D mayor	Bass line	CC 40 (B)	–	105.2	207.4	-0.421 ± 0.021	*
		Melody	CC 40 (M)	2	31.7	24.4	0.581 ± 0.029	
CC 46	D minor	Bass line	CC 46 (b)	3	30.5	27.9	0.581 ± 0.029	
		Melody	CC 46 (m)	1	21.2	42.8	0.581 ± 0.029	
CC 65	E minor	Bass line	CC 65 (b)	3	28.2	26.9	0.719 ± 0.044	
		Melody	CC 65 (m)	1	14.4	19.4	0.581 ± 0.029	
CC 69	F mayor	Bass line	CC 69 (B)	–	56.3	127.5	0.581 ± 0.029	
		Melody	CC 69 (M)	2	23.3	18.9	0.581 ± 0.029	
CC 77	F mayor	Bass line	CC 77 (B)	3	28.3	40.1	1.258 ± 0.068	
		Melody	CC 77 (M)	2	17.7	12.2	0.581 ± 0.029	
CC 86*	G mayor	Bass line	CC 86 (B)	4	69.5	30.0	-0.152 ± 0.012	*
		Melody	CC 86 (M)	2	33.6	27.8	0.177 ± 0.002	
CC 93	G minor	Bass line	CC 93 (b)	3	22.3	26.0	0.749 ± 0.045	
		Melody	CC 93 (m)	1	19.7	18.5	0.581 ± 0.029	

Fig. 14. Allemandes from the A. Curtis edition of the keyboard music by L. Couperin. CC stands for Couperin and Curtis. The numbers are the same as in the edition. B and M stands for bass and melody in major keys and correspondingly b and m for minor keys.

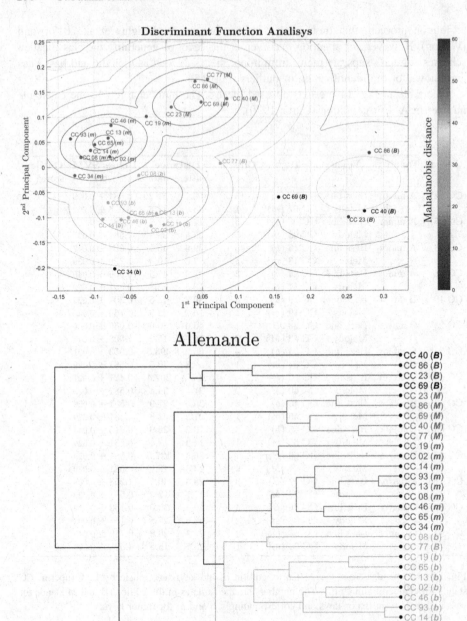

Fig. 15. Classification techniques for the Allemandes mentioned in the text.

Outlier detection via 1-class SVM

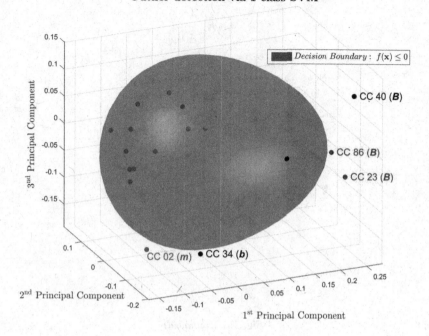

Allemande

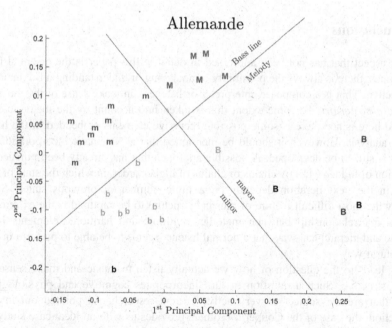

Fig. 15. (*continued*)

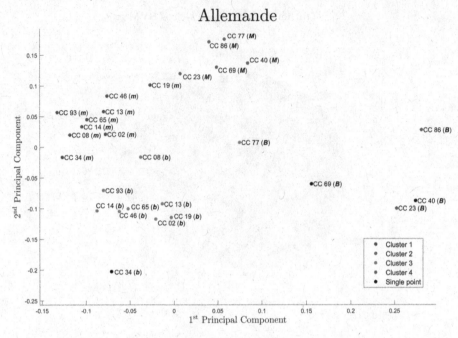

Fig. 15. (*continued*)

3 Conclusions

Another aspect that has not been addressed in detail in this paper is the essential fact that in music there is always the possibility of analyzing or understanding a fragment in retrospective. That is a composer, interpreter or listener can make sense of the musical structure *a posteriori*. To some extent this is taken into account by the methodology proposed here, since there is some possible recursive elements embedded in the hierarchical analysis. However, it should be mentioned that a systematic incorporation of this fact is still to be developed. A possible approach that has already been pursued is application of hidden Markov chains or chains of higher order, in which the state of the system in the next iteration depends on a finite number of previous states. Still, probably the most difficult feature of musical structure to be captured is the extremely complex interrelationship between melodic, rhythmic and harmonic elements in a recursive and hierarchical way that a normal listener seems to be able to perform in the most natural way.

This leads to the question of how we actually listen to music and make sense of musical structure. Such a question in turn incorporates cognitive and physiological aspects that are the subject of very active current research. As pointed out in the introduction, the case of the Couperin brothers provides us with an ideal case study to which apply the methodology presented here and this is the subject of a forthcoming work (Knights et al. 2017a). We also refer to (Knights et al. 2017b) for an application to elucidating joint attribution to Taverner and Tye of the motet *O splendor gloriae*.

However, there are countless possibilities of application. Even in the work of a composer like J. S. Bach, that has been thoroughly studied, stylistic issues are still far from being settled (Jones 2015).

Finally, we also mention that the extraction of modules and motifs provides a starting point to generate music in a standard way. Once a motif structure has been defined, one can use simulation techniques to generate patterns with similar characteristics (see (Hardoon et al. 2014) for a recent work in which this technique is used. We would like to stress the fact that this methodology can provide the basic recombination material that was proposed by Cope, which we referred to in the introduction. We also leave for a later work the more detailed study of automatic music generation using these techniques, since it constitutes a subject by itself.

References

Bassiou, N., Kotropoulos, C., Papazoglou-Chalikias, A.: Greek folk music classification into two genres using lyrics and audio via canonical correlation analysis. In: 9th International Symposium on Image and Signal Processing and Analysis (ISPA), pp. 238–243. IEEE (2015)

Collins, T., Laney, R.: Computer-generated stylistic compositions with long-term repetitive and phrasal structure. J. Creative Music Syst. **1**(2) (2017)

Cope, D.: Virtual Music. Computer Synthesis of Musical Style. MIT Press, Cambridge (2004)

Curtis, A., (ed.): L. Couperin, Pièces de clavecin (Montrouge), vol. 2 (1970)

Domek, R.: Some Aspects of Organization in Schoenberg's Book of the Hanging Gardens, Opus 15 (1979). http://symposium.music.org/index.php?option=com_k2&view=item&id=1837: some-aspects-of-organization-in-schoenbergs-book-of-the-hanging-gardens-opus-15&Itemid=124

Farbood, M., Schoner, B.: Analysis and synthesis of Palestrina-style counterpoint using Markov chains. In: Proceedings of International Computer Music Conference. International Computer Music Association, San Francisco (2001)

Jones, D.P.R.: The Creative Development of Johann Sebastian Bach, vol. I. Oxford University Press, Oxford (2015)

Hardoon, D.R., Saunders, C., Shawe-Taylor, J.: Using Fisher Kernels and Hidden Markov Models for the Identification of Famous Composers from their Sheet Music (2014). https://www.researchgate.net/publication/253897785_Using_Fisher_Kernels_and_Hidden_Markov_Models_for_the_Identification_of_Famous_Composers_from_their_Sheet_Music

Jeppesen, K.: Counterpoint. The Polyphonic Vocal Style of the Sixteenth Century. Dover, Mineola (1992)

Knights, F., Padilla, P., Tidhar, D.: Identification and evolution of musical style II: a statistical approach (2017a). formal-methods-in-musicology.webnode.com

Knights, F., Padilla, P., Tidhar, D.: A question of attribution in *O splendor gloriae*: Taverner and Tye (2017b). formal-methods-in-musicology.webnode.com

Loy, G.: Musimathics. The Mathematical Foundations of Music. MIT Press, Cambridge (2006)

Nierhaus, G.: Algorithmic Composition. Springer, Heidelberg (2009). https://doi.org/10.1007/978-3-211-75540-2

Stamatatos, E.: A survey of modern authorship attribution methods. J. Am. Soc. Inform. Sci. Technol. **60**(3), 538–556 (2009)

Tidhar, D.: A hierarchical and deterministic approach to music grammars and its application to unmeasured preludes. Dissertation.de (2005)

Tolviainen, P., Eerola, T.: Midi Toolbox 1.1. Github (2016). https://github.com/miditoolbox/1.1

Wilson, G.: The other Mr. Couperin. https://www.academia.edu/30474468/The_Other_Mr._Couperin

A Fuzzy-Clustering Based Approach for Measuring Similarity Between Melodies

Brian Martínez[1] and Vicente Liern[2(✉)]

[1] Conservatorio Superior de Música, Valencia, Spain
brian.martinez@csmvalencia.es
[2] Dep. Matemáticas para la Economía y la Empresa,
Universidad de Valencia, Valencia, Spain
vicente.liern@uv.es

Abstract. Symbolic melodic similarity aims to evaluate the degree of likeness of two or more sequences of notes. In this work, we propose the use of fuzzy c-means clustering as a tool for the measurement of the similarity between two melodies with a different number of notes. Moreover, we present an algorithm, FOCM, implemented in a computer program written in C♯ able to read two melodies from files with MusicXML format and to perform the clustering to calculate the dissimilarity between any two melodies. In addition, for each iteration step in the convergence process of the algorithm, a family of intermediate states (transition melodies) are obtained that can be used as new thematic material. This last feature, could be especially useful in the near future, as a complement in computer-aided composition.

Keywords: Fuzzy clustering · Symbolic melodic similarity
Computer-aided composition

1 Introduction

Symbolic melodic similarity is fundamental in the field of computer-aided composition [1,13]. The measure of the similarity/dissimilarity between melodies is a key factor both in defining transitions between two different melodies and to generate new melodic material from an already preexisting melody [11]. In this paper, we propose a procedure to measure the similarity between two melodies by using an algorithm based on fuzzy clustering.

Our starting point will be the characterization of musical notes as points in a metric space, where the coordinates represents musical characteristics. In this way, a melody would be an ordered sequence of notes. Measuring the dissimilarity between two monophonic melodies of equal number of notes can be made by comparing, one by one, each note of the first melody with a note of the same order in the second. However, for our purpose we would need to be able to establish a generic comparison mechanism allowing the measurement of the dissimilarity

© Springer International Publishing AG 2017
O. A. Agustín-Aquino et al. (Eds.): MCM 2017, LNAI 10527, pp. 279–290, 2017.
https://doi.org/10.1007/978-3-319-71827-9_21

of two (monophonic or polyphonic) melodies of different number of notes. For this, we will use an algorithm based on fuzzy c-means clustering.

The purpose of that clustering would be to establish to what extent the notes of a first melody are related to the notes of a second one. The purpose of that clustering would be to establish to what extent the notes of two different melodies are related. After the clustering, we will be able to calculate a global difference between the melodies (dissimilarity) aggregating the partial distances weighted by their corresponding membership coefficient. In the fuzzy logic context, the membership functions are the extension of the characteristic set functions [16]. While characteristic functions take values 0 or 1, membership functions can take any value between 0 and 1. Therefore, the membership coefficients express the membership degree of an element to a cluster [15, 16].

Subsequently, in the comparison of the general dissimilarity, we will take into account the order of the notes in each melodic sequence. For this, we will use neighborhood functions. These functions will allow us to define a comparison in which the clustering of the notes is influenced by their position within the sequence defining the melody.

In order to verify the utility of our proposal, we present an algorithm, FOCM, implemented in a computer program written in C$\sharp$. This algorithm will allow us to read two melodies from files in MusicXML format and to perform the clustering to calculate the dissimilarity between them. In addition, for each iteration step in the convergence process of the algorithm, a family of intermediate melodies will be obtained that can be used as new thematic material. This last feature could be especially useful in the near future, as an aid in computer composition. For this reason, in the last section we provide an example, in which the number of intermediate melodies created by our method is shown. As an instance, we present one of the intermediate melodies obtained from the measure of dissimilarity between two passages.

2   Preliminary Concepts

A musical note determined by k characteristics (picht, intensity, duration, timbre, etc.) can be expressed as a vector in $\mathbb{R}^q$, where $q \leq d$. Of course, each characteristic does not have to correspond to a single coordinate. For example, in [7, 8] pitch is defined as a fuzzy set [16],

$$\tilde{P} = \{(f, \mu_{\tilde{P}}(f)), \quad f \in [F_0, F_1]\}, \tag{1}$$

where f represents the frequency in Hz. and $\mu_{\tilde{P}}(f) \in [0, 1]$ is the membership degree of f to a note in a given tunning system. In this case, the fuzzy pitch would be given by two coordinates $(f, \mu_{\tilde{P}}(f))$.

The most simple way to represent a musical note is by setting $q = 2$, the pitch and the duration, and establishing two bijections from the pitch and the duration of the note $\mathbf{x} \in \mathbb{R}^2$. However, it is possible to work with a higher number of dimensions in order to represent more accurately the characteristics of music. New properties belonging to the requirements of other kinds of music or styles [14], like *Non Western-tradition music*, *Computer-Generated Music* or *Electroacoustics* could be easily assimilated as extra dimensions.

As usual, the distance in cents between two notes whose frequencies are f_1 and f_2 can be easily calculated in cents [3,8] by means of the expression

$$d(f_1, f_2) = 1200 \times \left| \log_2 \left(\frac{f_1}{f_2} \right) \right| \text{ cents.} \tag{2}$$

According to [12], the MIDI protocol defines a midi-pitch of a note by a integer number comprised on a range $[0, 127]$, being central $C_4 = 60$ and reference $A_4 = 69$. For the equal temperament of 12 notes [3], there are 100 cents of difference between two notes separated by one midi-pitch number (semitone). If a concert pitch frequency f_{A_4} (usually 440 Hz) corresponds to the midi-pitch number 69 then, the midi-pitch number of a frequency f, is

$$\nu = 69 + 12 \log_2 \left(\frac{f}{f_{A_4}} \right). \tag{3}$$

Taking the figure of the whole note as the unit, it is easy to define a note's duration coefficient $\delta \in \mathbb{R}$. A half note has a coefficient $1/2$, a quarter note $1/4$, a quaver $1/8$, etc., i.e.

$$\alpha = \frac{1}{2^a}, \quad -1 \leq a \leq 7. \tag{4}$$

In addition, each dotted note multiplies its duration by the factor

$$\beta = \sum_{k=0}^{b} \frac{1}{2^k} = \frac{2^{b+1} - 1}{2^b}, \tag{5}$$

where b is the number of dots of the note. On the other hand, tuplets (described in [2] as reading c notes in the space of d) are notated by the expression $c : d$, and modify the duration of each note with the factor

$$\gamma = \frac{d}{c}. \tag{6}$$

Taking into account expressions (4), (5) and (6), a number of τ tied notes will have a duration

$$\delta = \sum_{i=1}^{\tau} (\alpha_i \cdot \beta_i \cdot \gamma_i) = \sum_{i=1}^{\tau} \left[\frac{1}{2^{a_i}} \cdot \frac{2^{b_i+1} - 1}{2^{b_i}} \cdot \frac{d_i}{c_i} \right]. \tag{7}$$

Once the concept of musical note has been defined we can express a melody as an ordered sequence of n notes, being each note of the melody a point in a *metric space*.

Definition 1. *A melody is a sequence, $\mathcal{M} = \{x_i\}_{i=1}^{n}$, where each $x_i \in \mathbb{R}^q$ is a musical note.*

For instance, let us consider the melody in Fig. 1. If we were only interested in the duration δ and in the midi-pitch number ν of each note, the fragment could be expressed as the following sequence of 14 notes:

$$\mathcal{M}_1 = \{(\delta_i, \nu_i)\}_{i=1}^{14} = \{(0.041667, 67), (0.041667, 69), (0.041667, 70), (0.166667, 69),$$
$$(0.166667, 67), (0.166667, 70), (0.250000, 72), (0.041667, 69), (0.041667, 72)$$
$$(0.041667, 70), (0.125000, 69), (0.187500, 67), (0.062500, 65), (0.750000, 67) \}.$$

If the melody is polyphonic, as in Fig. 2, the notes' pitch is represented by a vector $\bar{\nu} \in \mathbb{R}^k$, where k is the least common multiple of the number of voices appearing in the melody. In Fig. 2 notes with 1, 2 and 3 voices appear, then $k =$ l.c.m.$(1, 2, 3) = 6$. Consequently, if our only interest are the duration and pitch of the notes, these could be expressed by using 7 coordinates. The corresponding melody would be $\mathcal{M}_2$.

$$\mathcal{M}_2 = \{(\delta_i, \bar{\nu}_i)\}_{i=1}^{5} = \{(0.125; 67, 67, 67, 71, 71, 71), (0.0625; 64, 64, 64, 67, 67, 67),$$
$$(0.0625; 67, 67, 67, 71, 71, 71), (0.125; 64, 64, 67, 67, 71, 71),$$
$$(0.125; 71, 71, 71, 71, 71, 71)\}.$$

Fig. 1. Example of a melodic line to be represented into the plane duration-pitch.

Fig. 2. Example of polyphonic melody.

3   Comparison of Melodies

If we consider two melodies $\mathcal{M}^A$ and $\mathcal{M}^B$, both belonging to a q-dimensional metric space, it is only possible to measure a well-defined distance between them if they have the same number of notes. This does not necessarily mean that both melodic lines have the same duration expressed in units of time; however, they need to have the same number of points.

In order to compare those melodies we will first calculate a total distance by accumulating the partial distance between each couple of notes $\mathbf{x_i}$, $\mathbf{y_i}, i \geq 1$, respecting the established order of the sequence of the points of both melodies.

Definition 2. *Let $\mathscr{M}^A = \{\mathbf{x_1}, \ldots, \mathbf{x_n}\}$ and $\mathscr{M}^B = \{\mathbf{y_1}, \ldots, \mathbf{y_n}\}$ be two melodies with n notes, belonging to a q-dimensional space $\mathbb{R}^q$. Given $d : \mathbb{R}^q \times \mathbb{R}^q \to \mathbb{R}$ a distance function, the distance between $\mathscr{M}^A$ and $\mathscr{M}^B$ can be defined as*

$$D(\mathscr{M}^A, \mathscr{M}^B) = \mathscr{F}\{d(\mathbf{x_1}, \mathbf{y_1}), \ldots, d(\mathbf{x_n}, \mathbf{y_n})\}, \tag{8}$$

where $\mathscr{F}$ is a prefixed aggregation operator [15].

In this work, until the contrary is noticed, we will use as $\mathscr{F}$ operator the arithmetic mean, that is

$$\bar{D}(\mathscr{M}^A, \mathscr{M}^B) = \frac{1}{n} \sum_{i=1}^{n} d(\mathbf{x_i}, \mathbf{y_i}). \tag{9}$$

Nevertheless, regardless the operator chosen, it is easy to verify the following result:

Proposition 1. *Assuming the previous notation, the following inequalities hold*

$$\min_i d(\mathbf{x_i}, \mathbf{y_i}) \leq D(\mathscr{M}^A, \mathscr{M}^B) \leq \max_i d(\mathbf{x_i}, \mathbf{y_i}). \tag{10}$$

As well known [3], the Weber-Fechner Law approximates the psychological rules of human perception of intensity or pitch. This idea can be easily incorporated this into the calculation of the distance. For example, if we express the notes $\mathbf{x}$ with three coordinates (x_1, x_2, x_3) representing duration, pitch and intensity, respectively, in [9] the following distance is used

$$d(\mathbf{x}, \mathbf{y}) = \alpha \cdot |x_1 - y_1| + \beta \cdot |\log(x_2/y_2)| + \gamma \cdot |\log(x_3/y_3)|, \tag{11}$$

where α, β and γ are some prefixed constant values.

If we want to compare two melodies with different number of notes, Definition 2 has to be generalized. In fact, in the symbolic melody similarity literature it is possible to find several examples in which some definitions of distance between two melodies of different length are defined [5,10]. The objective of many of these works is to approximate as much as possible to human perception [11]. With this aim, different techniques have been proposed ranging from the geometric structure of the melodies [1] to fuzzy logic [11], for instance.

A definition of an average distance based on the clustering of two melodies $\mathscr{M}^A = \{\mathbf{x_1}, \ldots, \mathbf{x_n}\}$ and $\mathscr{M}^B = \{\mathbf{y_1}, \ldots, \mathbf{y_m}\}$, being $n > m$, allows us to estimate how far away melody $\mathscr{M}^A$ is from melody $\mathscr{M}^B$. Despite the fact that this measurement will not satisfy the requirements of a distance function, the result provides some useful information about to the degree of similarity between these two melodies.

When a classical clustering process, e.g. *c-means clustering*, is applied to a general data set X of information, the result is a Boolean partition of X into c clusters, so each element of X belongs only to one cluster. Related to the comparison of melodies, we can use this procedure to cluster the set of n notes of melody $\mathscr{M}^A$ into m subsets. Once this is finished, we will be able to associate each subset in $\mathcal{M}^A$ to a note in $\mathcal{M}^B$ and finally, calculate an average distance from every point of each subset in $\mathcal{M}^A$ to its corresponding note in $\mathcal{M}^B$.

The global dissimilarity of the two melodies would be calculated by aggregating the partial average distance. However, while carrying out with this procedure we have to accept two arguable assertions:

1. It is assumed that comparing each note of $\mathcal{M}^A$ only to one note of $\mathcal{M}^B$ has musical sense.
2. In the process of comparing notes the order information is omitted. This is a key question in musical terms.

In what follows, a new proposal based on fuzzy logic will be presented. Real features of musical fact can be better represented by this new approach. With this objective, we will use fuzzy clustering applied to the calculation of a dissimilarity measure between two melodies of different number of notes.

3.1 Fuzzy C-Means Clustering (FCM)

The fuzzy c-means is a clustering method initially developed by Dunn [6] in 1973, based on the statement that any element of a given set is able to belong to more than one cluster. Thus, the fuzzy clustering method will provide a membership function that describes the belonging degree of each element to any centroid. As it is explained in [4], the generalization of fuzzy c-means algorithms comes from the iterative minimization of an objective functional.

Definition 3. *Let the data set* $X = \{x_1, x_2, \ldots, x_n\} \subset \mathbb{R}^q$. *Let* v *be a set of cluster centers* $v = (v_1, v_2, \ldots, v_m)$, *with* $v_i \in \mathbb{R}^q$ *and* $m < n$. *Fuzzy c-means functionals are defined as*

$$J_\lambda = \sum_{i=1}^{n} \sum_{j=1}^{m} (u_{ij})^\lambda (d_{ij})^2, \tag{12}$$

where $d_{ij}^2 = \| x_i - v_j \|^2$, *being* $\| \cdot \|$ *any inner product induced norm on* $\mathbb{R}^q$, $\lambda \in [1, \infty)$ *is the weighting exponent (degree of fuzzyness of the process), and* u_{ij} *is the membership coefficient of* x_i *to the cluster* j.

The fuzzy clustering is achieved through an iterative optimization of J_λ, updating, at each iteration, the membership coefficients u_{ij} as well as the *cluster centers* v_j by using the following expressions

$$u_{ij} = \frac{1}{\sum_{k=1}^{m} \left[\frac{\|x_i - v_j\|}{\|x_i - v_k\|} \right]^{\frac{2}{\lambda-1}}}, \quad v_j = \frac{\sum_{i=1}^{n} u_{ij}^\lambda \cdot x_i}{\sum_{i=1}^{n} u_{ij}^\lambda}. \tag{13}$$

The matrix U is a fuzzy partition of X, formed by the membership coefficients u_{ij}

$$U_{ij} = \begin{pmatrix} u_{11} & \cdots & u_{1n} \\ \vdots & \ddots & \vdots \\ u_{m1} & \cdots & u_{nm} \end{pmatrix}. \tag{14}$$

As convergence condition of any fuzzy clustering we have

$$\sum_{j=1}^{m} u_{ij} = 1, \quad 1 \leq i \leq n. \tag{15}$$

3.2 Fuzzy C-Means Algorithm

In what follows we will show the implementation of the Fuzzy c-Means Clustering Algorithm proposed by Bezdek in [4].

FCM-Algorithm

STEP 1. Fix a number of clusters m, $2 \leq m < n$. Choose any inner product norm metric for $\mathbb{R}^q$; fix λ, $1 \leq \lambda < \infty$. Initialize $U^{(0)}$.

STEP 2. Calculate the fuzzy cluster centers $\{v_j^{(k)}\}$ with $U^{(k)}$ and expression (13).

STEP 3. Update $U^{(k)}$ using expression (13) and $\{v_j^{(k)}\}$.

STEP 4. Compare $U^{(k)}$ to $U^{(k+1)}$ using a convenient matrix norm, being $\epsilon \in (0,1)$ and arbitrary termination criterion. If $\| U^{(k+1)} - U^{(k)} \| \leq \epsilon$ then stop, otherwise set $k = k + 1$ and return to STEP 2.

4 Measuring Dissimilarity by Means of Fuzzy Clusters

Let us consider two melodies $\mathcal{M}^A$ and $\mathcal{M}^B$ with different number of notes. We will now make a fuzzy partition of the notes from $\mathcal{M}^A$ with the initial cluster centers given by $\mathcal{M}^B$, and apply the FCM algorithm k times until the termination criterion is satisfied. Once the partition process is complete, we can define a dissimilarity function between $\mathcal{M}^A$ and $\mathcal{M}^B$ by using the final membership coefficients and the original cluster centers.

Definition 4. Let $\mathcal{M}^A = \{\mathbf{x_1}, \ldots, \mathbf{x_n}\} \subset \mathbb{R}^q$ and $\mathcal{M}^B = \{\mathbf{y_1}, \ldots, \mathbf{y_m}\} \subset \mathbb{R}^q$ be two melodies, where $n > m$. Let $d : \mathbb{R}^q \times \mathbb{R}^q \to \mathbb{R}$ be a distance function. Let u_{ij} be the final membership coefficients calculated with FCM algorithm. The average dissimilarity $\mathcal{D}$ from $\mathcal{M}^A$ to $\mathcal{M}^B$ is defined by

$$\mathcal{D}(\mathcal{M}^A, \mathcal{M}^B) = \frac{1}{n \cdot m} \sum_{i=1}^{n} \sum_{j=1}^{m} u_{ij} \cdot d(\mathbf{x_i}, \mathbf{y_j}). \tag{16}$$

By construction, $\mathcal{D}$ does not consider the natural order of the sequence of notes within each melody. Thus, the partition that FCM algorithms calculate does not weight in any special way the notes whose degree of neighbourhood is stronger. As an illustrative example of this fact, in Fig. 3 it is possible to see three different melodies. Since Melody B is a complete retrogradation of Melody A, average dissimilarity $\mathcal{D}$ between melodies A and C have exactly the same value than average dissimilarity between melodies B and C,

Fig. 3. Three example melodies, where Melody B is a retrogradation of Melody A.

$$\mathscr{D}(\mathscr{M}^A, \mathscr{M}^c) = 0.23354, \qquad \mathscr{D}(\mathscr{M}^B, \mathscr{M}^c) = 0.23354.$$

This example shows that a comparison of different melodies without taking into account the order of the notes does not completely reflect musical reality. To avoid this, we will introduce a dependence with the order in the algorithm. In this way, higher weights will be given to the pair of notes that share closer positions in the order of each melody, reducing the contribution to the global dissimilarity of the pair of notes that are far away from an ordinal point of view. Neighbourhood functions will provide the information related to the order in which the pair of notes must be compared.

Definition 5. *A continuous function* $f : \mathbb{R}^2 \to \mathbb{R}$ *is a neighbourhood function between two melodies* $\mathscr{M}^A, \mathscr{M}^B$ *if*

$$\int_1^n f(i,j)di < \infty, \quad \forall j \in \{1, 2, \dots, m\}, \tag{17}$$

where n *is the number of notes of* $\mathscr{M}^A$ *and* m *the number of notes of* $\mathscr{M}^B$.

If a correct setting for the neighbourhood function is defined, neighbourhood values of $i \in (j - \varepsilon, j + \varepsilon)$ will be assigned to higher coefficients and the rest of values will be assigned lower coefficients.

The procedure will be following: Once the fuzzy partition U has been calculated, we will assign a weight to any element u_{ij} by means of a specific neighbourhood function $f(i,j)$. In order to accomplish with the FCM convergence criterion, we will normalize U as follows

$$\tilde{u}_{ij} = \frac{u_{ij} \cdot f(i,j)}{\sum\limits_{k=1}^{m} u_{ik} \cdot f(i,k)}. \tag{18}$$

Example 1. Gaussian neighbourhood function

$$f_G(i,j) = A e^{-\frac{1}{2\sigma^2}\left[i+1-\frac{(n-1)\cdot(j-1)}{(m-1)}\right]^2}. \tag{19}$$

In this function it is easy to see how the original μ value has been replaced by $1 - \frac{(n-1)\cdot(j-1)}{(m-1)}$. This expression has been obtained from the equation of a line

$\mu = f(j)$ that crosses through points $(1,1)$ and (m,n). Given the fixed values $n, m \in \mathbb{N}$, the shape of $f(i,j)$ will change for each pair of values i, j. When $j = 1$, the Gaussian will be centered at $i = 1$, but when $j = m$ it will be centered on $i = n$.

Our proposal is to modify the algorithm FCM in such a way that the order of the sequences of the notes $\mathscr{M}^A = \{\mathbf{x_1}, \ldots, \mathbf{x_n}\}$ and $\mathscr{M}^B = \{\mathbf{y_1}, \ldots, \mathbf{y_m}\}, n < m$, is taken into account. With this objective we propose the following algorithm, named fuzzy ordered c-means (FOCM).

4.1 FOCM-Algorithm

STEP 1. Set $\{v_j^{(0)}\} = \{y_j\}$. Let m, n be the number of notes of $\mathscr{M}^B$ and $\mathscr{M}^A$, respectively. Choose any convenient neighbourhood function.

STEP 2. Choose any inner product norm metric for $\mathbb{R}^q$, and fix $\lambda \geq 1$. Calculate the initial $\widetilde{U}^{(0)}$ using (13), (18) and $\{v_j^{(0)}\}$.

STEP 3. Calculate the fuzzy cluster centers $\{v_j^{(k)}\}$ with $\widetilde{U}^{(k)}$ and the equation (13).

STEP 4. Update $\widetilde{U}^{(k)}$ using the Eqs. (13), (18) and $\{v_j^{(k)}\}$.

STEP 5. Compare $\widetilde{U}^{(k)}$ to $\widetilde{U}^{(k+1)}$ using a convenient matrix norm; being $\epsilon \in (0,1)$ and arbitrary termination criterion. If $\| \widetilde{U}^{(k+1)} - \widetilde{U}^{(k)} \| \leq \epsilon$ then stop; otherwise set $k = k + 1$ and return to STEP 3.

Once the melodies have been compared by taking into account all the described characteristics, we can establish the following definition.

Definition 6. *Let $\mathscr{M}^A = \{\mathbf{x_1}, \ldots, \mathbf{x_n}\} \in \mathbb{R}^q$ and $\mathscr{M}^B = \{\mathbf{y_1}, \ldots, \mathbf{y_m}\} \in \mathbb{R}^q$ be two melodies of different number of notes. Let $d : \mathbb{R}^q \times \mathbb{R}^q \to \mathbb{R}$ be a distance function. Let $\tilde{u}_{ij}$ be the final membership coefficients calculated with FOCM algorithm. The average ordered dissimilarity $\widetilde{\mathscr{D}}$ from $\mathscr{M}^A$ to $\mathscr{M}^B$ is defined by*

$$\widetilde{\mathscr{D}}(\mathscr{M}^A, \mathscr{M}^B) = \frac{1}{n \cdot m} \sum_{i=1}^{n} \sum_{j=1}^{m} \tilde{u}_{ij} \cdot d(\mathbf{x_i}, \mathbf{y_j}). \tag{20}$$

In what follows we show the utility of expression (20). For this, we will calculate the dissimilarity between different melodies.

4.2 Computational Examples

Example 2. We are now going to compare the melodies appearing in Fig. 4 using expressions (16) and (20).

The dissimilarity values are $\mathscr{D}(\mathscr{M}^A, \mathscr{M}^B) = 0.68938$ and $\widetilde{\mathscr{D}}(\mathscr{M}^A, \mathscr{M}^B) = 3.93483$. The reason behind the disparity in the obtained results is that when the order of the notes is not taken into account, we use $\mathscr{D}$, sharp notes in $\mathscr{M}^A$

are compared with sharp notes in $\mathcal{M}^B$ and flat notes in $\mathcal{M}^A$ are compared with flat notes in $\mathcal{M}^B$. In fact, when we give importance to the order, $\widetilde{\mathscr{D}}$, both melodies are quite different (dissimilarity is almost six times greater with $\widetilde{\mathscr{D}}$ than with $\mathscr{D}$). In Fig. 5 we show a screenshot appearing in our implementation of algorithm FOCM. We can observe how, for example, the two first notes in $\mathcal{M}^B$ are associated to the first time measure in $\mathcal{M}^A$, showing the above mentioned differences.

Fig. 4. Melodies of Example 1.

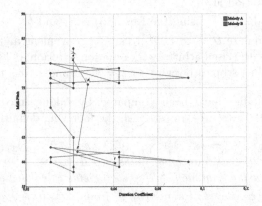

Fig. 5. Final result of clustering algorithm FOCM of $\mathcal{M}^A$ and $\mathcal{M}^B$.

Example 3. Using the melodies displayed in Fig. 6, we will now show the capacity of our proposal to measure the dissimilarity in polyphonic melodies.

Fig. 6. Melodies of Example 2.

The obtained values are $\widetilde{\mathscr{D}}(\mathscr{M}^A, \mathscr{M}^C) = 0.30695$, $\widetilde{\mathscr{D}}(\mathscr{M}^B, \mathscr{M}^C) = 0.46356$. As it was expected, melodies in this example are more similar than melodies in Example 1 and the differences between them increase when one of them is polyphonic and the other is not.

Example 4. In Table 1 we provide an example showing how our proposal functions. We have selected four passages of very well-known musical works: (1) = W.A. Mozart. Symphony No. 40. First movement. Measures 1–4, (2) = L.V. Beethoven. Symphony No. 6. First movement. Measures 1–4, (3) = J. Brahms. Symphony No. 3. Second movement. Measures 1–8, and (4) = B. Bartók. Music for strings, percussion and celesta. First movement. Measures 1–4.

Table 1. Measurement of dissimilarity between melodies from Example 4.

$\mathscr{M}^A$	$\mathscr{M}^B$	$\widetilde{\mathscr{D}}$	♯States	$\widetilde{\mathscr{D}}$	♯States	$\widetilde{\mathscr{D}}$	♯States
		Fuzzy coefficient = 2		Fuzzy coefficient = 2.5		Fuzzy coefficient = 3	
1	2	0.2395874272	389	0.1820221001	389	0.1835055288	17
3	1	0.1144880336	56	0.1079920868	375	0.1094903790	584
4	1	0.4231256333	508	0.3294539142	672	0.3609358015	636
3	2	0.0910976980	10	0.1231091704	20	0.1233827179	12
4	2	0.7071511697	81	0.6642747436	194	0.6666523284	353
3	4	0.0692844111	51	0.1451223554	429	0.1383288250	310

In Table 1 we display the dissimilarity measures between melodies from Example 4, as well as the number of intermediate compositions (♯States) generated by the algorithm with different values of the fuzzy coefficient used in the FOCM (Fig. 7).

Fig. 7. One of the intermediate melodies obtained by the algorithm when measuring the dissimilarity between passages (1) and (3) from Example 4.

5   Conclusions

The evaluation of the degree of likeness of two melodies is nowadays a topic of great interest. By comparing the similarity of different melodies it is possible to find patterns, to extract rules and to identify structures, all key questions in the study of musical styles. In this work, we have proposed a fuzzy logic tool, fuzzy c-means clustering, for the measurement of the similarity between two melodies with different number of notes.

The proposed FOCM algorithm allow us to define a measurement of the symbolic melodic dissimilarity between two different melodies, taking into account the order of the sequences of notes that each melody contains. To a certain extend, the definition of *fuzzy c-means average ordered dissimilarity* offers a geometric way to compare very different melodic lines that can be used with several purposes, like classification of melodies, computer-aided composition or musical-styles recognition.

Our proposal could also be applied to other fields of research in which is necessary to estimate the degree of closeness of two different sequences of ordered information.

References

1. Aloupis, G., Fevens, T., Langerman, S., Matsui, T., Mesa, A., Nuez, Y., Rappaport, D., Toussaint, G.: Algorithms for computing geometric measures of melodic similarity. Comput. Music J. **30**(3), 67–76 (2006)
2. Apel, W.: Harvard Dictionary of Music, 2nd edn. The Belknap Press of Harvard University Press, Cambridge (1994)
3. Benson, D.: Music: A Mathematical Offering. Cambridge University Press, Cambridge (2006)
4. Bezdek, J.C.: Pattern Recognition with Fuzzy Objective Function Algoritms. Plenum Press, New York (1981)
5. Downie, J.S.: Evaluating a simple approach to musical information retreival: conceiving melodic N-grams as text. Ph.D. Thesis, University of Western Ontario, Ontario (1999)
6. Dunn, J.C.: A fuzzy relative of the ISODATA process and its use in detecting compact well-separated clusters. J. Cybern. **3**, 32–57 (1973)
7. Haluska, J.: The Mathematical Theory of Tone Systems. Marcel Dekker Inc., Bratislava (2005)
8. Liern, V.: Fuzzy tuning systems: the mathematics of the musicians. Fuzzy Sets Syst. **150**, 35–52 (2005)
9. Liern, V.: La música y sus materiales: una ayuda para las clases de matemáticas. Suma **14**, 60–64 (1994)
10. Mongeau, M., Sankoff, D.: Comparison of musical sequences. Comput. Humanit. **24**, 161–175 (1990)
11. Müllensiefen, D., Frieler, K.: Cognitive adequacy in the measurement of melodic similarity: algorithmic vs. human judgments, Comput. Musicology **13**, 147–176 (2004)
12. Selfridge-Field, E.: Beyond MIDI: The Handbook of Musical Codes. MIT Press, Cambridge (1997)
13. Velardo, V., Vallati, M., Jan, S.: Symbolic melodic similarity: state of the art and future challenges. Comput. Music J. **40**(2), 70–83 (2016)
14. Xenakis, I.: Formalized Music: Thought and Mathematics in Composition. Pendragon Press, Launceston (1992)
15. Yager, R.R.: On ordered weighted averaging aggregation operators in multi-criteria decision making. IEEE Trans. Syst. Man Cybern. **18**, 183–190 (1988)
16. Zadeh, L.A.: Fuzzy sets. Inf. Control **8**, 338–353 (1965)

The Evolution of Tango Harmony, 1910–1960

Bruno Mesz[1](✉), Augusto Paladino[1], Juan Pérgola[1], and Pablo Amster[2,3]

[1] Departamento de Artes Electrónicas, UNTREF, Buenos Aires, Argentina
bruno.mesz@gmail.com
[2] Departamento de Matemática, FCEyN-UBA, Buenos Aires, Argentina
[3] CONICET, Buenos Aires, Argentina

Abstract. In this article, we look at the diachronic changes in tango harmony with the methods of network science. We are able to detect some significant tendencies of harmonic discourse in the first half of the 20th century, among them an enrichment of harmonic transitions and power law frequency distribution of triadic chords with exponents compatible with a quite small rate of accretion of the vocabulary.

1 Introduction

Tango is undoubtedly the most transcendent collective cultural creation of the Río de la Plata region. Several texts give account of its history, spanning from the last decades of the 19th century to the present day [1, 2]. In spite of this, to the extent of our knowledge, no computational musicological study has focused specifically on tango and its diachronic evolution.

The availability of big corpora of musical data has fostered quantitative evolutionary studies on American popular music [3], jazz harmony [4], electronic art music [5], musical influence of songs [6], to mention some examples. Recently, complex networks methods have been employed to analyse pitch and timbre transitions both in individual works [7], and large collections [8,9].

We consider chord transition networks built from sampling whole decades of a corpus of tango recordings. To this end, we assembled a database of 510 recordings of tangos, composed between 1910 and 1960, by downloading all the tangos from the Web archive Todo Tango [10], and discarding those that contained extramusical elements such as speech or clapping. Some of the recordings were denoised using Adobe Audition. In case several recordings of the same tango were available, we preferred the one with the earliest recording date. The median number of years between composition and recording is 0.

We built different dictionaries of pitch class chroma chords, which became the networks' nodes, as follows: we extracted the chromagram with Mirtoolbox for Matlab [11], using a frame size of 0.2 s without overlapping, keeping only the chroma with energy level above the average over all files, and circularly shifted them according to an estimation of the tonality of each tango to transpose them

This work was supported by project 'Evolución musical' UNTREF.

O. A. Agustín-Aquino et al. (Eds.): MCM 2017, LNAI 10527, pp. 291–297, 2017.
https://doi.org/10.1007/978-3-319-71827-9_22

to the tonality of C, in order to have a common tonal framework. Borrowing the terminology of [8], we call the resulting chroma vectors *codewords*.

The links of our pitch networks represent harmonic transitions between these codewords. Specifically, for the purpose of studying the evolution of harmonic discourse, we formed 5 collections of codewords, one for each decade in the year span 1910–1960. Two codewords are connected by a directed edge if they appear in consecutive analysis frames. In this way we are left with 5 networks corresponding to the periods 1910–1919 (16 tangos in the collection), 1920–1929 (176 tangos), 1930–1939 (135 tangos), 1940–1949 (139 tangos) and 1950–1959 (64 tangos).

Proceeding in this way, many of the generated codewords do not correspond to the standard harmonic vocabulary of Western tonal music: beyond usual triadic chords, all kinds of chromatic harmonies, including the 12-note chromatic cluster, are obtained. For this reason, we considered two kinds of networks:

a Unfiltered networks, containing all codewords.
b Triadic networks, that is, filtered networks generated by only the triadic code-words of at most 4 chroma classes, including single chromas and dyads. We call a codeword *triadic* if it can be obtained, modulo 12, from one (or more) of its pitches by stacking consecutive minor or major thirds over it (or them). In these reduced networks, two triadic codewords are connected if the second chord is the next triad appearing after the first, ignoring non-triadic codewords in between. In this way we aim to representing a core harmonic skeleton, ignoring noisy frames and non-triadic chords arising from passing notes.

2   Results

Based on the results of Serrà et al. [8] and the models of vocabulary frequency of [15], we essayed fitting the frequencies of codewords, sorted in decreasing order (that is, ordered by rank r, where $r = 1$ for the most frequent codeword and so forth), with a Zipf law of the form $z = Cr^{-\alpha}$. For our fitting procedure, we used the approach of Clauset et al. [13,14]. In the case of unfiltered networks, we found, for all decades, nice fits with truncated power laws (see Fig. 1a). The scaling exponents obtained vary very little over the years, ranging from $\alpha = 1.81$ to $\alpha = 1.94$. They are larger than those found in [8], pointing to a comparatively more compact and less innovative vocabulary [15], a fact which is to be expected since the corpus of Serrà et al. is much more varied and massive, consisting on a million themes of popular music of many different genres. These exponents are also somewhat smaller than those found in [15] for the distribution of notes in classical music.

For triadic networks, however, we did not find good fits with pure truncated power laws. One reason for this could lie in the limited vocabulary considered here. A more appropriate model in this case is a shifted power law $z = (a + br)^{-\alpha}$, with coefficients adjusted to the vocabulary size. This law is derived partly from the

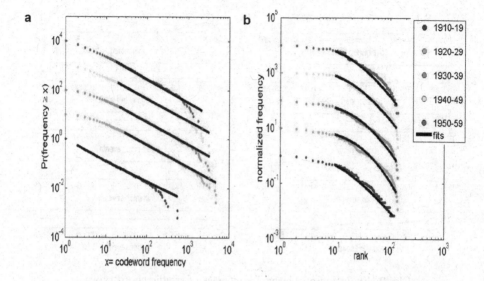

Fig. 1. (a) Complementary cumulative distribution of codeword frequencies and their fits by power laws for unfiltered networks. Curves are chronologically shifted by a factor of 10 in the vertical axis for ease of visualization. (b) Rank-frequency distribution of normalized codeword frequencies (respect to maximum frequency) and their fits by shifted power laws for triadic networks. Curves are chronologically shifted by 1 in the vertical axis.

hypothesis that, as the musical corpus grows in time, the frequency of harmonic innovations goes as a power $1/\alpha$ of the pre-existing language size [15].

We found nice fits of this model with triadic codewords frequencies, with exponents now varying between 2.48 and 6.05 (Fig. 1b). A tentative explanation of the unusually large exponents, in the context of the aforementioned Zipfian shifted power law, is that there is a very slow innovation rate going on in the basic triadic vocabulary as we consider the whole collection of tangos from a given decade (hence very small innovation exponent $1/\alpha$), and that the changes occur, instead, mostly at the level of nontriadic chords. In order to see if the codeword ranking remains stable across the years, we compute the Spearman rank correlation coefficients of triadic codewords for all pairs of decades. They are all high, with a minimum of 0.81. So frequent codewords continue to be so along the history of tango. Tracking the relative frequencies of each triadic chord type between 1910 and 1960, we observe some steady changes: augmented triads, half diminished sevenths have a twofold increase, minor sevenths also grow, although in lesser proportion; there is a small transitory drop in minor triads in 1920–1930 while major triads show a long term falling tendency (Fig. 2).

Beyond codeword frequencies, harmonic networks give us a panorama of how musical discourse transits between the elements of the vocabulary of codewords, creating stylistic patterns that can be learnt by repeated listening experiences and subsequently lead to the formation of expectancies and surprise [16,17].

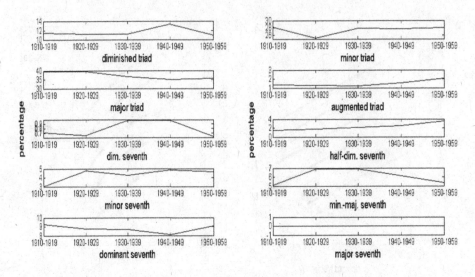

Fig. 2. Evolution of relative frequencies of triadic chord types.

Usual network measures and metrics can be easily interpreted in our context in terms of their musical meaning. In the following, we consider several such typical network coefficients

Density is defined as the fraction of edges present, compared with all possible $n(n-1)/2$ edges (where n is the number of nodes of the network). All our harmonic networks are sparse in this sense. For triadic networks density is at most 0.21, while unfiltered networks are much sparser, with densities below 0.006. Phrased in terms of predictability, this sparseness makes accessible the statistical learning of transition rules, involving around 2000 different transitions between the 140 possible triads.

Degrees. Node out-degree k is the number of neighbors following a codeword. For unfiltered networks, degree distribution is nicely fit with a truncated power law $P(k) = k^{-\gamma}$ for $k > k_{min}$ for the period 1920–1929, with exponent 2.42, while in the other periods a better fit is a truncated power law with exponential cutoff, with exponents in the interval [1.93, 2.12]. These values are similar to those obtained by Serrà et al. [8]. For triadic networks, also good fits are obtained with truncated shifted power laws, with exponents ranging from 2.92 to 6.17. While in unfiltered networks the median degree varies little between 2 and 6, for the triadic ones there is a big increase of degree connectivity from a median of 5 in 1910–1919 to values above 19 in the decades from 1920 to 1950, dropping somewhat in 1950–1959 to 13. This indicates a strong tendency towards greater freedom of harmonic discourse, and is also correlative with an increase of the size of the vocabulary, from 122 codewords in 1910–1919 to 139 codewords in 1920–1929, with a gradual and small decay to 133 nodes in the '50s. (Note that the total number of possible triadic codewords is 151).

From now on, we focus on networks of triads, where results are more easily interpreted in the framework of classical harmonic analysis. Codeword frequency and codeword degree are almost perfectly monotonically correlated, with Spearman rank coefficients above 0.99 for all decades. So the most frequent chords, among which there are the main triads defining tonality, are also the most connected. A notable symmetry emerges here, that also has been observed by Serrà et al. [8] Contrasting the out-degree of the major and minor triads over all chromatic scale degrees (in the musical sense of the word), with their similarly defined in-degree (the number of different chords that lead to a given one), their values are extremely similar, with their mean ratio over all triads between 0.99 and 1.02, and standard deviations between 0.01 and 0.1, for all decades.

Clustering measures the transitivity of the network. The local clustering coefficient $c_i = \frac{2T_i}{k_i(k_i-1)}$ gives the number of closed triangles among the nodes connected to node i. Harmonically, if we interpret the network as giving the stylistically permissible chord transitions, a high c_i implies that a transition between codeword i and another codeword that could be done directly, also could often be realized with an intermediate linking chord, adding to variety of harmonic conduction. Here we measure local connectivity by C, the average of c_i. A global measure of connectivity is the *average shortest path* length l. This gives the average of the minimum number of intermediate chords that are necessary to go between two given codewords. For instance, the appearance of bold and abrupt harmonic progressions that link tonally distant chords side by side would tend to reduce the value of l. High levels of clustering and small values of l define a small-world network [18]. Finally, *assortativity by degree* Γ is a coefficient measuring the tendency of nodes with similar degree to connect to each other. It is positive if this effectively occurs and negative if nodes of high degree tend to connect with nodes of low degree and vice versa. To interpret these coefficients, they are to be compared with the same coefficients computed from a random network with the same degree distribution, which we construct with the rewiring method described in [19]. For our networks, there is a marked increase of C (the average of c_i) from a value of 0.35 in 1910–1919 to values in the range [0.47, 0.58] for the following decades. Corresponding random networks have clustering coefficients in [0.1, 0.18]. At the same time, l decreases from 2.57 to 1.86 between the first two decades, and then slightly increases to 2.11 in the '50s; these values are smaller than the ones obtained by randomizing links. So globally, the small-worldness increases along time, implying a trend towards relatively more rich and daring harmonic progressions, with more different choices and also shorter ways to go from a chord to another (Fig. 3). Assortativity remains negative, in the range [−0.09, −0.21], with a slight increase to −0.17 in the '50s. Keeping in mind the direct correspondence between frequent and connected chords, this means an increasing tendency to avoid direct connections between the most common triads. However, while assortativity values are more negative than random in 1920–1950, they are less negative than for the randomized networks in 1910–1919 and 1950–1959.

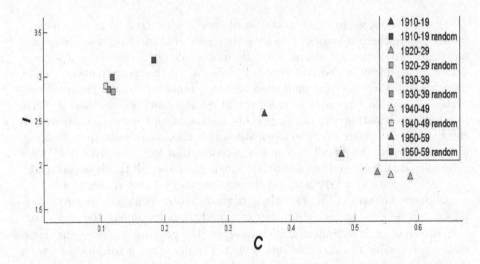

Fig. 3. Average shortest path length l versus clustering coefficients for actual (triangles) and randomized (squares) triadic networks.

3 Conclusions

Looking at tango from the network science perspective, we are able to single out some clear trends in tango evolution, and to compare them with the changes in other genres of music described in [8]. Tango appears to have a relatively limited harmonic vocabulary (even if we consider unfiltered networks), and data are compatible with an innovation model exhibiting a slow rate of appearance of novelties. But in the period considered here, inversely to the tendencies shown in [8], progressively richer and more complex chord transitions emerged within this universe, which increased its small world features.

References

1. Kohan, P.: Estudios sobre los estilos compositivos del tango: (1920–1935). Gourmet Musical (2010)
2. Ferrer, H.A.: El tango: su historia y evolución, vol. 12. A. Peña Lillo (1960)
3. Mauch, M., MacCallum, R.M., Levy, M., Leroi, A.M.: The evolution of popular music: USA 1960–2010. Roy. Soc. Open Sci. **2**(5), 150081 (2015)
4. Broze, Y., Shanahan, D.: Diachronic changes in Jazz harmony. Music Percept. Interdisc. J. **31**(1), 32–45 (2013)
5. Collins, N.: The UbuWeb electronic music corpus: an MIR investigation of a historical database. Organised Sound **20**(01), 122–134 (2015)
6. Shalit, U., Weinshall, D., Chechik, G.: Modeling musical influence with topic models. In: ICML, vol. 2, pp. 244–252, May 2013
7. Gomez, F., Lorimer, T., Stoop, R.: Complex networks of harmonic structure in classical music. In: Mladenov, V.M., Ivanov, P.C. (eds.) NDES 2014. CCIS, vol. 438, pp. 262–269. Springer, Cham (2014). https://doi.org/10.1007/978-3-319-08672-9_32

8. Serrà, J., Corral, Á., Boguñá, M., Haro, M., Arcos, J.L.: Measuring the evolution of contemporary western popular music. arXiv preprint arXiv:1205.5651 (2012)

9. Liu, L., Wei, J., Zhang, H., Xin, J., Huang, J.: A statistical physics view of pitch fluctuations in the classical music from Bach to Chopin: evidence for scaling. PloS ONE **8**(3), e58710 (2013)

10. http://www.todotango.com/

11. https://www.jyu.fi/hum/laitokset/musiikki/en/research/coe/materials/mirtoolbox

12. Stark, A.M., Plumbley, M.D.: Real-time chord recognition for live performance. In: Proceedings of the 2009 International Computer Music Conference, ICMC 2009, pp. 85–88, August 2009

13. Clauset, A., Shalizi, C.R., Newman, M.E.: Power-law distributions in empirical data. SIAM Rev. **51**(4), 661–703 (2009)

14. http://www.santafe.edu/aaronc/powerlaws/

15. Zanette, D.H.: Zipf's law and the creation of musical context. Musicae Scientiae **10**(1), 3–18 (2006)

16. Dubnov, S.: Information dynamics and aspects of musical perception. In: Argamon, S., Burns, K., Dubnov, S. (eds.) The Structure of Style, pp. 127–157. Springer, Heidelberg (2010)

17. Huron, D.B.: Sweet Anticipation: Music and the Psychology of Expectation. MIT press, Cambridge (2006)

18. Watts, D.J., Strogatz, S.H.: Collective dynamics of 'small-world' networks. Nature **393**(6684), 440–442 (1998)

19. Maslov, S., Sneppen, K.: Specificity and stability in topology of protein networks. Science **296**(5569), 910–913 (2002)

Determination of Compositional Systems Through Systemic Modeling

Liduino Pitombeira[✉]

Grupo de Pesquisas MusMat, Universidade Federal do Rio de Janeiro,
Rio de Janeiro, Brazil
pitombeira@musica.ufrj.br
http://musmat.org

Abstract. In this paper we propose the systemic modeling of Camargo Guarnieri's *Ponteio No.1* with the aim of identifying a hypothetical compositional system that gave rise to this work. From this compositional system we will plan a new work for woodwind trio. The model, specifically related to the harmonic syntax and the melodic gestures, is encoded into two algorithms written in Python and MATLAB.

Keywords: Systemic modeling · Compositional system
Compositional planning · Guarnieri's *Ponteios*

1 Introduction

This paper describes the methodological procedures for the systemic modeling of *Ponteio No.1*, from the *First Book of Ponteios*, by Brazilian composer Camargo Guarnieri(1907–1993). The purpose of this methodology is to propose a hypothetical compositional system that gave rise to this work, such that, from this system it is possible to plan and compose a new work for an instrumental set distinct from the original one (piano). The modeling will be achieved with the use of a technique we call parametric generalization, which is defined below and explained in more detail during the course of the modeling itself. Initially, we will formally define the terms systemic modeling, compositional system, and parametric generalization, and then accomplish the systemic modeling of Guarnieri's *Ponteio No.1*. From the resulting system of this modeling, we will plan and compose a piece for woodwind trio (oboe, clarinet, and bassoon). The results of the modeling will provide data for the creation of two computational algorithms in MATLAB and Python, which have the function of generating materials within the scope of the pitch parameter with the same profile as the original work.

2 The Fundamentals of Systemic Modeling

A model is a "simplified representation of a real system with the aim of studying this system" [1]. In the field of engineering, modeling can offer a physical model

O. A. Agustín-Aquino et al. (Eds.): MCM 2017, LNAI 10527, pp. 298–311, 2017.
https://doi.org/10.1007/978-3-319-71827-9_23

and a mathematical model that represents the characteristics of a system with high accuracy, in order to test its operating conditions and in particular its limits. In the musical domain, the model of a particular work may be proposed from analytical tools that can describe its structural relationships. However, from a compositional perspective, it is not our interest to propose a comprehensive and multidimensional modeling from which one can replicate the original work in all its aspects, since it is not our intention to rebuild the analyzed work, but to build a different one, which only keeps certain degree of kinship with the original. Thus, systemic modeling is intentionally partial and only focuses on some aspects of a work. The degree of relationship between the analyzed work, which can be considered in some respects an intertext, and the new work happens in the realm of a deep structure called compositional system, extensively studied in Lima [2]. A compositional system is determined from a series of guidelines that act directly on the construction of a musical vocabulary and syntax, that is, the building of materials and the relationships between them. These guidelines may be originally designed or modeled from another work, as it is the case in this study. In the latter case, it works as a kind of abstract intertextuality, in contrast with the kind of literal intertextuality, in which the surface levels of the intertexts reveal themselves more straightforwardly. These systemic guidelines (or definitions) can be expressed in a written language or translated into computer algorithms.

The systemic modeling consists methodologically of three stages: parametric[1] selection, analysis, and parametric generalization. In the first stage one selects, through a prospective analysis, the parameters that can render the best analytical result. In the analytical phase, we describe the behavior of the selected parameters of the analyzed work, for example, the syntax of the harmonic structure, the profile of melodic contours, the rhythmic patterns, etc. In the last stage, the values associated with these parameters obtained in the analysis are generalized, that is, are emptied of particular values. Thus, for example, if the analysis reveals that the intrinsic structure of a work is built from the parsimonious connection of set classes [012], [013], and [014] one can generalize this information simply stating that this structure is consolidated through the parsimonious connection of pitch-class sets. The parametric generalization is the methodological key that allows us to envision a hypothetical compositional system related to a musical work. Such a system is hypothetical because it disregards the author's intention, that is, it is not our interest to examine how the composer of the original work designed its structure, much less identify a compositional system suitable to encompass the entire set of works of a particular author.

Once determined the compositional system, one can perform the reverse path, which consists of assigning particular values to the generalized parameters. This phase is called compositional planning and is in a way opposite to the analysis. It is through the compositional planning that a composer assigns values to the

---

[1] It is important to mention that we are considering here an expansion of the concept of parameter: instead of being associated to surface level elements, which are closely related to a specific aesthetic profile, a parameter can be as abstract as an inversional axis, for example, which disregards tonal or atonal biases.

parameters described in the compositional system and also takes free compositional decisions about undeclared parameters. Thus, one arrives at a structure
that has, under certain perspectives, the same systemic lineage of the original
work but it is still in the raw state. In a final stage, this raw structure is refined
in order to set the work in line with certain aesthetic inclination.

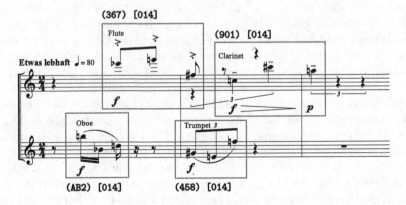

Fig. 1. Initial gestures of Webern's *Konzert, Op.24.*

In order to clarify the methodological steps of systemic modeling, we will take
the initial gestures of Webern's *Konzert, Op.24*, shown in Fig. 1. As one can see,
it consists of four three-note fragments for Oboe, Flute, Clarinet, and Trumpet.
Each fragment has its own information on the following parameters: pitch (in
terms of pitch-class and register), rhythm (in terms of duration and time-point),
dynamics, articulation, timbre, and tempo. In the first phase of systemic modeling, parametric selection, we will select the parameters that will be the focus
of the procedure. As we have previously mentioned, we are not interested in an
exhaustive modeling of the piece, since our purpose is only to capture some of
its deep features, from which we will plan and create a new original work. In this
short example, we will only select the pitch parameter. The first consequence of
this selection is that we loose all the information regarding the other parameters. Examining Fig. 2, we can verify that the pitches of Webern's excerpt were
transferred to a musical staff with no other additional parametric information.
The numbers inside parenthesis indicate the normal form and the numbers inside
brackets indicate the prime form for each fragment[2].

The second phase of systemic modeling—analysis—will reveal to us the relationships amongst the three-note fragments. It consists of three pitch-class sets
related by transposition and inversion, which means that they belong to the same
set class. Furthermore, they constitute a twelve-tone series, i.e., we have twelve
distinct pitch classes forming a series derived from trichordal class [014]. At this

---

[2] Numbers 10 and 11 are represented by their hexadecimal equivalent, A and B, to
avoid ambiguity.

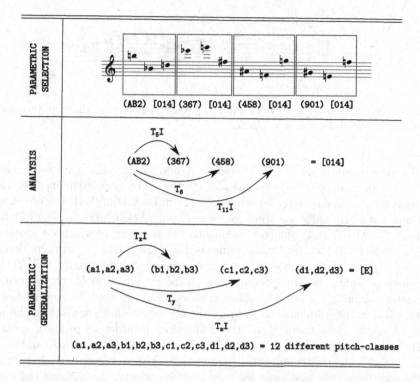

Fig. 2. The methodological cycle of systemic modeling applied to the pitch parameter in the first three-note fragments of Webern's *Konzert, Op.24*.

point we still have objects (pitch-class specification for the fragments) and relationships. The last phase of systemic modeling—parametric generalization—will consider only relationships between potential objects. This phase is fundamental to the modeling process, as we have laid down here, since it makes it possible to replace the original objects with entirely different new ones following the same relationships. This last phase is also important in the methodology because it allows us to define the compositional system. In this case, the system can be defined by the only rule: "Choose a trichordal class and build a derived series".

The compositional system can be used to plan a new work. In the compositional planning, we will execute the reverse process: we build a series derived from a single trichordal class and complete the parametric information that was removed as a result of the systemic modeling. If we choose, for example, the trichord (047), which is a member of trichordal class [037], we can apply transpositional and inversional operations in a such way that it will yield a derived series. One possibility for such a series would be: 0-4-7-9-5-2-3-8-B-A-6-1. This series, the only mandatory connection with Webern's fragment through the rule expressed in the compositional system, can be musically realized as shown in Fig. 3, with the other parameters (rhythm, timbre, dynamics, articulations, and tempo) freely chosen by the composer.

Fig. 3. New fragment created from the compositional system of the first three-note fragments of Webern's *Konzert, Op.24*.

The methodology of system modeling developed through a convergence of the theory of compositional systems and the theory of intertextuality. The theory of compositional systems loosely derives from Bertalanffy's theory of general systems. For Bertalanffy a system consists of "sets of elements standing in interaction" [3]. Music and language belong to the category of symbolic systems, which are formally defined by the "rules of the game". Klir [4], in turn, defines a system (S) as a set of objects (O) and relations (R), or formally $S = (O, R)$. Meadows [5] enhances Klir's definition with the introduction of the functionality factor. Lima [2], inspired by those authors, proposes that: "a compositional system is a set of guidelines to form a coherent whole that coordinates the use of musical parameters and their interconnection, in order to produce musical works". We suggest an update to this definition by adding the terms "and musical materials" right after "musical parameters". This is particularly important in the cases when the materials are used in their entirety, i.e., without the parametric fragmentation, such as in the works that combine intertextual materials in a literal fashion (Berio's *Sinfonia* or Rochberg's *Music for the Magic Theater*, for example).

The theory of intertextuality is another vital piece in the definition of systemic modeling. Kristeva [6] states, "all text is constructed as a mosaic of quotations, every text is absorption and transformation of another text". Kristeva [7] also highlights the relations between language and music, consequently bringing the intertextual thinking to the domain of music composition, a resource already employed in the past, as clearly demonstrated by Korsin [8] and Klein [9]. The methodology of system modeling employs intertextuality in a more abstract manner. As Lima [2] observed, the theoretical and artistic references in this field reveal that "the production of new texts can be obtained both through the literal use of intertext as through modified version of them". This latter can be called abstract intertextuality and parametric intertextualization, when applied to a set of specific musical parameters. The effectiveness and functioning of the methodology of systemic modeling can already be observed in the studies of xxxxxx and his graduate and undergraduate students [10–16].[3]

---

[3] Peer-reviewed papers blindly evaluated by researchers in the fields of composition and theory. All these papers contain at least one piece created with the systemic modeling of another piece. Some of the new pieces were already premiered.

3   Systemic Modeling of Guarnieri's *Ponteio No.1*

Figure 4 shows the first six bars of Guarnieri's *Ponteio No.1*, which has a slow tempo, sorrowful, with 32 bars. The macrostructure is a loose ABA+coda. In our analytical methodology we consider that the work can be divided into two layers. The first layer, corresponding to the right hand of the piano, consists of seven melodic gestures (the last being a slightly varied recapitulation of the first). The second layer, corresponding to the left hand of the piano, is divided into two sub-layers: (1) a rhythmic figuration that is repeated throughout the entire work and (2) long notes in a lower register. As the rhythmic feature is not being considered in this analysis, the rhythmic figurations are compressed such that the second layer will be seen as a single block of four voices.

Fig. 4. Six first bars of Guarnieri's *Ponteio No.1*. Copyright by Universal Music Publishing Group. Printed with permission.

The modeling will be accomplished separately for each layer. For the top layer, we were inspired, to some extent, by the theory of developing variation, especially as proposed by Almada [17], since we seek to identify a generator set for all the work's melodic gestures, which are obtained from a series of operations applied to this hypothetical generator set. The analytical methodology for the second layer, also inspired by the developing variation, consists in describing the parsimonious relations between the harmonic structures and the subsequent proposition of two generating structures: a chord and an interval set.

Figure 5 shows an analysis for the entire lower layer indicating the chromatic sets in the order in which they appear (and not in normal form), and the intervals separating each element of these sets in relation to the adjacent sets. We have noticed that these intervals configure parsimonious movements, here defined as intervals of major second at the most. These intervals of parsimonious movements form sets indicated in Fig. 2 inside brackets and labelled with letters of the alphabet. It is noteworthy that these sets of intervals are subsets of arrangements with repetitions of five elements, taken four at a time $(AR_{5,4})^4$, resulting in a

---

[4] The formula for the calculation of arrangements with repetitions is: $A_{n,p} = np$ [18].

total of 625 possibilities. Figure 6 displays the relationships between all sets of intervals and the first set, which will be considered here as the generator set: [+2, +2, +1, +2].

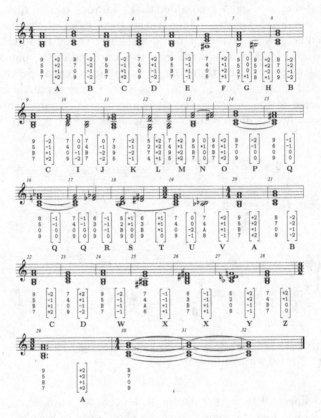

Fig. 5. Modeling of the upper layer of Guarnieri's *Ponteio No.1*.

In Fig. 6, the first column (left to right) indicates the position of the parsimonious movements intervals within the universe of the 625 above-mentioned possibilities; the second column shows the number of intervals; the third, the movement type according to the analytical labels shown in Fig. 2; the fourth, the type of operation that relates this set of intervals with the generator set (the first one)[5] ; and the last column indicates the number of times that the set of intervals appears in the analysis. To complete the modeling of the lower layer,

---

[5] $INV(C)$, Inversion: inverts the sign of each element of C; $RET(C)$, Retrogradation: realizes the retrogradation in C; $ROT(C,n)$, Rotation: rotates the set Cn times; $SUBROT(C,n)$, Subrotation: rotates the last three elements of C, n times; $COMP(C,n)$, Compression: subtracts n from each element of C; $MULT(C,n)$, Multiplication: multiplies n to each element of C; and $SOMA(C,D)$, Concatenation: concatenates the sets C and D.

Arr.	Sets of intervals	Type	Operation	Qty.
620	['+2', '+2', '+1', '+2']	A		3
6	['-2', '-2', '-1', '-2']	B	INV(A)	3
45	['-2', '-1', '+1', '+2']	C	SOMA((SOMA((MULT((SOMA(ROT(COMP (A,2),1),ROT(COMP (A,2),2))),3)), ROT(COMP (A,2),2))),A)	3
581	['+2', '+1', '-1', '-2']	D	INV(C)	2
44	['-2', '-1', '+1', '-1']	E	SOMA(SOMA(SOMA(MULT(ROT(COMP(A,2),3),3),B),C),A)	1
599	['+2', '+1', '+2', '+1']	F	SOMA((ROT(A,1)), (ROT(COMP(A,2),3)))	1
314	['0', '0', '0', '+1']	G	INV(ROT(COMP(A,2),3))	1
604	['+2', '+2', '-2', '+1']	H	SOMA(ROT(A,3),MULT(COMP(A,2),4))	1
306	['0', '0', '-1', '-2']	I	SOMA(INV(A),MULT(SOMA(ROT(INV(COMP(A,2)),1),ROT(INV(COMP(A,2),2)),2)	1
276	['0', '-1', '-2', '-2']	J	SOMA(ROT(INV(A),1),MULT(ROT(INV(COMP(A,2)),2),2))	1
27	['-2', '-1', '-2', '-1']	K	INV(F)	1
624	['+2', '+2', '+2', '+1']	L	ROT(A,3)	1
600	['+2', '+1', '+2', '+2']	M	ROT(A,1)	1
338	['0', '+1', '0', '0']	N	ROT(INV(COMP(A,2)),1)	1
595	['+2', '+1', '+1', '+2']	O	SUBROT(F,2)	1
38	['-2', '-1', '0', '0']	P	RET (I)	1
163	['-1', '-1', '0', '0']	Q	SOMA(ROT(COMP(A,2),1),ROT(COMP(A,2),2))	3
158	['-1', '-1', '-1', '0']	R	ROT(INV(COMP(A,1)),3)	1
463	['+1', '+1', '0', '0']	S	INV(Q)	1
468	['+1', '+1', '+1', '0']	T	INV(R)	1
302	['0', '0', '-2', '-1']	U	ROT(RET(I),2)	1
592	['+2', '+1', '+1', '-1']	V	SOMA(O,MULT(ROT(COMP(A,2),3),3))	1
32	['-2', '-1', '-1', '-1']	W	SOMA(ROT(R,3),MULT(ROT(COMP(A,2),2),2))	1
169	['-1', '-1', '+1', '+1']	X	SOMA(RET(S),Q)	2
607	['+2', '+2', '-1', '-1']	Y	SOMA(SOMA(MULT(ROT(COMP(A,2),3),2),U),A)	1
588	['+2', '+1', '0', '0']	Z	INV(P)	1

Fig. 6. Operations that relate all the sets of intervals of parsimonious movements with the generator set (A), in the lower layer of Guarnieri's *Ponteio No.1*.

these operations become part of a script written in Python, which will allow us to propose several generative values for both the initial chord and the initial set of intervals. This will enable us to generate the entire set of chords, which will be available in PDF format via Lilypond[6].

We discuss now the modeling of the upper layer. The first melodic gesture shown in Fig. 7 (upper part) can be segmented into two trichords [013], whose normal forms (B02) and (9B0) are mutually related by $T_{11}I$, where 11 is the first pitch-class of the first normal form. The juxtaposition of these two trichords defines the entire contents of the first gesture, which consists of a tetrachord [0235] in the normal form (9B02). This tetrachord, in turn, has two tricordal subsets: the generator trichord [013] and the trichord [025], which is used in the third and fourth gestures. Generalizing, we can say that the juxtaposition of the generator set $C_{x,0}$, consisting of pitch-class set $\{c_1, c_2, ...c_n\}$[7], with one of its transposed inversions—except for operations that would result in the generator

---

[6] Open-source application for editing music scores, available in http://www.lilypond. org/, visited in 02.22.2015.

[7] The generator set is identified as $C_{x,0}$, in which $x = 3, 4, 5, 6$, i.e., the set can be a trichord, a tetrachord, a pentachord, or a hexachord. The first value (x) indicates the set's cardinality (how many elements has the set) and the second value (0) is simply a label to differentiate the set from the other sets used in the system.

set itself[8]—produces a set with greater cardinality[9], which is the material of the first melodic gesture of the work. Formally we have that this set is $C_{x+1,1} = C_{x,0} + +T_iI(C_{x,0})$, and $i = c_1 \in C_{x,0}$, as long as $C_{x,0} \neq TiI(C_{x,0})$. Additionally, the derived set ($C_{x+1,1}$), generates a series of subsets of cardinality x. One is the generator set. The other set, $C_{x,1}$, is used in the third and fourth gestures.

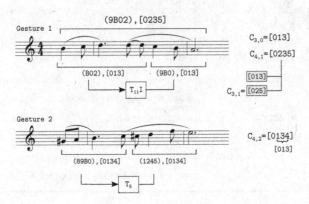

Fig. 7. Segmentation of the first and the second gestures of Guarnieri's *Ponteio No.1*.

The second melodic gesture, shown in Fig. 7 (lower part), is formed by the juxtaposition of two tetrachords [0134], whose normal forms (89B0) and (1245) are related by T_5, in which 5 is the last pitch-class of the second normal form, (1245). This tetrachord, labeled in Fig. 7 as $C_{4,2}$, i.e., the second tetrachord detected in the segmentation, is a superset of the generator trichord $C_{3,0} = [013]$. A fundamental feature of this tetrachord is that the prime form of its first three pitch-classes equals the prime form of its three last pitch-classes. One can generalize the constructional principle of this second gesture considering the principle of formation of its pitch-class sets. Thus, we consider that the generator set is $C_{x,0}$ and has pitch-classes $\{c_1, c_2, ...c_n\}$, and the derived set is $C_{x+1,2}$ and has pitch-classes $\{c_1, c_2, ...c_{n+1}\}$ in such a way that $[c_1, c_2, ...c_n] = [c_2, c_3, ...c_{n+1}]$. The derived set ($C_{x+1,2}$) appears in two normal forms related by T_y, in which y is the last pitch-class of the second normal form.

The third gesture is formed by the melodic trichord [025], which is the prime form of one of the subsets of the tetrachord of the first gesture, used in the normal form (790) and framed by two generator trichords, whose normal forms

---

[8] For example, $T_4I(024) = (024)$.

[9] This cardinality depends on the number of common pitch-classes between the generator set and its transposed inversion. In the first gesture of Guarnieri's *Ponteio No.1* the concatenation of two trichords produced a tetrachord because there are two common pitch-classes (0 and 11). If there is no common pitch-classes the result is a hexachord, as we see in the compositional planning of the new work, in which the generator trichord (45A) will produce the hexachord (456AB0) from the same operations used in Guarnieri's first gesture.

(79A) and (457) are correlated by T_2I, where 2 is the second pitch-class of the prime form [025]. In a generalized way, the chromatic set of the third gesture is formed by $C_{x,0} + C_{x,1} + C_{x,0}$. $C_{x,0}$ is manifested in two normal forms related to each other by T_zI where z is the second pitch-class of the prime form $[C_{x,1}]$. This gesture is shown in Fig. 8.

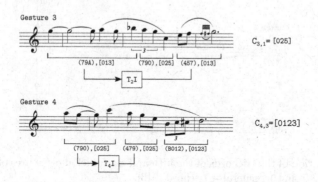

Fig. 8. Segmentation of the third and the fourth gestures of Guarnieri's *Ponteio No.1*.

The fourth melodic gesture (Fig. 8) is the juxtaposition of two trichords [025] juxtaposed to a tetrachord [0123]. The normal forms of the trichords, (790) and (479), relate to each other by T_4I, in which 4 is the first pitch-class of the second normal form. The first pitch-class of tetrachord [0123]'s normal form, that is, (B012) is the sum of the first pitch-classes of the trichords (790) and (479), i.e., $7 + 4 = B$. The procedure for the derivation of the tetrachord consists of the chromatic completion of the generator trichord by filling the spaces between pitch-classes. Thus, [013] becomes a chromatic tetrachord through the insertion of pitch-class 2 between 1 and 3. The chromatic completion in generalized situations, namely, in which the generator set will be chosen by the composer, can produce a pitch-class set with cardinality higher than 4, as in the case of the trichord [048], which in order to achieve chromatic completeness need to be transformed into the nonachord [012345678]. For the generalization of the fourth gesture's material we observe that it is formed by the juxtaposition of two sets $C_{x,1}$ whose normal forms are related by T_wI, in which w is the first pitch-class of the second normal form. To these two sets it is juxtaposed a third set consisting of the chromatic completion of the generator set $C_{x,0}$. The first pitch-class of this set's normal form is obtained by adding the first pitch-classes of the normal forms of $C_{x,1}$.

The chromatic materials for the fifth and sixth melodic gestures shown in Fig. 9 are also derived from the generator set. The material for the fifth gesture is the trichord [014] obtained by the unitary increment of the last pitch-class of the generator trichord [013], and the sixth gesture's material consists of the trichord [015], which is achieved by the unitary increment of the last pitch-class of the trichord corresponding to the fifth gesture. The normal forms (014)

and (015) coincide with the prime forms for both gestures. Generalizing, one
can derive the fifth gesture by the unitary increment of the last pitch-class of
the generator set and the sixth gesture by the unitary increment of the last
pitch-class of the fifth gesture.

Fig. 9. Derivation of the trichords of the fifth and sixth melodic gestures of Guarnieri's
Ponteio No.1, from the generator trichord [013].

The model for the upper layer was automated by a MATLAB function
(*ponteio1.m*) that contains all the relationships described in the modeling. The
composer inserts an initial set and the *ponteio1.m* function provides all the other
sets. Thus, for example, if we want to replicate the same material of the top layer
of Guarnieri's Ponteio No.1, we must enter the set {1120}.

4   Compositional Planning of *Germinación*

The compositional planning of the new work for woodwind trio (oboe, clarinet, and bassoon), entitled *Germinación*[10], started with the generation of the
set of chords for the bottom layer, through the insertion of a generator chord

Fig. 10. First eleven chords generated by the Python script, based on the systemic
modeling of Guarnieri's *Ponteio N.1*

[10] This is the first movement of a piece entitled *Vientos Tejanos*, Op. 203 (2016), dedicated to the Vientos Tejanos Trio, from Texas (USA). The other two movements—
Tejido and *Siluetas*—were also composed with the methodology of systemic modeling.

Generator set		(45A)	Prime form [016]
Gesture 1	g1_1	(45A)	(45A)
	g1_2	(6B0)	(6B0)
	g1_3	(456AB0)	Hexachord generated by the juxtaposition of g1_1 e g1_2. This hexachord will produce the pentachordal subset [01268], which will be used as the central material of the third gesture.
Gesture 2	g2_1	(56B0)	(56B0) e (1278) related to each other by T_8I, in which 8 is the last pitch class (1278)
	g2_2	(1278)	
Gesture 3	g3_1	(238)	(238) e (5AB) relate to each other by T_1I, in which 1 is the second pitch-class of [01268]
	g3_2	(78913)	The normal form of this pentachord resulted from the addition of its prime form [01268] by 7, the same value used in Guarnieri's *Ponteio No.1*(see endnote 12).
	g3_3	(5AB)	
Gesture 4	g4_1	(AB046)	This pentachord is obtained applying T_7I to the second pentachord. The value 7 is the first pitch-class of the second pentachord (g3_2).
	g4_2	(78913)	This pentachord is generated by applying T_9I to the pentachord's prime form [01268]. This relationship is the same one found in Guarnieri's *Ponteio N.1* (see endnote 13).
	g4_3	(56789AB)	This heptachord is generated by the chromatic completion of the generator trichord [45A], which is transposed in order to start with pitch-class 5 (the addition of the first pitch-classes of g4_1 and g4_2).
Gesture 5	g5	(017)	Parsimonious ascending movement in the last pitch-class of the prime form of the generator set.
Gesture 6	g6	(018)	Parsimonious ascending movement in the last pitch-class of the prime form of g5.

Fig. 11. Gestures generated by MATLAB's function —*ponteio1.m*, based on the systemic modeling of Guarnieri's *Ponteio No.1*.

and a set of parsimonious intervals in the Python script. We chose the half-diminished chord (7A15) and the interval set [+2, +1, +1, −1], resulting in the entire chordal structure, of which the first eleven ones are shown in Fig. 10. In turn, for the generation of the melodic gestures we started from trichord (45A), which inserted into the *ponteio1.m* function generated all the material of the six gestures, shown in Fig. 11. From these data, generated from the systemic modeling of Guarnieri's *Ponteio No.1*, we began the composition of the new work, freely dealing with other parameters (rhythm, dynamics, articulation) and the macrostructure. However, the rhythmic profile present in the bottom layer throughout Guarnieri's *Ponteio No.1*, as well as the manner of connection of some notes by non-harmonic tones in the upper layer (see Fig. 4)[11], sometimes appear in the new work. This is an attempt to relate the two works in the superficial level, reinforcing the already existing deep systemic connection built by the modeling. Figure 12 shows the first page of *Germinación*.

---

[11] See in Fig. 4 the passing G♭ connecting the G of measure 2 to the F of measure 3 (if one considers the structural harmony to be formed by the chord A-C-G-B), and the D♯ between G and E in the fourth measure (considering the harmony to be A-C-E-G).

Fig. 12. First page of the new work, entitled *Germinación* (first movement of *Vientos Tejanos*), based on the systemic modeling of Guarnieri's *Ponteio No.1*.

References

1. Mororó, B.: Modelagem Sistêmica do Processo de Melhoria Contínua de Processos Industriais Utilizando o Método Seis Sigma e Redes de Petri. Dissertation (Masters in Engineering). PUC (2008)
2. Lima, F.: Desenvolvimento de Sistemas Composicionais a partir de Intertextualidade. Dissertation (Masters in Music). UFPB (2008)
3. von Bertalanffy, L.: General System Theory: Foundation, Development, Application. George Brazillera, New York (1968)
4. Klir, G.: Facets of Systems Science. Plenum, New York (1991)
5. Meadows, D.: Thinking in Systems: A Primer. Earthscan, London (1991)
6. Kristeva, J.: Semiótica: Introdução à Semanálise. Perspectiva, São Paulo (2005)
7. Kristeva, J.: História da linguagem. Edições 70, Lisboa (1969)

8. Korsin, K.: Toward a new poetics of musical influence, music and analysis. In: Music and Analysis, vol. 10, no. 1/2, pp. 3–72, March–July, 1991
9. Klein, M.: Intertextuality in Western Art Music. Indiana University Press, Bloomington and Indianapolis (2005)
10. Moraes, P.M., Castro, G., Pitombeira, L.: Procedimentos Composicionais utilizados no Ponteio N.2 de Pedro Miguel a partir da modelagem do Ponteio N.12 de Camargo Guarnieri. In: Per Musi, vol. 27, pp. 61–74 (2013)
11. Moraes, P.M., Pitombeira, L.: Composição do Ponteio N.5 de Pedro Miguel a partir da Modelagem Sistêmica do Ponteio N.15 de Camargo Guarnieri. In: Música Hodie, vol. 13, pp. 8–33 (2013)
12. Moraes, P.M., Pitombeira, L.: Planejamento Composicional do Ponteio N.1 de Pedro Miguel a partir da Modelagem do Ponteio N.11 de Guarnieri. In: Revista Música, vol. 13, pp. 136–154 (2012)
13. Pitombeira, L.: Modelagem sistêmica do Ponteio N.2 de Camargo Guarnieri segundo a teoria dos contornos. In: Revista Brasileira de Musica, vol. 28, pp. 331–348 (2015)
14. Pitombeira, L., Kühn, M., Usai, C.: Modelagem sistêmica do primeiro movimento de Brinquedo de Roda, de Heitor Villa-Lobos, como uma metodologia para o planejamento composicional de Villa. In: Anais do XXVI Congresso da ANPPOM (2016). http://www.anppom.com.br/congressos/index.php/26anppom/bh2016/paper/view/3943
15. Castro-Lima, M., Pitombeira, L.: Composition of two works for woodwind quintet based on the systemic modelling of Guarnieri's Ponteio N.25. In: Anais do XXV Congresso da ANPPOM (2015). http://www.anppom.com.br/congressos/index.php/25anppom/Vitoria2015/paper/view/3454
16. Castro-Lima, M., Maddalena, G., Pitombeira, L.: Composição do primeiro movimento de Sonatina, para tuba e piano, de Marcel Castro-Lima, a partir da modelagem sistêmica do Ponteio 23 de Camargo Guarnieri. In: Anais do XXVI Congresso da ANPPOM (2016). http://www.anppom.com.br/congressos/index.php/26anppom/bh2016/paper/view/4063
17. Almada, C.: A Variação progressiva aplicada na geração de ideias temáticas. In: Anais do II Simpósio Internacional de Musicologia da UFRJ, pp. 79–90 (1991)
18. Iezzi, et al.: Matemática: 2a. Série, 2o. Atual Editora, Grau São Paulo (1976)

A Cluster Analysis for Mode Identification in Early Music Genres

Daniel C. Tompkins[✉]

Florida State University, Tallahassee, FL 32306, USA
dtompkins@fsu.edu

Abstract. This paper presents a corpus study that identifies the number of statistically distinct modes used in sacred and secular genres from 1400–1750. Corpora used for the study include Masses, motets, and secular songs from the Franco-Flemish School, works by Palestrina, secular Italian songs with alfabeto guitar tablature from the early seventeenth century, and works by J.S. Bach. A k-means cluster analyses of key profiles determine the number of distinguishable modes in each corpus. The results of this study show that the number of modes present in a corpus depend not only on date of publication but also on the genre of a composition.

Keywords: Music computation · Machine learning · Early music
Music genre · Cluster analysis · K-means

1 Introduction

It is often assumed that European music before common-practice tonality was built on a system of six, eight, or twelve modes and that music from the eighteenth century and onwards was built on two: major and minor. Historical notation supports this with the number of signatures and final cadences possible in any given system. However, the results of this study show that the modal framework of a musical corpus depends on its genre and not just its era of publication. K-means cluster analyses of key profiles to determine the number of statistically distinct modes in given corpora of music.

1.1 Historical Notation

Generally, music prior to the mid-seventeenth century was notated in a system that indicates several modes. Music from this period was often notated by two signatures—no flats (*durus*) and B flat (*mollis*)—with several possible final cadences for each signature, which can be seen in Fig. 1a. By the eighteenth century a system of several key signatures that were each associated with a major and minor key was well established, which can be seen in Heinichen's musical circle from 1711, reproduced in Fig. 1b. The *durus/mollis* system indicates a multi-modal[1] framework, as seen in Table 1a, while the later system implies a

---

[1] This paper uses the Greek modal names common today but with the understanding that these names were often not historically used.

© Springer International Publishing AG 2017
O. A. Agustín-Aquino et al. (Eds.): MCM 2017, LNAI 10527, pp. 312–323, 2017.
https://doi.org/10.1007/978-3-319-71827-9_24

two-mode framework. However, it is possible that the notated modes could cluster into only two based on the key profiles of their representative musical scores, as seen in Table 1b.

Signature	Final Cadence
♮ (*durus*)	C, D, E, F, G, A
♭ (*mollis*)	C, D, F, G, A B♭

(a) Mode possibilities in the *durus/mollis* system

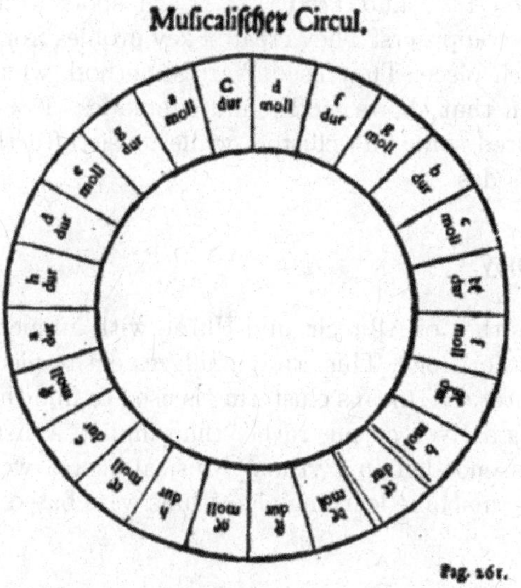

Muſicaliſcher Circul.

Fig. 161.

(b) Heinichen's musical circle (1711) in the common-practice system

Fig. 1. Comparison of modes and keys from the seventeenth and eighteenth centuries

Table 1. Two possible modal frameworks in early music

Mode	Signature:Final
Ionian	♮:C, ♭:F
Dorian	♮:D, ♭:G
Phrygian	♮:E, ♭:A
Lydian	♮:F, ♭:B♭
Mixolydian	♮:G, ♭:C
Aeolian	♮:A, ♭:D

(a) Notated modes

Mode	Signature:Final
Major	♮:C, ♮:F, ♮:G, ♭:F, ♭:B♭, ♭:C
Minor	♮:D, ♮:E, ♮:A, ♭:D, ♭:G, ♭:A

(b) Possible coalesced modes due to accidentals and/or *musica ficta*

1.2 Previous Approaches

Key profiles have been used by Temperley and Marvin [18,19] and are based on cognitive studies by Krumhansl and Kessler [14]. A key profile is a twelve-dimensional vector—representing pitch-classes 0–12. The value of each dimension is the percent the associated pitch class is used in a given selection of music. Temperley and Marvin created major and minor key profiles based on a corpus study of the string quartets by Mozart and Haydn. The key of a selection of music can be compared to the major and minor key profiles to determine the key and mode of the given selection.

Albrecht and Huron published a study modality and key profiles in fifty-year epochs between 1400 and 1750 [1,2]. Each epoch contained fifty scores from representative composers. They created key profiles from the first and final ten measures of each piece. They used Ward's method, which is a hierarchical clustering algorithm that shows modes and submodes. They found that music prior to 1700 featured some sub-clusters while music after 1700 clustered into only two distinct modes.

2 Methodology

This study expands that of Albrecht and Huron with a much larger corpus and uses a different methodology. This study analyzes entire pieces rather than the first and last measures.[2] K-means clustering is used to find the number of modes that best represents a given corpus rather than finding a hierarchy.

This study was encoded in Python. Musical scores were parsed through music21 [5], and the machine learning algorithms were based on the sci-kit learn module [15].

2.1 Corpora: Chronology, and Genres

To compare genres and chronology, the corpus is divided by genre and time period, which can be seen in Table 2. All corpora except for Bach were notated in the *durus/mollis* system. The Bach and Palestrina corpora were included with the music21 Python module [5], and the Franco-Flemish corpora were taken from the Josquin Research Project [20].

The alfabeto corpus is a collection of 529 secular Italian songs from the early seventeenth century that contain alfabeto tablature—letters that indicate specific hand shapes of chords for a Baroque guitarist to strum [3]. These chords all form either major or minor triads.[3] These letters, each paired with a bass note, provide a kind of continuos realization. The alfabeto chords and their associated bass notes were encoded by the author for this study.

---

[2] This method still provides clear and high-scoring clustering. However, Albrecht and Huron's method is necessary when extracting key profiles from later music where large sections may be in different keys.

[3] Christensen has explored the theoretical significance of this unique triadic practice [4].

Table 2. Corpora used for this project

Corpus	Source	Dates	Genre	Corpus size
Franco-Flemish	Josquin research project	c.1420-c.1520	Motet	175
			Mass	394
			Secular	151
Palestrina	music21	16th c.	Sacred	903
Alfabeto songs	Original	1610–1651	Secular	529
J.S. Bach	music21	18th c.	Mixture	352

2.2 K-means Clustering

To find the modal clustering of a corpus, a key profile for each musical score is created with the first dimension being the final bass note. The placement of other pitches are based on their distance in semitones (pitch-class interval) from the bass note. Each song is also labeled with its key signature and final bass note. The labels will be used to compare the key profile cluster results with the notated mode.

To cluster songs of a corpus, the k-means clustering algorithm is used, which is an unsupervised machine learning algorithm that clusters n points of data into k number of clusters. Key profiles are measured using the Euclidean distance. Songs that have similar key profiles will have small Euclidean distances. Given a large set of data, songs with similar key profiles should cluster together in a 12-dimensional space.

Given the Euclidean distances of the songs within a corpus, the k-means algorithm attempts to partition the data, a set of songs $(x_1, x_2, \ldots, x_n)$, into a number of clusters, k sets where $\mathbf{S} = \{S_1, S_2, \ldots, S_k\}$. The success of clustering is measured by finding the inertia (within-cluster sum-of-squares), which the algorithm tries to minimize [15]:

$$\sum_{i=1}^{k} \sum_{x \in S_i} ||x - \mu_i||^2,$$

where μ_i is the mean of points in S_i. After initiation, the partitions are adjusted and scored again. This process continues until the inertia reaches convergence, or a desired minimum.

To determine the number of modes (k) that most accurately represents a corpus, k 2–12 are tested. Two scoring metrics are used to determine the distinctiveness of the clustering and the degree to which the notated mode agrees with the clustered mode.

The silhouette coefficient, as defined by Rousseeuw [17], measures the distinctiveness of the clusters without considering the data labels and is based on two metrics: $a(i)$, the mean distance between a sample (i) and all other points in the same class (a), and $b(i)$, the mean distance between a sample (i) and all other points in the next nearest cluster ($b(i)$) [15]. Thus the silhouette score, where $-1 \leq s(i) \leq 1$, can be defined as follows:

$$s(i) = \frac{b(i) - a(i)}{max\{a(i), b(i)\}}.$$

The model with the highest silhouette score will be selected as the number of modes that best represents the corpus.

The completeness score (c), as defined by Rosenberg and Hirschberg [16, 411-2], measures the degree to which the notated modes (data labels) agree with the clustered modes by determining the success of notated modes belonging to the same clusters. A corpus is comprised of N data points, which is comprised of a set of classes, or labels, (C) and a set of clusters (K). The classes are the labeled key signatures, and the clusters are the number of modes. Rosenberg and Hirschberg define A as a contingency table produced by the clustering algorithm where $A = a_{ij}$ and where a_{ij} is the number of data points that belong to class c_i and cluster k_j.

The result, where $0 \leq c \leq 1$, can be defined as follows:

$$c = 1 - \frac{H(K|C)}{H(K)}$$

where $H(K|C)$ is the conditional entropy of the cluster assignments given the class:

$$H(K|C) = -\sum_{c=1}^{|C|} \sum_{k=1}^{|K|} \frac{a_{ck}}{N} \log \frac{a_{ck}}{\sum_{k=1}^{|K|} a_{ck}},$$

and where

$$H(K) = -\sum_{k=1}^{|K|} \frac{\sum_{c=1}^{|C|} a_{ck}}{n} \log \frac{\sum_{c=1}^{|C|} a_{ck}}{n}.$$

The model with the highest completeness score will show which number of modes best agrees with the key signature notation, if any.

To determine the mode that represents each cluster, the coordinates of each cluster's centroid is found. The seven dimensions with the highest values are extracted, which infers the mode. For example, Ionian mode (Major) would score highest in dimensions 1, 3, 5, 6, 8, 10, and 12 (scale degrees $\hat{1}$, $\hat{2}$, $\hat{3}$, $\hat{4}$, $\hat{5}$, $\hat{6}$, $\hat{7}$), while Mixolydian mode would score highest in dimensions 1, 3, 5, 6, 8, 10, and 11 (scale degrees $\hat{1}$, $\hat{2}$, $\hat{3}$, $\hat{4}$, $\hat{5}$, $\hat{6}$, $\flat\hat{7}$).

2.3   Visualizing Results

Silhouette and completeness scores of the uncompressed (12-dimensional) data will be shown for 2–12 clusters. However, to visualize the clustering, principal component analysis (PCA) is used, which decompresses the number of dimensions from twelve to two. The decompressed data can be plotted on a 2-dimensional graph. Each data point on a graph is the notated mode for a song in the corpus. The distances between data points approximate the proportional distance of the Euclidean distance of key profiles. Cluster centroids are numbered and later labeled with their modes. The partitions are visualized in as Voronoi diagrams where cluster membership is shown by color partitions in the background.

To determine the accuracy of the PCA-reduced graphs, the reduced data is processed through k-means clustering and scored using the silhouette coefficient and completeness score. If the compressed scores are similar to the uncompressed scores, the PCA-reduced graphs are an appropriate representation of the uncompressed data.

3   Results

The results of each corpus will be presented in reverse chronological order. There will be a line graph that shows the silhouette and completeness scores for 2–12 clusters and a PCA-reduced graph of the highest-scoring number of modes. The J.S. Bach corpus is the only corpus not notated in the *durus/mollis* system, and each song on the PCA graph is labeled with its key. All other corpora were notated in the *durus/mollis* system and are labeled with their signature and final cadence (ex. ♭:G for G mollis or "G dorian").

3.1   J.S. Bach

The songs from the J.S. Bach corpus show strong clustering for only two clusters, which can be seen in Fig. 2. The completeness score is a perfect 1, indicating that all notated modes belong to the same clusters, and the silhouette score is also quite high. Both the completeness and silhouette scores drop off significantly after two clusters. It is also important to note that the clusters are mostly compact, which indicates strong clustering. The minor cluster is somewhat more

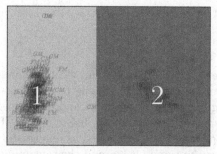

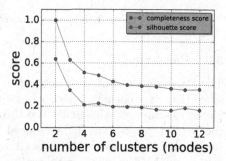

(a) *K*-means graph: distance is proportional to difference

(b) Silhouette and completeness scores of the Bach corpus

Fig. 2. *K*-means clustering of the Bach corpus

spread out, which is perhaps to be expected due to the flexibility of scale degrees $\hat{6}$ and $\hat{7}$. The cluster centroids are as follows:

1. Major (Ionian)
2. Minor (Aeolian)[4]

3.2 Alfabeto

Although the Alfabeto corpus was notated in the *durus/mollis* system, it clusters into two modes, which can be seen in Fig. 3. The silhouette and completeness scores, while not as high as in the Bach corpus, still show a strong peak at two clusters and quickly fall off. If the notated modes reflected the clustered modes, there should be a strong peak at six clusters. However, it is clear that a two-mode system is at work in the actual notes despite the music's notation. The cluster centroids are as follows:

1. Major (Ionian)
2. Minor (Aeolian)

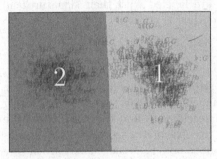
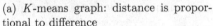
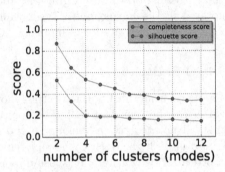

(a) *K*-means graph: distance is proportional to difference

(b) Silhouette and completeness scores of the alfabeto corpus

Fig. 3. *K*-means clustering of the alfabeto corpus

3.3 Palestrina

The Palestrina corpus, also notated in the *durus/mollis* system, clearly shows more than two clusters, which can be seen in Fig. 4. The silhouette and completeness scores are highest for two through five modes, with two and five being the highest peaks. The PCA-reduced graph in Fig. 4a shows five clusters. If two clusters were selected, clusters 2 and 5 would be partitioned as one cluster (although the data points would still be in the same place), and the remaining clusters would be another cluster—albeit a sprawling one. The cluster centroids are as follows:

---

[4] This natural minor mode centroid differs from Temperley and Marvin's Mozart and Haydn minor key profile, which is in harmonic minor. However, the raised seventh degree is only slightly higher than the lowered [18, 195].

1. Minor (Aeolian)
2. Major (Ionian)
3. Dorian
4. Phrygian
5. Mixolydian

The only notated mode that did not cluster to itself was Lydian. The songs notated in Lydian mode mostly fell into the Ionian cluster, which shows that the ♯4 was lowered more frequently than it was left raised. This is due in part to the notated accidentals and the moderate amount of editorial *ficta* in the digital corpus.

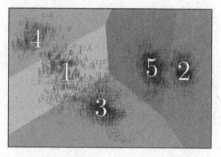

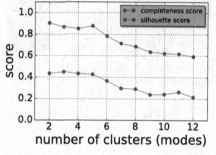

(a) *K*-means graph: distance is proportional to difference

(b) Silhouette and completeness scores of the Palestrina corpus

Fig. 4. *K*-means clustering of the Palestrina corpus

3.4 Franco-Flemish Genres

The Franco-Flemish corpus from the Josquin Research Project can be divided into Mass movements, sacred motets, and secular songs. The corpora provide an opportunity to test whether the multi-mode system is a product of its time or whether genre may also play a role.

Masses. The mass movements cluster well in two to six modes, which can be seen in Fig. 5. None of the clusters are as distinct as the Palestrina corpus, but the completeness score shows that the corpus can be divided into six modes, and the notated modes still mostly cluster together. The five cluster centroids in Fig. 5a are as follows:

1. Dorian
2. Major (Ionian)
3. Phrygian
4. Mixolydian
5. Minor (Aeolian)

The only mode that does not become a distinct cluster is again the Lydian mode. If six clusters are chosen, the small cluster of songs between centroids 3 and 5 become a cluster, but their centroid key profile and their notated mode both indicate Aeolian mode, like centroid 5. The songs notated in Lydian mode are mostly nested within cluster 2, Ionian.

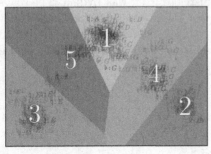

(a) K-means graph: distance is proportional to difference

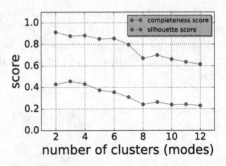

(b) Silhouette and completeness scores of the Franco-Flemish Masses corpus

Fig. 5. K-means clustering of the Franco-Flemish masses corpus

Motets. The results for the sacred motet corpus is quite similar to the Mass corpus, although there is a stronger peak at five modes in the completeness score, which can be seen in Fig. 6. The five cluster centroids in Fig. 5a are as follows:

1. Mixolydian
2. Phrygian
3. Dorian
4. Major (Ionian)
5. Minor (Aeolian)

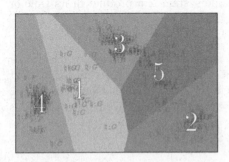

(a) K-means graph: distance is proportional to difference

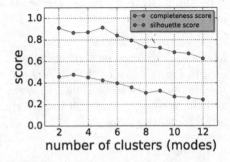

(b) Silhouette and completeness scores of the Franco-Flemish Motet corpus

Fig. 6. K-means clustering of the Franco-Flemish Motet corpus

Once again, the Lydian mode is not distinguishable, which shows that accidentals and/or *ficta* consistently lowers the ♯4̂. All songs notated in Lydian mode belong to cluster 4, Ionian.

Secular Songs. The secular song corpus, however, clusters best in two modes with a slight secondary peak at four modes (Figs. 7a and 6b). The clusters are not as distinct as later corpora, but it is clear that two modes better represents this secular song corpus than any other number of modes. Like the Bach and alfabeto corpora, the centroids of the Franco-Flemish secular songs are again major and minor:

1. Minor (Aeolian)
2. Major (Ionian)

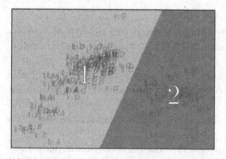

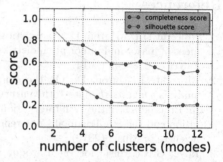

(a) K-means graph: distance is proportional to difference

(b) Silhouette and completeness scores of the Franco-Flemish secular songs corpus

Fig. 7. K-means clustering of the Franco-Flemish secular songs corpus

The differences between the secular songs and the sacred genres are quite significant, although the two sprawling clusters shown in Fig. 7a do show some modal drift. For example, the notated Phrygian songs are mostly at the bottom of cluster 1. Likewise, the notated Mixolydian songs are slightly to the left of the notated Ionian songs in cluster 2. However, it is quite possible that a larger data size would fill in those gaps. Despite the possible modal drift, the silhouette scores in Fig. 7b clearly shows that two modes best describes the secular song corpus.

4   Discussion

The results of this study show that different genres have different modal frameworks even if composed within the same time period, country, and even by the same composer. Secular genres cluster into only two modes despite the music's notation long before sacred genres. This leaves open speculation of the modal framework for vernacular genres that were not notated. Perhaps the alfabeto corpus is the closest we can come to vernacular music [4, 7, 12]. Further digitization of early music can extend this study, especially in the seventeenth century. Given enough music, comparisons could also be made geographically or perhaps by instrumentation.

Of course, understanding the harmonic structures of a corpus includes more than pitch class frequency but also how those pitches are grouped[5] and move from one to another,[6] but this study is an important first step. It gives a view of the modal framework in early music that provides a foundation for other ways of investigating harmonic practice—a foundation that recognizes the different harmonic practices of secular and sacred genres and that sometimes compositional practices were not always in line with notational or theoretical conventions of their time.

References

1. Albrecht, J.D., Huron, D.: A statistical approach to tracing the historical development of major and minor pitch distributions, 1400–1750. Music Percept. Interdisc. J. **31**(3), 223–243 (2014)
2. Albrecht, J., Huron, D.: On the emergence of the major-minor system: cluster analysis suggests the late 16th century collapse of the Dorian and Aeolian modes. In: Proceedings of the 12th International Conference on Music Perception and Cognition and the 8th Triennial Conference of the European Society for the Cognitive Sciences of Music (2012)
3. Boye, G.R.: Music for the Baroque Guitar: The Rasgueado (Strummed) Style (2013). http://applications.library.appstate.edu/music/guitar/strummed.html
4. Christensen, T.: The Spanish Baroque guitar and seventeenth-century triadic theory. J. Music Theor. **36**(1), 1–42 (1992)
5. Cuthbert, M.S., Ariza, C.: music21: A Toolkit for Computer-Aided Musicology and Symbolic Music Data (2010)
6. Dahlhaus, C.: Study on the Origins of Harmonic Tonality. Princeton University Press, Princeton (1990). Translated by Robert O. Gjerdingen
7. Dean, A.: The five-course guitar and seventeenth-century harmony: Alfabeto and Italian song. Ph.D. dissertation. University of Rochester (2009)
8. Eisenhardt, L.: Italian Guitar Music of the Seventeenth Century: Battuto and Pizzicato, vol. 130. Boydell & Brewer, Woodbridge (2015)
9. Eisenhardt, L.: Baroque guitar accompaniment: where is the bass? Early Music **42**(1), 73–84 (2014). Oxford University Press
10. Gavito, C.M.: The Alfabeto song in print, 1610-ca. 1665: Neapolitan roots, Roman codification, and 'Il Gusto Popolare'. Ph.D. dissertation, The University of Texas at Austin (2006)

---

[5] The author has also conducted a study of bass harmonization frequency and found that the clusters are mostly similar for the Bach and alfabeto corpora. The results for the Palestrina and Franco-Flemish corpora show very weak clustering, which is perhaps to be expected given the style.

[6] Another promising area of research is in n-gram clustering. White has explored n-gram similarity to compare composers of different time periods [21].

11. Gjerdingen, R.O.: 'Historically informed' corpus studies. Music Percept. Interdisc. J. **31**(3), 192–204 (2014)

12. Hudson, R.: The concept of mode in Italian guitar music during the first half of the 17th century. Acta Musicologica **42**(Fasc. 3/4), 163–183 (1970)

13. Kaufman, L., Rousseeuw, P.J.: Finding Groups in Data: An Introduction to Cluster Analysis, vol. 344. John Wiley & Sons, Hoboken (2009)

14. Krumhansl, C.L., Kessler, E.J.: Tracing the dynamic changes in perceived tonal organization in a spatial representation of musical keys. Psychol. Rev. **89**(4), 334 (1982)

15. Pedregosa, F., et al.: Scikit-learn: machine learning in Python. J. Mach. Learn. Res. **12**, 2825–2830 (2011)

16. Rosenberg, A., Hirschberg, J.: V-measure: a conditional entropy-based external cluster evaluation measure. In: EMNLP-CoNLL, vol. 7 (2007)

17. Rousseeuw, P.J.: Silhouettes: a graphical aid to the interpretation and validation of cluster analysis. J. Comput. Appl. Math. **20**, 53–65 (1987)

18. Temperley, D., Marvin, E.W.: Pitch-class distribution and the identification of key. Music Percept. Interdisc. J. **25**(3), 193–212 (2008)

19. Temperley, D.: Music and Probability. The MIT Press, Cambridge (2007)

20. The Josquin Research Project (2016). http://josquin.stanford.edu/

21. White, C.W.: Some statistical properties of tonality, 1650–1900. Ph.D. dissertation, Yale University (2013)

Cross Entropy as a Measure of Coherence and Uniqueness

Christopher Wm. White[(✉)]

The University of Massachusetts Amherst, Amherst, USA
cwmwhite@umass.edu

Abstract. Cross entropy, a measurement of the complexity/predictability of a series of observations given a probabilistic model, has been used in a variety of domains in music scholarship for decades. This paper presents a novel application of this metric to musical corpus analysis. Given a series of divisions to a larger corpus, a sub-corpus is relatively "unique" if a probabilistic model derived from its pieces better predicts its constituent pieces than do models derived from other sub-corpora. A sub-corpus is relatively "coherent" if its own model describes its pieces better than a model derived from the entire corpus. The Yale-Classical-Archives corpus was used to illustrate several strategies for sub-corpus division, each of which are tested for uniqueness and coherence. Some broader interpretive applications are also described.

Keywords: Computation · Corpus analysis · Cognitive modeling · Style

1 Introduction

Music researchers have been experimenting with concept of *entropy* almost since the field of informatics began in the mid 20th century [1, 2]. Since the 1950s, scholars have connected entropy, or the relative complexity of some signal, to musical style, communication, normativity, meaning, and compositional modeling [3–7]. As shown in Eq. 1, the entropy H of an observed series O measures the complexity of a signal by calculating the log-probability of an event o to occur within some observed series O (here, $\log P(o_i)$), weighting that value by the relative frequency with which that observation occurs in O (here, $P(o_i)$, and summing all such values. The negative sign turns the negative value resulting from the logarithm into a positive value, such that the higher the value, the more randomness – or more entropy – as series has. A very redundant signal – one in which a particular event happens most of the time – will have low entropy since those events are highly predictable given the rest of the signal, while a series of wildly unpredictable events would have a high entropy. (Here, the logarithm's base can be chosen as appropriate for the situation: this study uses base 2 in order to report entropy in bits.)

$$H(O) = -\sum_{i=1}^{n} P(o_i)\log P(o_i) \qquad (1)$$

© Springer International Publishing AG 2017
O. A. Agustín-Aquino et al. (Eds.): MCM 2017, LNAI 10527, pp. 324–334, 2017.
https://doi.org/10.1007/978-3-319-71827-9_25

In the past several years, work by David Temperley [8] has introduced a particular modification of this technology to music: *cross entropy*. In this framing of the general concept, the probability of each event is judged by some other model rather than by some probabilistic distribution drawn from the observation itself. As shown in Eq. 2, this formula takes the negative log-probability of an event o given some probabilistic model m, again weighting each value by its probability mass within the observation series O, and then summing for all n events. This essentially captures how well some probability distribution m accounts for the series of observations O.

$$H(O, m) = - \sum_{i=1}^{n} P(o_i) \log m(o_i) \qquad (2)$$

This paper proposes several novel ways of applying this modeling technique to the analysis of musical data. I will show how cross entropy can capture the *coherence* and the *uniqueness* of musical corpora. Because, in one sense, using a composer's identity to build a corpus creates an unassailably coherent and unique dataset: using this framework, the composer's identity provides the desideratum as to whether a piece is included in some corpus. But, one might also wonder whether a composer writes pieces that are distinct from their contemporary colleagues, or if a composer's style is basically interchangeable with that of their contemporaries. If the former were true, the composer's pieces would exhibit notably divergent statistical properties from those of their colleagues; but, if two composers have made virtually identical decisions surrounding some musical parameter, then the same statistical model could represent both corpora.

This paper agues that cross entropy can shed light onto these sorts of questions by manipulating which models are used to assess the corpus. Given some corpus with potential smaller divisions (or, *sub-corpora*), if the individual pieces within some sub-corpus are predicted by the overall corpus better than any other sub-corpus, that sub-corpus is *unique* as compared to other sub-corpora. If that sub-corpus contains pieces that are more statistically similar to one another than to the overall corpus, that sub-corpus is *coherent*. Below, I show a computational model that exploits these properties to test the coherences and uniqueness of several different divisions of a large corpus of Western-European common-practice MIDI files. I end by discussing the interpretive potentials that this modeling provides.

2   The Corpus, the Sub-corpora

This experiment relied on data from the Yale-Classical-Archives Corpus [9]. This corpus collects MIDI files from classicalarchives.com (a website of user-sourced MIDI files), each associated with metadata that specifies the file's opening key, meter, composer, date of composition, instrumentation, composer's nationality, genre, and so on. Given that this study was interested in dividing corpora by composer, the 19 composers listed as "The Greats" on the website were used: Bach, Beethoven, Brahms, Byrd, Chopin, Debussy, Handel, Haydn, Liszt, Mendelssohn, Mozart, Saint-Saens, Scarlatti, Schubert, Schumann, Tchaikovsky, Telemann, Vivaldi, and Wagner. The overall corpus was more than 5,000 pieces, and the average composer's dataset contained 231 pieces, with the smallest corpus – Wagner's – containing only 33, and the largest – Scarlatti's – containing 554. The corpus is divided into "salami slices" – every verticality where the pitch-class content changes. The average composer's sub-corpus had 339,185 such slices, with Wagner's again being by far the smallest (67,538), and Mozart's being the largest (1,322,716). The corpus also contains tonal annotations, which were used to convert the corpus's pitch material of each slice into scale degrees.

These scale-degree sets were used to create Markov (n-gram) chains designed to probabilistically model how surface harmonies progressed to one another. Different sizes of n-grams within these tonal passages were then tallied, and after initial experimentation it was determined that trigrams (i.e., $n = 2$) seemed to balance between precise and sparse data. (An n-gram model involves contiguous sequences of n items from a sequence of observations. When $n = 2$, the observation at the current timepoint is conditioned on the two previous observations. The model is therefore concerned with three-chord trigrams – the current and previous two chords – at every observed timepoint.) In order to remain as theory-neutral as possible, the meter metadata was used to gather trigrams at three metric levels; these three levels were then combined. Repeating data collection at several levels and agglomerating the resulting trigrams allows for patterns that recur at several durational or metric levels to become more dominant in a distribution while remaining agnostic as to the relative importance of different surface divisions. The three metric levels were (1) the salami slices themselves, (2) the contents of each beat as defined by the corpus's metric data (i.e., the quarter-note in 4/4), and (3) the contents of the beat's primary division (i.e., the eighth note in 4/4; this division is also recorded by the corpus). NB: this process recognizes not only traditional chords (like triads and seventh chords) but also less traditional chords (like passing chords and dissonances): this study therefore assumes that any surface structure is a legitimate "chord," following [10–12]. The tallying and organization of the YCAC's trigrams was implemented with Python version 2.6 using the music21 software package [13].

In order to compare the uniqueness and coherence of various different divisions of the larger dataset, several different divisions of the larger corpus were undertaken. Most basically, each individual composer's output will first be considered a sub-corpus. Next, chronological divisions were used, grouping pieces in the corpus by their date of publication, first arranged by the half-century beginning in 1650 and ending in 1900, and then by 30-year epochs (now beginning in 1680 because of the sparse data between 1650 and 1679). Finally, to introduce machine-learned groupings into the corpus, the groupings found in [14] were used. Here, the identical dataset and modeling as described above were used, and composers' trigram frequencies were submitted to a k-means cluster analysis to group composers whose surface harmonic progressions were statistically similar. The study used values $k = [0 \ldots 10]$; peaks in silhouette widths were used to identify optimal k values; and, such peaks values were identified for 7 and 10 clusters. The groupings – used here as sub-corpora are reproduced in Table 1.

Table 1. k-means clusters drawn from White (2014)

K-means clusters	
$k = 7$	$k = 10$
Bach	Bach
Byrd	Byrd
Beethoven, Mozart, Haydn, Schumann, Mendelssohn, Brahms, Schubert, Wagner	Beethoven, Mozart, Haydn, Mendelssohn, Schubert
Tchaikovsky, Liszt, Chopin, Saint-Saens	Tchaikovsky, Liszt, Chopin, Saint-Saens
Telemann, Vivaldi, Handel	Telemann, Vivaldi
Debussy	Debussy
Scarlatti	Scarlatti
	Wagner
	Brahms, Schumann
	Handel

3 Modeling Coherence and Uniqueness

The *coherence* and *uniqueness* of each division was corpus quantified by determining the cross entropy of each piece given every other piece (exclusive of the piece under question) in some sub-corpus. In terms of Eq. 2, for each piece, the observations O would be those trigrams within an individual piece within the sub-corpus, and the model m would be the probability distribution of all trigrams within the remaining pieces in that corpus. As a baseline, each piece within the sub-corpus was judged in relation to the entire corpus (here, the entire YCAC becomes the model m). The average and standard error of these cross entropies across the sub-corpora is tallied, as well as the pieces in each sub-corpus as judged by the entire corpus. A sub-corpus is *unique* if its standard error is sufficiently low to not overlap with the window of any other corpus's standard error ("does this sub-corpus predict its own pieces better than any other sub-corpora above chance?"). A sub-corpus is *coherent* if the standard error of its self-assessments is outside the standard error of the overall corpus ("does this sub-corpus predict itself better than it would be predicted by the entire corpus?").

NB: As cross entropy is itself a relative measurement, so too are uniqueness and coherence. Each of these numbers must only be judged in relation to other numbers: a piece is only unique *in relation to other sub-corpora* or only more coherent *than the overall corpus*.

Importantly, both these ideas have conceptual overlaps with the central idea of "entropy." When applied to a single dataset (i.e., using the format of Eq. 1), entropy rises when each event is more random in terms of the other events, and falls when each event it is more predictable. Uniqueness and Coherence manipulate these relationships by comparing a dataset's randomness not simply to the dataset itself, but to other potential datasets with which the original dataset has some relationship. In other words, these ideas capitalize on the original informatic structure of entropy to draw out additional relationships between datasets. Note also that the difference between uniqueness and coherence is not mathematical in nature (indeed, they are mathematically identical), but in the relationships between the models used, with uniqueness quantifying relationships between sub-corpora and coherence quantifying relationships between a corpus and its sub-corpora.

4 Sub-corpora

4.1 Dividing by Composer

By dividing the corpora by composer, 74% composer-by-composer comparisons were significantly unique. The median proportion of unique comparisons was 83%. 88% of these sub-corpora predicted themselves better than the overall corpus. Only two composers – Byrd and Handel – registered perfect results: the trials produced significantly lower cross entropy when comparing these composers' own pieces to their own corpora than when comparing them to any other composer's corpus. In other words, these results show the models to be "sure" these composers' pieces were significantly more likely on average to be composed by themselves than by someone else. Example 1a shows Handel's sub-corpus compared to that of each other composer. The cross entropy of the composer's pieces when compared to the other composers' sub-corpora are shown as the clear bar, other composers are shown by solid bars, the self-wise comparison is shown by the white bar, and the dashed bar shows Handel's average cross entropy judged by the entire corpus. Handel's own pieces are judged statistically significantly better than they are judged by other corpora– the corpus is therefore unique. The corpus also judges itself better than it is judged by the overall corpus– it is therefore coherent (Fig. 1).

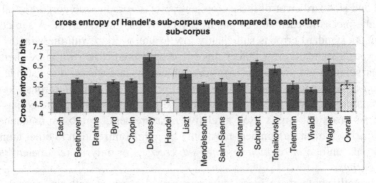

Fig. 1. Comparative cross entropies using Handel's sub-corpus model

However, more than a quarter of the time, these trials judged other corpora to predict a composer's pieces with either a lower or insignificantly different level of cross entropy when compared to the composer's own corpus. Mendelssohn's corpus, for instance, performed around the average with two non-unique comparisons: the cross entropies of the Brahms, Handel, and Schubert sub-corpora were not significantly different from the cross entropy resulting from a self comparison. However, the corpus does predict itself significantly better than the agglomerated corpus predicts its pieces. This result indicates that the Mendelssohn model is coherent insomuch as it predicts its own pieces well; however, it also shows that the model is not sufficiently unique, as other models predict Mendelssohn's corpus virtually identically to Mendelssohn's own model.

On the whole, it seems that these results suggest that grouping corpora by composer tends to create coherent corpus models, although these models are often not sufficiently unique from one another.

4.2   Dividing by Chronological Epochs

Here, pieces were divided into sub-corpora based on their date of composition, first into fifty-year epochs, and then in thirty-year epochs. The fifty-year model performed worse than that using composer-defined trials, the former returning a 68% uniqueness rate. However, the median success rate was higher, registering an 80% uniqueness rate. This rate stems from the fact that one time period, 1751–1800, did not have a single successful trial; this epoch also did not predict itself better than did the overall corpus. Example 2a shows the offending epoch's results. Not only can the late 18th-century corpus not be significantly distinguished from the late 17th-century corpus, but the other three corpora produce significantly *lower* cross entropies, indicating that these corpora predict the trigrams within the late 18th-century corpus better than they predict the trigrams of their own time periods. The fact that the overall corpus predicted its pieces better than did this sub-corpus also indicates this sub-corpus to not be coherent.

Example 2b shows the case of the 1801–1850 corpus, a relatively successful example representing this test's median. While a self-comparison yields the lowest average cross entropy, the average cross entropy when compared with the 1851–1900 corpus is not significantly different than the self-wise average. (Interestingly, 75% of the unsuccessful judgments throughout the 50-year-epoch test (i.e., incorrect/insignificant comparisons) involved time periods adjacent to one another; if one removes the late 18th-century results from the percentage, this number rises to a complete 100%. In other words, with the exception of the problematic late 18th-century, the models generally become "confused" as to a piece's time period only when comparing that piece to a chronologically adjacent corpus.)

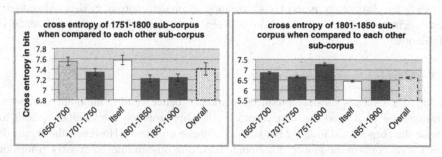

Fig. 2. Comparative cross entropies of (a) the 1751–1800 sub-corpus, and (b) the 1801–1850 sub-corpus

Dividing the corpora into 30-year segments produced similar results. The overall average success rate was 68.75%, and the median success rate was 75%. Half of the unsuccessful returns involved adjacent time periods, and 80% were within two time periods (i.e., within 60 years). As with the 50-year segments, the remaining 20% were not evenly distributed throughout the results, but centered in two particularly unsuccessful epochs. Two trials were not relatively coherent: the 1801–1830 and 1891–1920 trial. Example 8 shows a median example, the 1741–1770 corpus, while Example 9 shows the largely unsuccessful 1801–1830 results. (While it may be satisfying that the only significant positive results involve corpora that are maximally chronologically distant from the 1801–1830 corpus, note that the two adjacent corpora register a significant but lower cross entropy.)

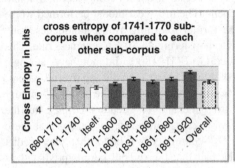 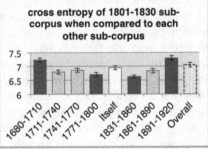

Fig. 3. Comparative cross entropies of (a) the 1741–1770 sub-corpus, and (b) the 1801–1830 sub-corpus

These results suggest that dividing a corpus by chronological epochs may be successful in some respects – it creates a high median success rate – but it also generates several corpora that are incoherent. From a modeling perspective, this incoherence could be explained by the presence of multiple and distinct chord-progression practices within a single corpus. For instance, the 1801–1830 corpus seems to have properties that are better modeled by the corpora surrounding it, perhaps indicating that this era contains practices that overlap those of its two surrounding epochs. If dividing corpora by composers seemed to create too many divisions, dividing by chronology is too broad, creating incoherent corpus models. Also, these tests seem to indicate a connection between chronological proximity and models' similarities.

4.3   Machine-Learned Sub-corpora

Using the *k*-means clusters produced markedly better results, although somewhat unsurprising as it used the same metric – chord-progression probabilities – both to divide the corpora and to judge the success of those divisions. (However, the results of this test do confirm the power of harmonic transition probabilities to classify groups of composers into unique and coherent corpora.) The seven clusters provide nearly perfect results, with only Debussy's corpus providing insignificant/non-unique comparisons, likely due to its small membership ($n = 60$). Figure 4a shows a typical perfect 7-cluster trial, using the "Romantic" (Tchaikovsky, Chopin, Liszt, Saint-Saens) sub-corpus. The ten clusters performed slightly worse, with an 88% success rate. If, however, one discounts the insignificant results of the two smallest corpora – now adding Wagner's corpus ($n = 32$) to Debussy's – the results rise to a 97.22% success rate. Figure 4b shows one of the two remaining insignificant results, the other being the average cross entropy of Vivaldi/Telemann's pieces given Handel's corpus.

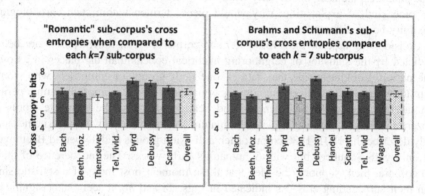

Fig. 4. Comparative cross entropies of (a) the Tchaikovsky, Chopin, Liszt, and Saint-Saens sub-corpus, and (b) the Brahms and Schumann sub-corpus

5   Applications

This type of modeling has various applications in how we think about and interpret musical corpora and the works contained therein. In what follows, I outline four potential applications of this kind of modeling, showing ways it can be used to interpret stylistic trends and compositional schools, how it can be used to identify points of innovation, how it can be used to broach the (admittedly thorny) topic of authorship, and how it might be used to formalize models of historical styles.

5.1   Describing Stylistic Trends and Compositional Schools

When the model identifies several composers whose sub-corpora and not unique, but – when grouped together – create a unique and coherent sub-corpora, this potentially identifies a compositional cohort operating within a similar compositional school. Here, we imagine that the compositional trends and norms used by these composers are sufficiently similar that the variation within their outputs makes them statistically indistinguishable (at least within the tested parameters). Furthermore, non-unique comparisons can show other potential avenues of influence. For instance, in Fig. 4, Brahms and Schumann's sub-corpus is coherent and unique in all but the comparison to the Tchaikovsky–Liszt–Saint-Saens–Chopin sub-corpus. This suggests that the output of these two composers comprise a distinct style that influences the output of these later Romantic composers. The fact that chronological adjacencies within the epoch-based models frequently accounted for non-unique comparisons also suggests stylistic trends. Here, this non-uniqueness captures the chronological developments of historical styles: historically proximate sub-corpora share statistical tendencies.

5.2   Moments of Innovations

Non-unique and incoherent findings also provide an opportunity for interpretation. At these junctures, the lack of similarity within the pieces constituting the sub-corpus begs for some kind of explanation: why would pieces written within such chronological proximity be so different?

Consider the case of the 1751–1800 sub-corpus: its constituent pieces are better predicted by the statistics of neighboring historical epochs than the pieces in its own time period. Looking inside that dataset, one finds groups of composers who would seem to be drawn from divergent compositional practices. It is not only a time period that saw the late works of Telemann and Scarlatti, but also the complete works of Mozart and Haydn, and ended with the mature works of Beethoven. One could similarly describe the 1801–1830 corpus: such a division groups middle-period Beethoven not only with Schubert, but with Schumann's early works. The incoherence of these sub-corpora, then, supports the idea that these moment host more of a stylistic shift than their surrounding eras. As indicated in Figs. 2a and 3b, it seems that significant portions of these groupings are better predicted by the surrounding epochs than contemporary compositions, further suggesting that these eras feature dramatic shifts between the previous and following styles.

5.3   Authorship

This modeling technique also allows for potential evaluations of reproductions, completions, or potentially spurious compositions. In each of these instances, one could take the piece(s) in question and treat them like their own sub-corpus, comparing its coherence and uniqueness with other sub-corpora in the piece(s) historical orbit. For instance, Fig. 5 compares a famous example of forgery to the 10-cluster sub-corpora. The forgeries here are those of Nicolas Chedeville publishing Vivaldi's fictitious "Opus 13" in 1737. The "X" above each of the bars shows the corpus' self-wise cross entropy, each constituent bar shows the forgeries' cross entropy compared to other sub-corpora, and the final bar again shows the agglomerated corpus's assessment of the forgeries. The sub-corpus is coherent, but it is not unique. Many other sub-corpora fall within the standard deviation of the average assessment: as before, these are shown as lined bars. Bars outside of the average standard deviation are shown as solid. There are two below the average: the Handel and Telemann-Vivaldi clusters. This means that these reproductions do indeed adequately imitate Vivaldi, but do so in a way that they could potentially also be passed off as composed by Handel!

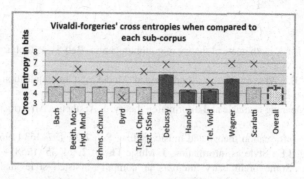

Fig. 5. Comparative cross entropies of Nicolas Chédeville's Vivaldi forgeries compared to the 10-cluster sub-corpora

5.4   Generative Modeling

Using these metrics to identify relatively unique and coherent statistical systems can potentially create well-formed generative models of some style. For instance, if one used the chord-progression (i.e., Markov-chain) probabilities embedded in the, say, Brahms-Schumann sub-corpus to generate sucessions of harmonies, one could reasonably argue that this models aspects of that style's compositional norms. The same cannot be said of, say, a model drawn from the 1751–1800 sub-corpus: because of its incoherence, it is not clear what such a generative model would capture outside of manifesting the era's stylistic heterogeneity. These metrics, then, can be imagined as ways to isolate statistical systems that can express some historically, culturally, or compositionally independent style.

6   Future Work

Of course, this work is incomplete. It relies entirely on simple Markov chains drawn from the very surface of a musical corpus. It is possible, for instance, that judging the similarity of two systems using something like a Context Free Grammar or at least some hierarchical system would better represent similarities and differences in chord progression usage. Additionally, other surface events rather than chord progressions may capture salient differences between sub-corpora: melodic figuration, recurrent bass lines, orchestration, or ornamentation may all contribute to stylistic differences better than (or in addition to) surface chord progressions. However, regardless of these potential avenues for future investigation, this study has identified a general method of using cross entropy to identify the uniqueness and coherence of various datasets, quantifying overlaps and consistencies within musical corpora.

References

1. Shannon, C.E.: A mathematical theory of communication. Bell Syst. Techn. J. **27**, 379–423, 623–656 (1948)
2. Shannon, C.E., Weaver, W.: A Mathematical Model of Communication. University of Illinois Press, Urbana (1949)
3. Meyer, L.: Meaning in music and information theory. J. Aesthetics Art Criticism **15**(4), 412–424 (1957)
4. Cohen, J.E.: Information theory and music. Behav. Sci. **7**(2), 137–163 (1962)
5. Youngblood, J.E.: Style as information. J. Music Theor. **2**, 24–35 (1958)
6. Mendel, A.: Some preliminary attempts at computer-assisted style analysis in music. Comput. Humanit. **4**(1), 41–52 (1969)
7. Duane, B.: Agency and information content in eighteenth- and early nineteenth-century string-quartet expositions. J. Music Theor. **56**(1), 87–120 (2012)
8. Temperley, D.: Music and Probability. The MIT Press, Cambridge (2007)
9. White, C., Quinn, I.: The yale-classical archives corpus. Empirical Musicology Rev. **11**(1), 50–58 (2016)
10. Quinn, I.: What's 'Key for Key': A Theoretically Naive Key–Finding Model for Bach Chorales. Zeitschrift der Gesellschaft für Musiktheorie 7 (2010)
11. Quinn, I., Mavromatis, P.: Voice-leading prototypes and harmonic function in two chorale corpora. In: Agon, C., Andreatta, M., Assayag, G., Amiot, E., Bresson, J., Mandereau, J. (eds.) MCM 2011. LNCS, vol. 6726, pp. 230–240. Springer, Heidelberg (2011). https://doi.org/10.1007/978-3-642-21590-2_18
12. White, C.W.: An alphabet-reduction algorithm for chordal n-Grams. In: Yust, J., Wild, J., Burgoyne, J.A. (eds.) MCM 2013. LNCS, vol. 7937, pp. 201–212. Springer, Heidelberg (2013). https://doi.org/10.1007/978-3-642-39357-0_16
13. Cuthbert, M.S., Ariza, C.: music21: a toolkit for computer–aided musicology and symbolic music data. In: Proceedings of the International Symposium on Music Information Retrieval, pp. 637–42 (2011)
14. White, C.: Changing styles, changing corpora, changing tonal models. Music Percept. **31**(2), 244–253 (2014)

Probability and Statistics in Musical
Analysis and Composition

Complementary Collections in Ligeti's *Désordre*

Clifton Callender[✉]

Florida State University, Tallahassee, USA
clifton.callender@fsu.edu

Abstract. This paper examines one aspect of Ligeti's approach to writing music that is neither tonal nor atonal—the use of complementary collections to achieve what Richard Steinitz has termed combinatorial tonality. After a brief introduction, the paper explores properties of the intervallic content both within and between complementary collections, which I term the intra- and inter-harmonies. In particular, the inter-harmonies are useful in understanding harmonic control in works based on complementary collections, as demonstrated by revisiting Lawrence Quinnett's analysis of Ligeti's first Piano Étude, *Désordre*.

Keywords: Ligeti · *Désordre* · Complementary collections
Combinatorial tonality

1 Introduction

Figure 1 shows the opening of Ligeti's first Piano Étude, *Désordre*. A remarkable feature of this étude is that the right hand plays only the white keys while the left hand plays only the black keys. The étude thus systematically divides the aggregate into two quite familiar complementary collections—diatonic and pentatonic. Due to the overlapping registers (proximity) and similarity of contour in the ascending scalar fragments (common fate), it can be difficulty to separate the two hands into independent psychological streams, thus making the diatonic and pentatonic collections difficult to hear separately. (See Bregman [3] for the importance of pitch proximity and common fate in the formation of independent auditory streams.) Instead, it is much easier to hear the *between* hand note-against-note harmonic intervals, which we might term the *inter-harmonies*.

As the étude progresses, the hands gradually drift apart, the accents in the two hands become desynchronized, and the durations between accents in each hand are gradually shortened, leading to a fragmentation of the scalar segments. Near the climax of the étude (see Fig. 6b), both the lack of pitch proximity and common fate in the melodic contours strongly encourages the formation of two independent streams, making it difficult to hear the harmonic relations between the hands, but relatively easy to hear the *within* stream intervallic relationships, which we might term the *intra-harmonies*. From the opening to the climax of *Désordre* there is thus a change in focus from the inter-harmonies to the intra-harmonies. Richard Steinitz [12] has referred to this interplay of complementary collections as "combinatorial tonality."

© Springer International Publishing AG 2017
O. A. Agustín-Aquino et al. (Eds.): MCM 2017, LNAI 10527, pp. 337–349, 2017.
https://doi.org/10.1007/978-3-319-71827-9_26

Fig. 1. Opening of *Désordre*

While there are clear precedents in the use of complementary collections, Ligeti's extensive use of opposed collections in his late works and, in particular, his exploration of the unfamiliar harmonic possibilities between collections and the ways in which a listener's attention can be focused on the relations within (inter) and between (intra) collections is something very different. Indeed, the technique of complementary collections may represent Ligeti's most systematic approach to achieving his goal of creating music that is neither tonal nor atonal in his late works. (See Ligeti's own comments in [2].) As such, our lack of understanding of the harmonic relations between complementary collections takes on greater significance. The current paper explores this aspect of Ligeti's combinatorial tonality, focusing on the relevant mathematical properties of complementary collections and the first part of *Désordre*, as a preliminary step to a greater understanding of Ligeti's exploration and realizations of the theoretical properties and corresponding possibilities of complementary collections.[1]

2   Properties of Intra- and Inter-harmonies

We begin examining the harmonic relationship between complementary collections by looking at the interval content (IC) from one pitch-class set to another using Lewin's [8] interval function (IFUNC)[2]:

$$\mathrm{IC}_{A,B}(k) = \mathrm{IFUNC}(A,B)(k) = |\{(a,b) \in A \times B, b - a = k\}|.$$

The interval function is a histogram of the pc intervals by which a member of one set can move to a member of the other set, yielding a 12-valued vector of pc interval multiplicities. For example, the interval content from $\{C, D\}$ to $\{C\sharp, D\sharp\}$ is $(0, 2, 0, 1, 0, 0, 0, 0, 0, 0, 0, 1)$, indicating that there are two ways to move by pc interval 1 ($C \rightarrow C\sharp$ and $D \rightarrow D\sharp$), one way to move by pc interval 3 ($C \rightarrow D\sharp$),

---

[1] My thanks to Nancy Rogers and an anonymous reviewer for comments that greatly improved this paper.

[2] The reader is strongly directed to Amiot [1] for an excellent and detailed presentation of the interval function and its relation to recent applications of the discrete Fourier transform in music theory.

one way to move by pc interval 11 (D→C♯), and no ways to move by any other interval.[3] In the special instance of the interval content from a set to itself, the interval content *within* a set, we will use the shorthand $\mathrm{IC}_{A,A} = \mathrm{IC}_A$.

For complementary collections A and $\bar{A}$, the *intra-harmonies* are a combination of the interval content within each collection separately:

$$A_{\mathrm{intra}} = \mathrm{IC}_A + \mathrm{IC}_{\bar{A}}.$$

For example, setting W to be the white-key diatonic collection ($W = \{0, 2, 4, 5, 7, 9, 11\}$), the intra-harmonies for the white-key/black-key complementary collections are given by

$$\begin{aligned}
W_{\mathrm{intra}} &= \mathrm{IC}_W + \mathrm{IC}_{\bar{W}} \\
&= (7, 2, 5, 4, 3, 6, 2, 6, 3, 4, 5, 2) + (5, 0, 3, 2, 1, 4, 0, 4, 1, 2, 3, 0) \\
&= (12, 2, 8, 6, 4, 10, 2, 10, 4, 6, 8, 2).
\end{aligned}$$

Similarly, for complementary collections, the *inter-harmonies* combine the interval content that obtains exclusively *between* the two collections:

$$A_{\mathrm{inter}} = \mathrm{IC}_{A,\bar{A}} + \mathrm{IC}_{\bar{A},A}.$$

For complementary collections, A and $\bar{A}$, $\mathrm{IC}_{A,\bar{A}} = \mathrm{IC}_{\bar{A},A}$. Thus,

$$A_{\mathrm{inter}} = 2 \cdot \mathrm{IC}_{A,\bar{A}}.$$

For example, again setting W to be the white-key diatonic collection, the inter-harmonies for the white-key/black-key complementary collections are given by

$$\begin{aligned}
W_{\mathrm{inter}} &= 2 \cdot \mathrm{IC}_{W,\bar{W}} \\
&= 2 \cdot (0, 5, 2, 3, 4, 1, 5, 1, 4, 3, 2, 5) \\
&= (0, 10, 4, 6, 8, 2, 10, 2, 8, 6, 4, 10).
\end{aligned}$$

For a given pair of complementary collections, any pc interval must occur either within or between the collections, and thus the intra- and inter-harmonies exhaust the set of possible pc intervals:

$$A_{\mathrm{intra}} + A_{\mathrm{inter}} = \mathrm{IC}_{\mathbb{Z}_n} = (n, \ldots, n).$$

Figure 2 shows graphs of the intra- and inter-harmonies for two different pairs of complementary collections. Note that the distributions of intra- and inter-harmonies for Fig. 2a are fairly uneven, while the distribution in b is nearly flat. Collections in which the interval content is highly uneven may be thought to be more distinctive, since the interval content is dominated by only a few (and therefore salient) pc intervals. Contrarily, when the interval content is very flat, the corresponding collections cannot be typified by a limited number of distinct pc intervals. In this sense, there is a strong correlation between the "distinctiveness" of a collection and the unevenness of its interval content.

---

[3] Multiplicity of pc interval i is indicated by the ith component of the interval function, which begins with pc interval 0.

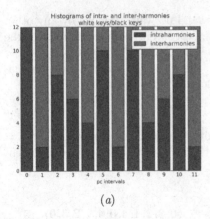
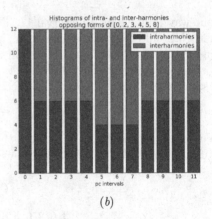

(a)   (b)

Fig. 2. Histograms of intra- and inter-harmonies for (a) white key/black key collections and (b) "flat" hexachord {C, D, D♯, E, F, G♯} and its complement

We can measure the unevenness of a collection's interval content by taking its standard deviation. For example, the interval content of the "flat" hexachord from Fig. 2b has a standard deviation of $\sigma(\mathrm{IC}_{\{0,2,3,4,5,8\}}) = 1.0$, while that of the whole-tone collection (with its maximally uneven interval content of all even intervals and no odd intervals) has a standard deviation of $\sigma(\mathrm{IC}_{\{0,2,4,6,8,10\}}) = 3.0$. The standard deviation of the white-key interval content lies between these two extremes: $\sigma(\mathrm{IC}_{\{0,2,4,5,7,9,11\}}) \approx 1.66.$[4]

As measured by the standard deviation of interval content, a collection and its complement are equally distinctive[5]:

$$\sigma(\mathrm{IC}_A) = \sigma(\mathrm{IC}_{\bar{A}}).$$

Moreover, the distinctiveness of the intra-harmonies is the same as the inter-harmonies:

$$\sigma(A_{\mathrm{intra}}) = \sigma(A_{\mathrm{inter}}).$$

Since Ligeti's favorite complementary collections, including diatonic, pentatonic, whole-tone, Guidonian and similar collections, are all highly distinctive, this ensures that the interval content between these collections and their complements will also be highly distinctive.

This does not, however, guarantee that the intra- and inter-harmonies will also be highly differentiated. In order to measure this differentiation, we can take the magnitude of the difference of the two vectors:

$$\|A_{\mathrm{inter}} - A_{\mathrm{intra}}\|_2.$$

---

[4] The distinctiveness of a collection, A, can also be measured in terms of the magnitude of its interval content, $\|\mathrm{IC}_A\|_2$. (See Callender [4].) For the present purposes, the standard deviation is preferable.

[5] This follows directly from the complement theorem. See Hanson [6] and Lewin [7].

For example, let W be the white-key collection and X be the "flat" hexachord $\{0,2,3,4,5,8\}$. The Euclidean distance between the intra- and inter-harmonies for the diatonic collection is nearly 24 ($\|W_{\text{inter}} - W_{\text{intra}}\|_2 \approx 23.98$), while the corresponding distance for the flat hexachord is nearly only 14 ($\|X_{\text{inter}} - X_{\text{intra}}\|_2 \approx 13.86$), reflecting the high differentiation of intra- and inter-harmony distributions in Fig. 2*a* and relatively low differentiation in Fig. 2*b*.[6]

More generally for complementary collections, there is a strong relationship between the differentiation of intra- and inter-harmonies and the distinctiveness of the collections. If complementary collections A and $\bar{A}$ are the same cardinality, then

$$\|A_{\text{inter}} - A_{\text{intra}}\|_2 \propto \sigma(IC_A).$$

(Specifically, for a chromatic universe of C pitch classes, $\|A_{\text{inter}} - A_{\text{intra}}\|_2 = 4\sqrt{C} \cdot \sigma(IC_A)$.) If the cardinalities are nearly equal, then the relation is nearly, though not exactly, proportional. Thus, highly distinctive collections indeed posses highly differentiated intra- and inter-harmonies. By working with highly distinctive collections, Ligeti ensures that there will be maximal variation between the melodic and harmonic components of the resulting combinatorial tonality.

3 *Désordre*

Returning to the opening of *Désordre* (Fig. 1), we would like to answer the following question: To what extent does Ligeti exert control over the note-against-note harmonies in the étude? The opening of *Désordre* consists of two layers that persist throughout the entire étude. The accented notes correspond to a highly complex isorhythmic structure, detailed in Kinzler [9], in which the left and right hands have very similar but non-identical *colores* (sequences of pitches) and *taleae* (sequences of durations). While the accents of the two hands are synchronized in the beginning of the étude, they quickly become misaligned, due to the slight difference in their *taleae*. Unaccented notes are not a part of the isorhythmic structure, but rather form a second layer consisting of generally ascending scalar fragments used to smoothly connect the accented notes. Perhaps the harmonic relations at any point are simply the result of the particular and temporary configuration of the two hands within the overarching isorhythmic structure. If this is the case, then over a large enough span of time the observed harmonies will be equivalent to the result of repeated random selection from the distribution of possible harmonies between the two collections. In other words, as the étude progresses, the distribution of observed harmonies will converge on the expected distribution of the inter-harmonies.

In the opening line of *Désordre* (Fig. 1), tritones and minor sixths predominate, while there are almost no minor second/major sevenths or perfect

---

[6] Measuring the distance between intra- and inter-harmonies using other metrics, such as angular (or cosine) distance, yields similar relative distances. (See Rogers [11].) The Euclidean metric is sufficient and advantageous for the present purposes.

fourths/fifths. The lack of interval class (ic) 5 is easily explained by the relative lack of this interval in the inter-harmonies. However, the relative lack of ic 1 may indicate some degree of control on the part of the composer, since there are plenty of minor seconds/major sevenths spanning the two collections. Does this favoring of tritones and minor sixths (major thirds) over minor seconds/major sevenths persist?

Figure 3 gives the normalized actual (observed) and expected interval distributions for the first section of the étude, concluding with the climax on the first downbeat on the sixth page of the published score.[7] (Intervals are reckoned between hands interpreted as pitch-class sets.) There is a noticeable emphasis on ic 6 and de-emphasis on ic 1. In his dissertation on harmony and counterpoint in Ligeti's *Études*, Quinnett [10] presents a comparison of observed and expected interval counts of the first section of *Désordre* divided into two parts, with the second part beginning where the accents of the two hands temporarily become (mostly) realigned (t_2, beginning just before the bottom system on page 2 of the score). (See Figs. 4 and 5.) Quinnett notes that the interval profile of the first section heavily favors tritones and minor sixths/major thirds over minor seconds/major sevenths, while the interval content for the second section is much more similar to the expected distribution.

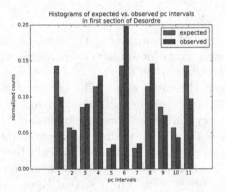

Fig. 3. Histograms of observed and expected intervals in the first section of *Désordre*

Why are the observed and expected distributions of the second part much more similar than in the first section? Quinnett suggests that this is due to the progressive rhythmic diminution of the isorhythmic structure that begins in the second part. As the durations of the *talea* decrease, the density of accented notes from the *color* increases, and the freedom that Ligeti had in his choice of pitches diminishes. Thus, as discussed above, we would expect the interval content to become increasingly governed by the distribution of intervals in the

---

[7] Statistical analysis of *Désordre* was greatly aided by Cuthbert's music21 [5], which is a Python toolkit for computer-aided musicology.

Fig. 4. Passage before and after the realignment of accents, marked by the vertical bar at t_2. Instances of interval class 1 are marked with asterisks.

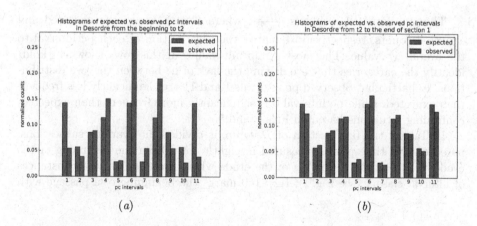

Fig. 5. Histograms of observed and expected intervals in the (*a*) first and (*b*) second part of the first section of *Désordre* (after Quinnett [10])

inter-harmonies. While the histograms of Figs. 3 and 5 and Quinnett's explanations are suggestive, in what follows we will briefly consider the statistical significance and size of the differences between the observed and expected interval distributions, look at the progression of the observed intervals at a finer level of detail, and consider alternative explanations for the convergence of observed and expected intervals toward the end of the first section.[8]

In order to compare the actual distribution of between-hand intervals in *Désordre* with the expected distribution based on the inter-harmonies, χ^2 goodness of fit tests were conducted for various time spans within the first section. In all cases the null hypothesis (H_0) is that the observed intervals are consistent with the distribution of the inter-harmonies. The alternative hypothesis (H_1)—that the observed intervals are not consistent with this distribution—implies that Ligeti is exerting control over the harmonic quality of a given time span in ways that cut against simple scalar connections between notes of the isorhythmic structure. Table 1 gives observed and expected interval counts in the first section of *Désordre* along with the corresponding χ^2 statistic and p-value. The test confirms the intuition that the differences between the two distributions are highly significant, though it does not address the size of this difference (see below).

Table 1. Contingency table of observed and expected intervals in the first section of *Désordre*, standardized residuals, and corresponding χ^2 statistic and p-value

Pc intervals	1	2	3	4	5	6	7	8	9	10	11
Observed	74	40	67	96	25	147	26	108	55	32	72
Expected	106.0	42.4	63.6	84.8	21.2	106.0	21.2	84.8	63.6	42.4	106.0
Std. residuals	**−3.11**	−0.37	0.43	1.22	0.83	**3.98**	1.04	2.52	−1.08	−1.60	**−3.30**

$\chi^2 = 50.05$, $p < 0.000001$

The standardized residuals ($\frac{O-E}{\sqrt{E}}$, where O and E are the observed and expected counts, respectively) quantify the contribution of each pc interval to the overall χ^2 value. The most significant values in this row (shown in bold) identify the categories that are driving the lack of fit between the two distributions. In particular, observed pc intervals 1 and 11 are significantly less frequent than expected, while pc interval 6 is significantly more frequent than expected, confirming intuitions based on Figs. 3 and 5.

In Table 2 the first section of *Désordre* is divided into various subsections, based on five time points measured in eighth notes from the beginning of the étude: $t_0 = 0$ is the beginning of the étude where there are very few instances of interval class 1 (see Fig. 1); $t_1 = 160$ marks the beginning of a passage with

---

[8] Comparison of interval counts with the inter-harmonies in Désordre in both Quinnett's treatise and the current paper stem from our conversations while Quinnett was a student at Florida State University.

Table 2. Comparison of observed and expected intervals for various time spans in the first section of *Désordre* with p-values and reduced phi coefficients (ϕ_ν) for χ^2 goodness of fit tests

begin	end	p-value	ϕ
t_0	t_4	<0.0001	0.26
t_0	t_2	<0.0001	0.60
t_2	t_4	0.91	0.10
t_0	t_1	<0.0001	0.72
t_1	t_2	0.01	0.49
t_2	t_3	<0.0001	0.79
t_3	t_4	0.96	0.10

Fig. 6. (*a*) section before and after time point t_3, (*b*) section immediately before the end of the first section (t_4). Asterisks indicate instances of interval class 1.

slightly increasing presence of ic 1 (see Fig. 4); $t_2 = 248$ marks the realignment of accents between the two hands, accompanied by a return to very few instances of ic 1 (see Fig. 4); at $t_3 = 316$ durations of the isorhythm are progressively shortened and ic 1 becomes much more prevalent (see Fig. 6a); and $t_4 = 634$ is the end of the first section, which includes accents in both hands on every pulse (see Fig. 6b). The final column of the table reports the phi coefficient, ϕ, which is a χ^2-based measure of effect size: $\phi = \sqrt{\frac{\chi^2}{n}}$, where n is the number of samples in the data. Larger values for ϕ indicate a greater difference between the two distributions.[9]

The results of Table 2 suggest that perhaps Ligeti exerted a finer degree of intervallic control than can be captured by dividing the opening section into two large parts as in Fig. 5. Row 1 of the table repeats the test from Table 1 of the entirety of the first section. Rows two and three divide the section into two parts. In first part, from t_0 to t_2, the difference between observed and expected intervals is significant and also has a larger effect size than the entire section. In the second part, from t_2 to t_4, the differences are not significant and the effect size is correspondingly very low. Time point t_1 divides the time span from t_0 to t_2 into two subparts in rows four and five, and time point t_3 similarly divides the span from t_2 to t_4 in rows six and seven. In both pairs of rows the first subpart differs strongly from the expected interval distribution, while the effect size is lower in the second subpart due to the increase in the prevalence of interval class 1. This is particularly true in the time span beginning at t_3. The upshot is that changes in the interval distribution support a division of the opening section into two parts, with a significant return to synchronized accents and avoidance of ic 1 at t_2. This sense of return is enhanced by the slight increase in ic 1 in the span from t_1 to t_2.

These changes in the intervallic distribution over the course of the first section can be seen more clearly in Fig. 7, which plots ϕ for a moving window of 65 eighth notes. Here, ϕ is based on a χ^2 goodness of fit test consisting of only two categories of intervals: those that belong to interval class 1 and those that do not. This graph demonstrates that the changes in harmonic content noted in Table 2 happen abruptly rather than gradually. (This is evident even though the transitions between regions of higher and lower values for ϕ are smoothed by the moving-window analysis.) Regions of lower values for ϕ are either mostly or almost entirely below the lines indicating various significance levels, whereas regions of higher values are almost entirely above these lines. These abrupt transitions as well as the return at t_2 of the intervallic content of the opening complicate the earlier explanation of the changing harmonic distribution over the course of the section. If these changes were simply the result of the progressive rhythmic diminution of the isorhythmic structure (beginning at t_3) and the consequent lack of harmonic freedom, the values for ϕ in Fig. 7 would remain consistently high until t_3 and then gradually decrease. Ligeti appears to be exerting control in switching from one distribution to the other.

To the extent that the quickening of the isorhythms plays a role in the differing intervallic distributions of the two parts, might there be other factors involved? Perhaps as the hands drift apart toward the registral extremes of the piano, Ligeti became less concerned with note-against-note harmonies, since the increased distance between the two hands encourages the perception of two

---

[9] Note that because there are more than a single degree of freedom in the data, ϕ is not normalized to a maximum of 1.

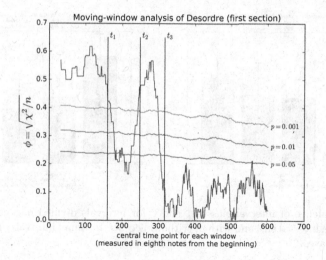

Fig. 7. Moving-window analysis of observed vs. expected frequency of interval class 1 in the first section of *Désordre*. Larger values for ϕ correspond to a greater difference between observed and expected frequencies. Window size is 65 eighth notes. Expected frequencies are based on the inter-harmonies. Time points t_1, t_2, and t_3 are indicated with vertical lines. Lines running across the graph indicate values for ϕ corresponding to various levels of significance.

independent streams and makes it difficult to perceive the quality of the between-hand intervals. The challenge in assessing the relative strengths of these two explanations is that interval size and *talea* durations are strongly (inversely) correlated.

One approach to separating these factors is to divide intervals for each time point by size and whether or not an accent is present. (Recall that accents always and only accompany elements of the isorhythm, so the presence of accents can be used as an indicator for the presence of the isorhythm.) In Fig. 8 intervals throughout the first section of *Désordre* are divided into categories based on small (S) or large (L) interval size and presence (T) or lack (F) of accents. (Small intervals are no larger than two octaves. Pitch intervals involving octaves are reckoned from the lower note of the octave.) For example, the interval in the first eighth note of the étude belongs to the category 'S-T', since it is less than two octaves and the time point contains at least one accent. The interval of 30 semitones on the unaccented time point immediately before t_3 (Fig. 6a) belongs to the category 'L-F'.

The plot on the left of Fig. 8 shows the ratio of interval class 1 for each category, while the plot on the right shows the ratio of interval class 6. (Recall that interval classes 1 and 6 had the highest standardized residuals in Table 1 and were most responsible for the divergence between observed and expected interval frequencies.) As the categories progress from 'small intervals without accents' to 'large intervals with accents' there is a clear trend in the ratios of both

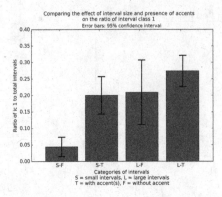 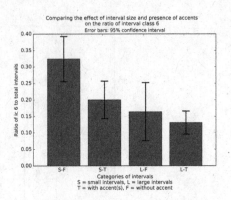

Fig. 8. Prevalence of interval classes 1 and 6 for intervals divided into categories of small or large interval size and with or without accent(s). (Small intervals are no larger than two octaves.)

Table 3. Significance and effect size for prevalence of interval classes 1 and 6 depending on interval size and presence of accents

Categories	ic 1	ic 6
Small size, accent varies	$p < .0001, \phi = .23$	$p < .01, \phi = .14$
Large size, accent varies	$p = .34, \phi = .05$	$p = .60, \phi = .03$
No accent, size varies	$p < .001, \phi = .24$	$p < .02, \phi = .15$
1 or 2 accents, size varies	$p = .07, \phi = .08$	$p < .05, \phi = .09$

interval classes from their (de-)emphasis at the beginning of the étude toward their expected ratios corresponding to the inter-harmonies, though not all of the differences between categories are significant. Table 3 summarizes the significance and effect size for interval classes 1 and 6 when holding interval size constant while varying presence of accents and vice versa. For both interval classes there is a significant and moderate effect of the presence of accents when the interval size is small and of interval size when no accents are present. There is a small and borderline significant effect of interval size when accents are present and a notable lack of effect of the presence of accents when the interval size is large. Taken together, the prevalence of these two interval classes differs noticeably from the inter-harmonies when both the size of the interval is not greater than two octaves and no accents are present; when neither of these conditions is present, the prevalence of these interval classes does not differ strongly from the inter-harmonies. Thus, both explanations seem to be justified—the isorhythmic structure limited Ligeti's freedom in controlling note-against-note harmonies and Ligeti exerted less control over harmonic intervals as the hands drifted apart, forming independent streams, and focusing the listener's attention on the intra- rather than inter-harmonies.

References

1. Amiot, E.: Music Through Fourier Space. Springer, Switzerland (2016). https://doi.org/10.1007/978-3-319-45581-5
2. Bossin, J.: György Ligeti's new lyricism and the aesthetic of currentness: the Berlin festival's retrospective of the composer's career. Current Musicol. **37**(8) (1984)
3. Bregman, A.: Auditory Scene Analysis: The Perceptual Organization of Sound. MIT Press, New York (1994)
4. Callender, C.: Continuous harmonic spaces. J. Music Theor. **51**(2), 277–332 (2007)
5. Cuthbert, M.: music21: a toolkit for computer-aided musicology. http://web.mit.edu/music21/
6. Hanson, H.: The Harmonic Materials of Twentieth-Century Music. Appleton-Century-Crofts, New York (1960)
7. Lewin, D.: Generalized Music Intervals and Transformations. Oxford University Press, Oxford (1987)
8. Lewin, D.: Forte's interval vector, my interval function, and regener's common-note function. J. Music Theor. **21**(2), 194–237 (1977)
9. Kinzler, H.: Decision and automatism in 'Désordre': 1re Étude. Premier Livre. Interface **20**(2), 89–124 (1991)
10. Quinnett, L.: Harmony and counterpoint in Ligeti Études. In: Book I: An Analysis and Performance Guide. Doctoral Treatise, Florida State University (2014)
11. Rogers, D.: A geometric approach to PCSet similarity. Perspect. New Music **37**(1), 77–90 (1999)
12. Steinitz, R.: The dynamics of disorder. Musical Times **137**(1839), 7–14 (1996)

Probabilistic Generation of Ragtime Music from Classical Melodies

Joel Michelson, Hong Xu, and Phillip B. Kirlin[✉]

Department of Mathematics and Computer Science, Rhodes College,
Memphis, TN 38112, USA
{micjp-17,xuho-17,kirlinp}@rhodes.edu

Abstract. This paper examines the computational problem of taking a classical music composition and algorithmically recomposing it in a ragtime style. Because ragtime music is distinguished from other musical genres by its distinctive syncopated rhythms, our work is based on extracting the frequencies of rhythmic patterns from a large collection of ragtime compositions. We use these frequencies in two different algorithms that alter the melodic content of classical music compositions to fit the ragtime rhythmic patterns, and then combine the modified melodies with traditional ragtime bass parts, producing new compositions which melodically and harmonically resemble the original music. We evaluate these algorithms by examining the quality of the ragtime music produced for eight excerpts of classical music alongside the output of a third algorithm run on the same excerpts; results are derived from a survey of 163 people who rated the quality of the ragtime output of the three algorithms.

Keywords: Algorithmic composition · Ragtime · Corpus-based study

1 Introduction

Ragtime is a musical genre that is best described by its syncopated, or ragged, rhythms. Studies of ragtime compositional techniques have concluded that syncopation is the single unifying characteristic of the genre. Though widely assumed to be a term only applied to piano music, ragtime encompasses a wide variety of instrumentations, techniques, and styles [1]. Widespread conceptions and misconceptions regarding the ragtime genre have recently led to the creation of a corpus of digitized ragtime compositions in the MIDI file format [10] to enable large-scale studies of the ragtime music. Not surprisingly, concurrent with the development of this corpus and others like it is the rise of data-driven studies in music informatics, with large-scale data sets being used to discover or confirm trends and tendencies about classical and pop music alike [3,6]. In particular, it is tempting to apply corpus techniques to the field of algorithmic composition in order to automatically extract the unifying characteristics of a musical genre and apply those patterns in a new composition. This is the problem we

© Springer International Publishing AG 2017
O. A. Agustín-Aquino et al. (Eds.): MCM 2017, LNAI 10527, pp. 350–360, 2017.
https://doi.org/10.1007/978-3-319-71827-9_27

examine here: we test the feasibility of composing ragtime music based solely on probabilistically applying rhythmic patterns extracted from a corpus of ragtime music to existing classical compositions. Specifically, we develop two databases of rhythmic patterns derived from a corpus of roughly 5,000 ragtime pieces and propose two algorithms that alter the rhythms of existing classical melodies to sound more like the ragtime rhythms in the databases. We evaluate these algorithms through a survey of 163 people who rated the quality of the ragtime music produced by these algorithms.

2   Methodology and Algorithms

The goal of this research is to study the feasibility of algorithmically composing ragtime music based solely on realigning existing classical music melodies to fit into ragtime rhythms. We algorithmically discover what rhythms are appropriate in ragtime by using a previously-created corpus of ragtime compositions, known as the RAG-collection, or the RAG-C data set. This data set is a collection of 11,591 MIDI files of ragtime music originally introduced by Volk and de Haas [10] as a first effort in putting together a large-scale database of ragtime music. These ragtime compositions were originally compiled and organized by a group of ragtime music enthusiasts collaborating over the internet, and are sourced from various original ragtime scores, from piano rolls translated into the MIDI format, and from recordings of performances as well. Though the data set contents vary in quality in terms of the MIDI translations, it is probably the most comprehensive collection of ragtime music in a symbolic digitized format known.

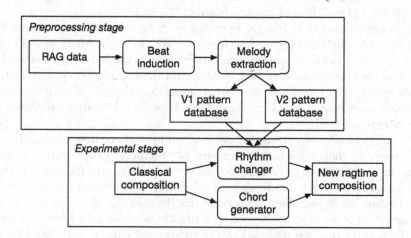

Fig. 1. Methodological setup

Figure 1 illustrates the components in our research setup, which broadly consists of a preprocessing stage and an experimental stage. The goal of the preprocessing stage is to create two databases of rhythmic patterns. We do this by processing the RAG-C data set with a beat induction algorithm; this is a necessary step as the MIDI files in the corpus do not contain enough information to automatically discover rhythmic information such as time signatures or measure boundaries. After the beat induction algorithm aligns the music metrically, we extract the melodies as monophonic sequences of notes using a standard skyline algorithm. This leaves us with a series of measures which we further process into two rhythmic pattern databases.

The goal of the experimental stage is to allow one to transform an existing classical music composition into a ragtime composition. Beginning with a classical music composition, we identify the monophonic main melody of the composition and its harmonic chordal structure. We send the main melody to the rhythm changer algorithms, which alter the metrical placement and duration of the notes in the main melody—but not their ordering—to make the melody sound more like ragtime. This is accomplished with the help of the information in the rhythmic pattern databases. At the same time, we use the harmonic structure of the classical music to compose a prototypical ragtime bass line, which is combined with the altered melody into a final composition in a ragtime style.

We give further details of each step of this process below.

2.1 Beat and Measure Detection

Though MIDI files can be encoded to include information such as time signature and tempo, the files in the RAG-C data set are derived from a variety of sources, including live performances, some of which do not encode these data. Therefore, we use a version of Dixon and Cambouropoulos's beat detection algorithm [4] to estimate the locations of musical beats in the MIDI data.

The algorithm's beat inducer operates by computing inter-onset intervals, or IOIs—times between pairs of note onsets—for the input MIDI file and then clustering them in the hope that small differences in the IOIs will be smoothed out. The clusters are then ranked by size, and the top-ranked clusters usually correspond to the inter-beat interval or a fraction thereof. With a correctly-predicted inter-beat interval, measure boundaries can be easily calculated for the entire piece.

In practice, however, the inter-beat interval predictions may be slightly miscalculated. We noticed a certain amount of "temporal drift" in the measure boundary predictions for the RAG-C data set. Specifically, as one predicts measure boundaries further and further ahead in a MIDI file, the predicted boundaries deviate more and more from their true locations, most likely due to the accumulation of small errors caused by a slightly miscalculated inter-beat interval. To remedy this, we returned to the top-ranked clusters calculated by the beat induction algorithm, and examined every possible inter-beat interval within 12 MIDI ticks of the cluster's interval, calculated to the tenth of a tick. For every potential inter-beat interval, we calculated the predicted measure boundaries for that interval over the entire MIDI file, then binned all the notes of the MIDI file according to which 16th note of the predicted measure they would fall into.

For correctly-predicted measure boundaries, we would expect this frequency distribution of notes across the 16 bins to be weighted more heavily towards the bins corresponding to strong beats, simply because it is more common—even in ragtime—for notes to occur on strong beats. For incorrectly-predicted measure boundaries, we would expect this distribution to be flatter. Therefore, we chose the predicted set of measure boundaries that produced a frequency distribution with the highest standard deviation as our correct measure boundaries.

2.2   Melody Extraction and Pattern Database Construction

We use an adapted version of Temperley's streamer algorithm [8] to split the RAG-C MIDI files into streams of notes. In order to isolate the main melodic voice in each file, a skyline algorithm is used as described in [9, 10] to select a set of notes from these streams using the average pitch of all the notes in a given stream as its height.

Though the RAG-C data set contains over 11,000 MIDI files, we used a specific set of 5,176 for this project. We omitted all files with changing tempos—mostly from live performances—because our algorithm for detecting measure boundaries assumed a fixed tempo. Additionally, there were many excessively long MIDI files—containing many repeats of sections of the music—that could not be processed by the melody extraction algorithm due to memory limitations.

Recall that our ultimate goal is to produce new ragtime music by realigning classical music melodies to fit ragtime rhythms. In order to choose ragtime rhythms appropriately during the algorithmic composition phase, we analyze the rhythms of the melodies in the 5,176 MIDI files. We do this by assuming all the ragtime compositions are in 2/4 or 4/4 time (a reasonable assumption for ragtime), and segment each piece at the level of a 4/4 measure (merging consecutive measures of 2/4 pieces). We represent the rhythm of each 4/4 measure using the method described in [5, 10]: the rhythm of a measure is described by a pattern "I"s and "O"s specifying the locations of the note onsets at the granularity of a 16th note: an "I" standing for an onset and "O" standing for no onset at that time. For instance, a 4/4 measure with a quarter note on every beat would be represented by the string "IOOOIOOOIOOOIOOO." We refer to these strings as *binary onset patterns* because their contents can be represented by 1s and 0s instead of Is and Os.

Once every ragtime composition is converted into a sequence of binary onset patterns, we create two rhythmic pattern databases, Version 1 (V1) and Version 2 (V2). The V1 database simply records the frequencies of every possible binary onset pattern observed in the melodies of the 5,176 ragtime MIDI files. The V2 database records the frequency of *transitions* between binary onset patterns corresponding to every pair of adjacent measures in the corpus. In Sect. 3, we describe some noteworthy facts that can be learned from examining the information in the rhythmic pattern databases.

2.3   Experimental Phase

The experimental phase is designed to harness the information in the pattern databases in order to produce new ragtime compositions from classical music files. For testing and evaluation, we use a set of eight excerpts of classical music. These excerpts are taken from "Dance of the Sugar Plum Fairy" by Tchaikovsky; "Eine kleine Nachtmusik," by Mozart; Concerto No. 1 in E major, Op. 8, RV 269, "Spring" by Vivaldi; three Christmas carols: "Deck the Halls", "Hark! The Herald Angels Sing", "Jingle Bells"; and two traditional tunes: "Old MacDonald Had a Farm" and "Yankee Doodle." We chose these pieces for their easily-identified melodies and duple meters.

We encoded the melody and harmony separately for these eight testing files and used them as input to the chord generation and rhythm changing algorithms, described next.

2.4   Chord Generation

The chord generation algorithm generates a ragtime-style bass line consisting of a sequence of chords. The input to the algorithm is a sequence of chord symbols, in this case from one of the eight classical music excerpts which have had their harmonies manually labeled. We turn these chord symbols into ragtime-style chord progressions based on a subset of the guidelines prescribed in [2]. We use a straightforward algorithm: we choose octaves or single bass notes on the first and third beats of a measure and chords on the second and fourth beats. The first beat is always the root of the current harmony, and the third beat is always the fifth. Additionally, we stochastically change some of the second- or fourth-beat chords into passing tones if the surrounding harmonic structure allows for this transformation. We found a probability of 1/6 works well for choosing whether or not to insert a passing tone.

As an example, Fig. 2 shows the bass line generated from the first eight measures of the fourth movement of Beethoven's Ninth Symphony ("Ode to Joy"). Notice how there is a passing tone generated in the transition from the end of measure 4 into measure 5.

Fig. 2. Illustration of chord generation for "Ode to Joy." The chord symbols above the staff are used as input, and the notes displayed are the output. Note the passing tone in measure 4.

2.5   Rhythm Changing Algorithms

At the heart of this algorithmic composition system is the rhythm changing algorithm. Recall that our goal is to adjust the rhythm of a classical melody to fit a ragtime rhythm. We do this by identifying, for every measure of the classical input composition, a corresponding ragtime measure with the same number of notes, and altering the classical measure to fit the ragtime measure's rhythm. For example, consider Fig. 3, which shows (on the top left) a measure of music taken from the Christmas carol "Deck the Halls," and also (on the top right) a measure taken from the ragtime composition "The Entertainer," by Scott Joplin. The rhythm changer would combine these measures into the new measure of music at the bottom of the figure, which aligns the notes of the "Deck the Halls" melody with "The Entertainer"'s rhythm.

Fig. 3. An example of the rhythm-changing algorithm. We combine a classical melody (top left) with a ragtime rhythm (top right), producing a new measure of ragtime-sounding music (bottom)

We develop and evaluate two different strategies for using the rhythm changer in conjunction with the V1 and V2 rhythmic pattern databases.

Recall that the V1 rhythm database simply stores the frequency of every binary onset pattern in the corpus of ragtime MIDI files. Given a piece of classical music as input, our goal is to probabilistically generate a set of rules that map every binary onset pattern in the input music to a ragtime binary onset pattern, to which we then apply the rhythm changer algorithm as described above. We generate this set of rules by enumerating all the onset patterns in the input classical composition on a measure-by-measure basis, and sorting them in order of descending frequency. For every one of the classical onset patterns, we choose a corresponding ragtime onset pattern proportionally to its frequency in the V1 database (keeping in mind that the number of onsets in the two patterns must be equal), and then create a rule associating those two binary onset patterns. There are two caveats. First, because un-syncopated rhythms (e.g. I000I000I000I000) are so common in the data, rules that map an onset pattern to itself are never permitted. Additionally, rules that would shift onsets by a total of more than eight positions (where a position is an individual 16th note shift) are rewritten to prevent drastic changes, such as a note at the beginning of a half measure being shifted to the end of the measure. This technique is presented as Algorithm 1.

The strategy for using the V2 database is similar to that of the V1 database, except we examine *pairs* of binary onset patterns in the classical input and in the ragtime corpus. This technique is presented as Algorithm 2.

Algorithm 1. Version 1

for each unique binary onset pattern X in the input song **do**
    Count the number of onsets in X
    while a rule has not yet been generated **do**
        Choose a random binary onset pattern Y with the same number of onsets in
        the data set, weighted by its frequency
        if $Y \neq X$ **and** onset shifting distance from X to $Y \leq 8$ **then**
            Generate new rule $X \rightarrow Y$
        end if
    end while
end for
for each measure in the input song with binary onset pattern X **do**
    Generate measure in output song with note positions Y from the rule $X \rightarrow Y$ and
    pitches from the original measure
end for

Algorithm 2. Version 2

for each unique binary pair of onset patterns X and its subsequent measure Y in
the input song **do**
    Count the number of onsets in X and the number of onsets in Y
    while a rule has not yet been generated **do**
        Randomly select a transition Z from the transition table in which a measure
        with count of X transitions into a measure with count Y
        if $Z \neq Y$ **and** onset shifting distance from Y to $Z \leq 8$ **then**
            Generate new rule $Y \rightarrow Z$
        end if
    end while
end for
for each measure in the input song with binary onset pattern Y **do**
    Generate measure in output song with note positions Y from the rule $Y \rightarrow Z$ and
    pitches from the original measure
end for

2.6   Syncopalooza Rhythm Changer

To serve as a baseline algorithm, we implemented the Syncopalooza algorithm as described in [7]. This algorithm is neither data- nor corpus-driven, but rather manipulates the syncopations in a composition by shifting note onsets to stronger or weaker metrical positions individually, rather than by rewriting an entire measure of rhythm at once.

3   Results, Survey, and Evaluation

Some interesting results can be gleaned from the V1 pattern database which records the frequency of every binary onset pattern in the 5,176 ragtime compositions. In general, the V1 database frequencies display a long-tailed distribution as can be seen in Fig. 4; the correspondence between the frequency of a pattern and its rank in the list follows a power law relationship. Furthermore, the most commonly-occurring binary onset patterns in ragtime music do not correspond to syncopated rhythms at all, but rather to simple rhythms such as a measure of two regularly-spaced half notes (the most frequent pattern), a measure with one whole note (the second-most frequent pattern), or a measure of four regularly-spaced quarter notes (the third-most frequent pattern). Though the data set contains 8,803 different binary onset patterns, these three patterns account for roughly 11% of all the measures in the corpus. It is noteworthy that the fourth-most common rhythm, with onsets on beats 1, 2, and 4 (but not 3), displays the characteristic "short-long-short" pattern of note durations which is especially prevelant in ragtime [5].

In order to evaluate the quality of the rhythm-changer algorithms presented earlier, we conducted two separate surveys. The surveys differed in length and in the participant demographics, but contained the same basic type of question. Each question in the survey asked the participant to listen to three different algorithmically-produced ragtime excerpts, one each derived from the V1 and V2 databases paired with the rhythm-changer algorithm, and the third from the Syncopalooza algorithm. The order of the three excerpts was randomized for every question. After listening to each excerpt as many times as the participant desired, they were asked how much they agreed with the statement "This excerpt sounds like ragtime" for each of the three excerpts. Their answers were recorded on a five-point Likert scale, with the choices of strongly disagree (1), disagree (2), neither agree nor disagree (3), agree (4), and strongly agree (5).

The first survey was taken by 33 college undergraduates with some familiarity with ragtime music. Each participant was asked to evaluate 6 sets of excerpts (listening to 18 excerpts in total) according to the schema above. The second survey was taken by 130 different people solicited from internet message board about piano music.

The college students' average responses to the questions on the Likert scale were 3.50, 2.83, and 2.99 for Syncopalooza, V1, and V2, respectively; while the internet users' average responses were 3.36, 2.55, and 2.64, respectively. These values indicate that while the college students rated the output of all three algorithms slightly higher than the general internet population did, both groups preferred Syncopalooza to the algorithms presented in this study, though by less than one point. Furthermore, because even the best-performing algorithm—Syncopalooza—did not surpass a 3.50 rating, there is clearly plenty of room for improvement in the algorithms.

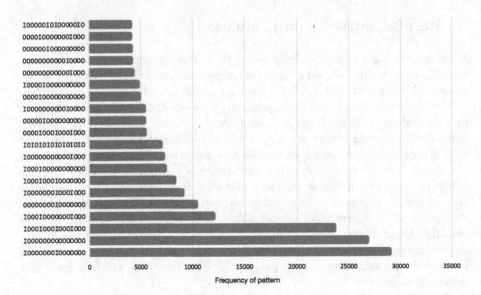

Fig. 4. This graph illustrates the frequencies of the 20 most common binary onset patterns found in the ragtime corpus.

Figure 5 illustrates the survey results grouped by the classical music piece used as input. We can see that Syncopalooza consistently outperforms both the V1 and V2 algorithms, though there are cases where all three scores are clustered closely together. It is also noteworthy that neither V1 nor V2 consistently outperforms the other; their ratings are usually close together.

Anonymous comments solicited from the survey participants revealed some of the reasons for their ratings. Multiple users noted some inconsistencies and mismatches between the beats of the melody and the bass notes of the accompaniment. We believe this may be due to (1) the algorithms moving notes that were originally consonant with their underlying harmonies to new metrical locations that then become dissonant with the underlying harmonies; (2) the algorithms choosing ragtime rhythmic patterns that, while technically common, do not "flow" with the preceding or following music; (3) beat induction errors. Multiple users also commented on the low register of the accompaniment chords; some thought this rendered the audio muddy and distracted from the overall sound. We plan on investigating these issues fully in the next iteration of this project.

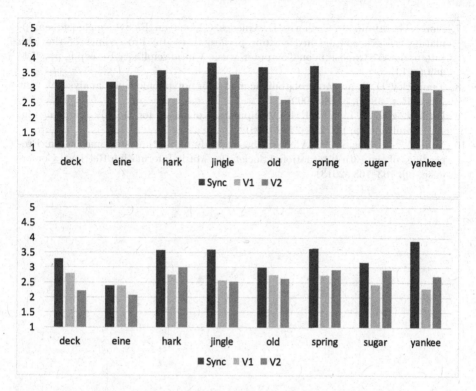

Fig. 5. Survey responses from college undergraduates (top), and internet respondents (bottom), separated by classical input composition and by ragtime algorithm (Syncopalooza, rhythm shifter version 1, rhythm shifter version 2). Participants were asked how much they agreed that the output music sounded like ragtime.

References

1. Berlin, E.A.: Ragtime. In: Grove Music Online, Oxford Music Online. Oxford University Press, March 2017. http://www.oxfordmusiconline.com/subscriber/article/grove/music/A2252241
2. Bradlee, S.: Ragtimify: How to Turn Any Song into Ragtime and Stride Piano. Scott Bradlee (2013)
3. de Clercq, T., Temperley, D.: A corpus analysis of rock harmony. Pop. Music **30**(1), 47–70 (2011)
4. Dixon, S., Cambouropoulos, E.: Beat tracking with musical knowledge. In: Proceedings of the 14th European Conference on Artificial Intelligence, pp. 626–630 (2000)
5. Koops, H.V., Volk, A., de Haas, W.B.: Corpus-based rhythmic pattern analysis of ragtime syncopation. In: Proceedings of the 16th International Society for Music Information Retrieval Conference, pp. 483–489 (2015)
6. Rohrmeier, M., Cross, I.: Statistical properties of tonal harmony in Bach's chorales, pp. 619–627 (2008)

7. Sioros, G., Miron, M., Cocharro, D., Guedes, C., Gouyon, F.: Syncopalooza: manipulating the syncopation in rhythmic performances. In: Proceedings of the 10th International Symposium on Computer Music Multidisciplinary Research, pp. 454–469 (2013)

8. Temperley, D.: A unified probabilistic model for polyphonic music analysis. J. New Music Res. **38**(1), 3–18 (2009)

9. Uitdenbogerd, A.L., Zobel, J.: Manipulation of music for melody matching. In: ACM Multimedia, pp. 235–240 (1998)

10. Volk, A., de Haas, W.B.: A corpus-based study on ragtime syncopation. In: Proceedings of the 13th International Society for Music Information Retrieval Conference, pp. 163–168 (2013)

Using Probabilistic Parsers to Support Salsa Music Composition

Brayan Rodríguez[1]([⊠]), Raúl Gutiérrez de Piñérez[1], and Gerardo M. Sarria M.[2]

[1] Escuela de Ingeniera de Sistemas y Computación,
Universidad del Valle, Cali, Colombia
{brayan.rodriguez.rivera,raul.gutierrez}@correounivalle.edu.co
[2] Departamento de Electrónica y Ciencias de la Computación,
Pontificia Universidad Javeriana Cali, Cali, Colombia
gsarria@javerianacali.edu.co

Abstract. Salsa is a long-established music genre. It has been used as a way to define, identify and express social beliefs. Due to the limited computational study of this genre, we consider relevant to identify and analyze the musical features of this music genre. Thus, we train a corpus with Grupo Niche songs for generating the production rules for an induced probabilistic context-free grammar through a probabilistic parser. In addition, we implement a web-based tool to support musical composition and generate automatic Salsa songs. In this work, we also compare three automatic songs using cross-validation on the corpus. We show the stability of the grammar because the precision of the generated songs compared to corpus' songs is close to those that are not in the corpus.

Keywords: Salsa · Treebank · Probabilistic context-free grammar
Rules · Probabilistic parser · Musical composition · Web-based tool
Precision · Recall · Automatic songs

1 Introduction

Salsa is a long-established music genre in modern Latin American culture and it has been used as a way to define, identify and express all social beliefs. For this reason, ethnomusicologists have argued that Salsa has a unique set of features that distinguishes it from other rhythms [1]. Despite the popularity of Salsa music, this music genre has not been formally analyzed in order to understand the components that define it and make it different from other rhythms. In the context of computer music, we are aware of only one research work (refer to [1]) developed in 2015 in which a dataset of around 25,000 Salsa songs was worked on without annotated musical information. In our work, we annotate a corpus with harmonic information of Grupo Niche and propose a set of rules that makes up the song structure very similar to the guideline bracketing for treebanks.

In this work, we aim to the integration of computational linguistic in the art of music generation. We want to identify and analyze the features of this music

© Springer International Publishing AG 2017
O. A. Agustín-Aquino et al. (Eds.): MCM 2017, LNAI 10527, pp. 361–372, 2017.
https://doi.org/10.1007/978-3-319-71827-9_28

genre that is part of the Latin American folklore. Composers of Salsa music usually want to follow the patterns promulgated by popular bands and they can support their process by this system that generates automatic melodies. We like to make possible to generate Salsa music for supporting musical composition through probabilistic parsers. We are particularly interested in automatizing the Grupo Niche music which is a Salsa band founded in the seventies in Cali, Colombia [2].

In this paper, we attempt to tackle several subjects of linguistic computational such as treebanks, probabilistic parsers, and induced grammars. Musical composition can be proposed like a problem of rules rewrite in context-free grammars. Other approaches include the use of logic, hierarchical structures, constraint programming, Markov models, L-systems, and concurrency theories.

This document is organized as follows: Sect. 2 presents related work, Sect. 3 describes the Salsa music treebank, Sect. 4 explains music generation using probabilistic context-free grammars, and finally, we present experimental setting and results in Sect. 5 and conclude with some remarks and future work in Sect. 6.

2   Related Work

Stochastic processes may be applied to musical analysis, sound synthesis, and composition. Specifically, the n-dimensional property of the probabilistic grammars can model the four properties of sound: pitch, tone, volume, and rate [3]. The best type of grammar to represent music due to their ability to represent multi-leveled syntactic formations is the context-free grammar [4]. The similarity between Natural Language Processing (NLP) and music processing allows techniques from NLP (e.g. probabilistic context-free grammars) to be applied to music processing through hierarchical structures by fragments of melody [5].

On the other hand, general patterns in musical composition for generating new automatic music have been used [6,7] and particularly an implementation of a machine learning system from a treebank of monophonic melodies is described in [7]. In this connection, how to manually make the production rules based on standards of the Bach's music with probabilities assigned and how to create a stochastic grammar to generate new melodies is described in [8]. These production rules use concepts such as musical areas to group the chords which are the terminals of the grammar. In order to know what information to annotate in a treebank of songs, three papers are found that use different ways to annotate a song through music theory concepts [9–11].

3   Salsa Music Treebank

The musical treebank consists of twenty-eight syntax trees each one representing a song restricted, naturally, by the composer (Grupo Niche) and the date it was composed (since 1980 to 1999), this is because of the evolution of Salsa music. For this purpose, it is necessary to carry out a process of gathering piano and bass scores in that the following compositions are obtained: Ana Milé,

A Prueba De Fuego, Busca Por Dentro, Cali Ají, Cali Pachanguero, Canoa Rocha, Caso Social, Cómo Podré Disimular, Del Puente, Digo Yo, El Coco, Ese Día, Etnia, Hagamos Lo Que Diga El Corazón, Han Cogido La Cosa, La Cárcel, La Culebra, La Danza De La Chancaca, La Magia De Tus Besos, La Negra No Quiere, Listo Medellín, Me Sabe A Perú, Mi Pueblo Natal, Miserable, Nuestro Sueño, Se Pareció Tanto A Ti, Sin Sentimiento and Solo Un Cariño. It is necessary to create a process, through concepts of music theory, to draw explicitly the sequence of chords for each song from the scores.

3.1 Syntactic Functions of Nonterminal Symbols

In music, every song can be divided into several parts concerning each section as popular songs have been traditionally split. These parts could be, for instance, introduction, verse, chorus, etc. In this way, every song is annotated dividing it into those parts [9]. This idea is taken and adapted to the context of Salsa songs. So, the first nonterminal symbols defined are the different elements of a Salsa song, that means anacrusis, introduction, verse, chorus, instrumental, pregon (that is particular and the most representative in Salsa) and coda. Introduction, verse, instrumental and pregon are found in all the songs that make up the treebank. On the other hand, the remaining (anacrusis, chorus, and coda) are not present in all those songs. All or some of these parts composes the initial symbol S (which symbolizes the whole song) in a specific order.

In the same way, following an example of annotation proposed by Weyde and Wissmann [11], each element of the song is composed of musical cadences or areas in case that these do not appear in order of a common (predominant, dominant, tonic or dominant, tonic) or an Andalusian cadence (tonic, dominant, predominant, dominant) [12]. The musical areas in which a chord could be classified are, precisely, predominant, dominant or tonic and in the case of cadence order, those would be grouped in a nonterminal symbol representing it. Those areas, in turn, are composed of one or more nonterminal symbols immediately preceding the terminal symbol (chord). The nonterminal of the area defines the number of chords that are in line continuously. Each chord is contained by a nonterminal symbol of the area corresponding to that chord.

Besides this, new nonterminal symbols become necessary to represent the end of each element of the song in the sequence of chords. Thus, these nonterminal symbols are represented by the string "FIN" following the symbol of the element that is finishing. Each one leads to a period ('.') as the terminal symbol.

One particularity of the anacrusis is that it always consists of one area dominant or tonic, as shown in Table 1. It also shows that each element finishes with the nonterminal symbol related to its end, which in turn always leads to a period.

3.2 Terminal Symbols

The terminal symbols are: 'i', 'ii', 'III', 'iv', 'IV7', 'V7', 'VI', 'VII', '.'.

With this in mind, the terminal symbols are the chords in roman numerals, due to music theory, indicating the degrees of a scale. Some in uppercase (if it is

Table 1. One-level depth examples of each nonterminal symbol

Symbol	Description	Example
S	Initial symbol	(S (INTRO ...) (VERSO ...) (CORO ...) (INS ...) (INS ...) (PREGON ...)(CODA ...))
ANA	Anacrusis	(ANA (AD (D ...) (D ...)))
INTRO	Introduction	(INTRO (AT ...) (CD ...) (AT ...) (AD ...) (FININTRO ...))
VERSO	Verse	(VERSO (CD ...) (CD ...) (CD ...) (CD ...) (CD ...) (CD ...) (AD ...) (FINVERSO ...))
CORO	Chorus	(CORO (CD ...) (CD ...) (CD ...) (CD ...) (ASD ...) (FINCORO ...))
INS	Instrumental	(INS (AT ...) (CD ...) (AT ...) (AD ...) (FININS ...))
PREGON	Pregon	(PREGON (CD ...) (CD ...) (CD ...) (CD ...) (CD ...) (ASD ...) (FINPREGON ...))
CODA	Coda	(CODA (AT ...) (ASD ...) (AD ...) (FINCODA ...))
CD	Common cadence	(CD (ASD ...) (AD ...) (AT ...))
CF	Andalusian cadence	(CF (AT ...) (AD ...) (ASD ...) (AD ...))
ASD	Predominant area	(ASD (SD ...) (SD ...))
AD	Dominant area	(AD (D ...) (D ...) (D ...) (D ...) (D ...))
AT	Tonic area	(AT (T ...) (T ...) (T ...) (T ...) (T ...))
SD	Predominant chord	(SD 'ii')
D	Dominant chord	(D 'V7')
T	Tonic chord	(T 'i')

a major chord) and some in lowercase (if it is minor), following the same theory. It is always a minor key because most Salsa songs are in that key. If a song is in a major key, it is translated to its relative minor. The IV and V chord have the seventh explicit because the scores have that chord so. In addition to this, the period has been added as a terminal symbol for the reasons set out above.

Figure 1 shows the syntax tree related to the fragment shown in Fig. 2 of the song Cali Ají which is in the treebank transcribed in A minor key the relative of C major key. This is an example of the annotation that has been given to every song concerning the following bracket annotation:

(S (ANA (AD (D VII) (D VII))) (INTRO (CD (AD (D VII) (D VII) (D VII) (D VII)) (AT (T III) (T III) (T III) (T III))) (CD (AD (D VII) (D VII) (D VII) (D VII)) (AT (T III) (T III) (T III) (T III))) (AD (D VII) (D VII) (D VII) (D VII)) (FININTRO .)) (CORO (CD (AD (D VII) (D VII) (D VII) (D VII)) (AT (T III) (T III) (T III) (T III)))))

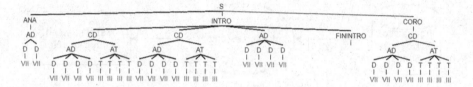

Fig. 1. Syntax tree of a piece of the song Cali Ají.

Fig. 2. Piano score of a piece of the song Cali Ají.

4 Composing Salsa Using Probabilistic Context-Free Grammars

In this section, we present a music generation model based on induced probabilistic context-free grammars (PCFG). We take the chords of the induced trees by the grammar and automatically generate a song's melody. Figure 3 shows a diagram of the music generation model. First, we train an induced grammar on the Grupo Niche Treebank and produce a set of production rules which is a PCFG. Second, we implement a chords generation algorithm which is provided by PCFG and obtain the sequence of chords. Finally, we implement a melodies generator using the sequence of chords considering tempo and tonality features. Following we explain in detail each element of the model.

4.1 Computational Model Based on PCFG

Before you explain induced grammars, we define a context-free grammar (CFG) as a set of production rules which describes strings that belong to a language and are syntactically valid. A CFG is a tuple $G = \langle N, \Sigma, P, S \rangle$, where P is a set of the production rules, N is a non-terminal symbol, Σ is a set of terminal symbols and finally, S is the initial non-terminal symbol. The production rules follow

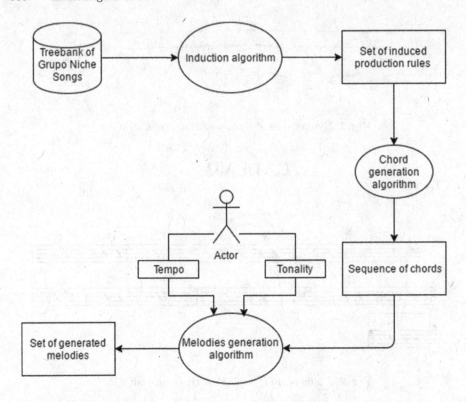

Fig. 3. Diagram of the development of the computational model

the form $A \rightarrow X$, where $A \in N$, and $X \in (\Sigma \cup N)^+$. A PCFG is a context-free grammar that has associated a probability distribution in P, where each production rule has associated a probability $q(\alpha \rightarrow \beta)$ for each rule $\alpha \rightarrow \beta \in P$. For any $X \in N$, we have the constraint:

$$\sum_{\alpha \rightarrow \beta \in P : \alpha = X} q(\alpha \rightarrow \beta) = 1$$

In addition we have $q(\alpha \rightarrow \beta) \geq 0$ for any $\alpha \rightarrow \beta \in P$.

When we have a PCFG induced from a treebank, the probability of each production rule $\alpha \rightarrow \beta$ is estimated using the maximum-likelihood estimation:

$$p(\alpha \rightarrow \beta) = \frac{Count(\alpha \rightarrow \beta)}{Count(\alpha)}$$

Where $Count(\alpha \rightarrow \beta)$ is the number of times that the rule $\alpha \rightarrow \beta$ is seen in the treebank, and $Count(\alpha)$ is the number of rules that the non-terminal α is seen on the rule left-hand side in the treebank.

In more detail, the algorithm that describes that process of estimating the probabilities of occurrence is:

```
pcount = {}
lcount = {}
for each prod in productions:
    lcount[prod.lhs()] = lcount[prod.lhs()] + 1
    pcount[prod]       = pcount.get[prod] + 1
prods = []
for each p in pcount:
    prods = prods + ProbabilisticProd(p.lhs(), p.rhs(),
                        prob=pcount[p] / lcount[p.lhs()])
return prods
```

First, given the treebank of Grupo Niche songs, the rules are induced from it by a top-down cyclic method through all the syntax trees. Once we have the production rules, parsing is carried out by estimating the probabilities of occurrence from the list of productions as mentioned above.

The induced grammar fulfills the characteristics of a PCFG. Thus, the probabilities of all choices of a nonterminal symbol must sum 1.0. Some induced production rules are shown below. According to assigned probability, the algorithm can choose one particular nonterminal symbol of a production rule (for example, in the production rule S shown below) or another nonterminal symbol, and a similar process with the other symbols.

```
S -> INTRO VERSO CORO INS VERSO CORO INS PREGON INS PREGON INS  CODA
[0.0357143]
PREGON -> CD CD CD CD CD FINPREGON [0.0434783]
CD -> AD AT [0.516279]
CD -> ASD AD AT [0.483721]
AD -> D D D D [0.233561]
D -> 'V7' [0.531322]
FINPREGON -> '.' [1.0]
```

4.2   From the Model to the Music

As shown in Fig. 3, given the production rules, we do a walkthrough top-down by the grammar until generating a sequence of terminal symbols. Namely, through a recursive method that chooses a single rule among all the options of a nonterminal symbol, starting with the initial symbol S. If the choice leads to a sequence of nonterminal symbols, this process is repeated with each symbol to choose a rule among its options. On the contrary, if the choice leads to a terminal symbol, this branch is finished and the terminal symbol is appended to a list which is finally returned. In more detail, the algorithm is:

```
generate_chords(grammar, items, terminals):
    for each item in items:
        if item is NonTerminal:
            prods = grammar.productions(lhs=item)
            probs = []
            for each prod in prods:
                probs = probs + prod.prob()
            chosen_prod = choose_prod(prods, probs)
            generate_chords(grammar, chosen_prod.rhs(), terminals)
        else:
            terminals = terminals + item
    return terminals
```

Following the diagram of the development, the sequence of chords requires a process to be the melodies of an automatic song. To achieve this goal, it is necessary to choose the instruments that are going to take part in this process. The representative non-percussion instruments of Salsa music genre chosen are electric bass and piano because these are considered, with the percussion instruments, the basis of the music genre.

Specifically, to generate the piano melodies, the rhythm is divided into four variations which are cyclically assigned to every chord in the sequence. The chords are set from the second one of them to the last one. Before this, the first chord is assigned a rhythm variation that is specially designed for this. It should be noted that every rhythm variation represents a half measure. On the other hand, to generate the electric bass melodies, the same process is done but only with two variations that are assigned to the sequence of chords without including the first chord. In order to generate these melodies, it is necessary to previously set the tonality and the tempo, as shown in Fig. 3.

In order to complement the song (i.e. makes it sound more like Salsa) we add loops of some percussion instruments (cowbell, congas, clave, and maraca) to the melodies mentioned above.

The process of getting the melodies from the sequence of chords is implemented in Python language programming by the GNU software Lilypond, which allows translating the musical language to a programming language and generates in MIDI format the melodies.

4.3   Practical Tool for Supporting Composers

In order to visualize the possible outcomes of the model, we implement a web-based tool to make it intuitive to users. We are convinced that music composers may be interested in the patterns that the Grupo Niche used in their songs. We want to highlight the Python library that we use for supporting the NLP processes called Natural Language Toolkit (NLTK).

As shown in Fig. 4 the web-based tool needs the tonality and the tempo of the composition. These are parameters that the user must input. After the user clicks on "Generate" button, the model built a song based on the parameters.

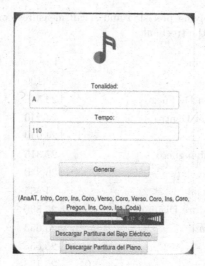

Fig. 4. Graphic user interface of the web-based tool

The song structure is shown in a web-player and the user can download the bass and piano scores of the song.

5   Experiments and Results

In this section, we explain the experiments performed consisting of three generated automatic songs in order to analyze them in terms of their chord progressions and their hierarchical structures. These songs are different to the annotated songs in the treebank. It is important to note the lack of music scores in Salsa music makes it difficult to gather of scores for the experiments.

5.1   Precision and Recall

We use Precision and Recall measures, as defined in [13]. There is no a formal technique which allows interpreting these measures in the context of large sized syntax trees (around 300 leaf nodes and 700 interior nodes in each tree). The syntax trees used in NLP generally have up to 30 leaf nodes. In our case, performance measures are adapted to evaluation of automatically generated songs and we use precision for looking at the human-composed songs, instead of evaluating the parsing of a sequence of terminals. We evaluate three sets of human-composed songs: Songs composed by Grupo Niche in the treebank, songs composed by Grupo Niche which are not in the treebank and Salsa songs that are not composed by Grupo Niche.

Tables 2, 3 and 4 show the results of the process of contrasting taking one automatic song. Taking the highest number between precision and recall because it only depends on the tree of which song the automatic or the human-composed

Table 2. Results of applying precision and recall measures comparing an automatic song with every song in the treebank

Song's name	Precision	Recall
Ana Milé	20,190	11,756
A Prueba De Fuego	22,565	13,085
Busca Por Dentro	24,940	15,395
Cali Ají	24,940	16,587
Cali Pachanguero	27,315	17,215
Canoa Rocha	26,840	14,790
Caso Social	24,465	13,419
Cómo Podré Disimular	25,890	18,136
Del Puente	25,653	10,577
Digo Yo	25,653	12,705
El Coco	26,128	18,425
Ese Día	19,477	9,692
Etnia	20,665	10,369
Hagamos Lo Que Diga El Corazón	23,752	16,393
Han Cogido La Cosa	23,515	12,547
La Cárcel	20,427	10,449
La Culebra	23,040	14,741
La Danza De La Chancaca	26,365	14,471
La Magia De Tus Besos	22,565	15,422
La Negra No Quiere	28,266	13,134
Listo Medellín	28,741	13,781
Me Sabe A Perú	22,802	14,814
Mi Pueblo Natal	20,902	17,886
Miserable	24,940	15,601
Nuestro Sueño	21,378	13,081
Se Pareció Tanto A Ti	21,852	13,294
Sin Sentimiento	18,289	12,520
Solo Un Cariño	22,565	12,195

Table 3. Results of applying precision and recall measures comparing an automatic song with Grupo Niche songs that are not in the treebank

Song's name	Precision	Recall
A Mi Medida	15,914	11,571
Las Mujeres Están De Moda	23,040	15,695
Mi Valle Del Cauca	27,078	18,269
Ni Como Amiga	23,040	15,299

Table 4. Results of applying precision and recall measures comparing an automatic song with Salsa songs that are not composed by Grupo Niche

Song's name	Precision	Recall
Ahora Quién	13,438	8,959
Arrepentida	18,281	25,167
Te Voy A Enseñar	16,223	19,198

has more interior nodes, the general results with three automatic songs shows that the numbers are in a range from 17 to 30 and on the whole, the numbers of the tables related to the songs that are not composed by Grupo Niche are lower than the numbers of the other tables, but not much.

Based on the small range of the numbers of the tables related to the songs in the treebank, it follows that the nodes of the syntax tree of the automatic songs are built in almost equal proportions from every song in the treebank. That means a harmonic similarity between these two groups. Further, because there is a small difference between both the numbers of the tables related to compositions of Grupo Niche and the numbers of the remaining tables, there is a little difference in their harmonic structure but most harmonic patterns remain in both groups. Due to the similarity of the numbers of the songs in the treebank and the songs composed by Grupo Niche that are not in it, the grammar is well-formed because that means that it covers a large number of the harmonic patterns promulgated by Grupo Niche.

In the context of the results obtained by performance measures, we deduce that the song structure of an automatic song is always an order of one or more songs in the treebank, which means that the automatic songs have an order used by Grupo Niche. Similarly, the cadences and the musical areas are organized inside every part like in one or more songs in the treebank. Thus, the annotated musical patterns in the automatic songs are the same that this band used in one or more songs. This is why the musical analysis of the experiments shows that the chord progression of each experiment follows the general musical rules which are established in this music genre and specifically in the Grupo Niche songs. The rhythm variations of the instruments are selected from several songs in the treebank.

6   Concluding Remarks and Future Work

We implemented a musical formal system that, through NLP formal techniques, generates music. Because it is not known a formal procedure to assess syntax tree with such width as described above, the data was analyzed in a way that allowed us to get information from it. Precision and recall measures were used and analyzed for that purpose.

The absence of the background documents about computational models in the context of Salsa music genre resulted in the lack of an annotated treebank or a standard of music annotation. Furthermore, the lack of head rules affected the quality of the induction process because of the difference between the function of a chord inside a song and the function of a word in a sentence.

Inducing a grammar from a treebank has resulted in an accurate way to generate music following the musical patterns that are written in the treebank and it is also subject to analysis and evaluation. Broadly this project does three important contributions in the academic field: A treebank with Salsa songs, a generative grammar of Salsa music genre and the analysis of Grupo Niche songs and its support to musical composition.

We plan to continue this work in three directions: Extend the range of the instruments considering brass instruments because of their melody and their contribution to Salsa music; head rules implementation on musical grammars will offer a very high degree of sophistication; and the adaptation of computation techniques such as evolutionary computation and artificial intelligence which will always be suitable for this work.

References

1. Sarria, G.M., Mora, M.J., Arce-Lopera, C.: Salsa dataset: primera base de conocimiento de música salsa. Ricercare **5**, 63–72 (2016)
2. Historia Grupo Niche. http://www.gruponiche.com/es/historia/
3. Jones, K.: Compositional applications of stochastic processes. Comput. Music J. **5**(2), 45–61 (1981)
4. Roads, C., Wieneke, P.: Grammars as representations for music. Comput. Music J. **3**(1), 48–55 (1979)
5. Bod, R.: Probabilistic grammars for music. In: Belgian-Dutch Conference on Artificial Intelligence (2001)
6. García Salas, H.A., Gelbukh, A., Calvo, H., Galindo Soria, F.: Automatic music composition with simple probabilistic generative grammars. Polibits **44**, 59–65 (2011)
7. McCormack, J.: Grammar based music composition. Complex Syst. **96**, 321–336 (1996)
8. Perchy, S., Sarria, G.: Musical composition with stochastic context-free grammars. In: 8th Mexican International Conference on Artificial Intelligence, MICAI 2009 (2009)
9. Temperley, D., Clercq, T.: Statistical analysis of harmony and melody in rock music. J. New Music Res. **42**(3), 187–204 (2013)
10. Kaneko, H., Kawakami, D., Sagayama, S.: Functional harmony annotation database for statistical music analysis. In: Proceedings of the International Society for Music Information Retrieval Conference (ISMIR): Late Breaking Session (2010)
11. Weyde, T.: Automatic semantic annotation of music with harmonic structure. In: 4th Sound and Music Computing Conference (2007)
12. Granados, M.: Teoría musical de la guitarra flamenca. Casa Publicacions Beethoven, Barcelona (1998)
13. Carroll, J., Minnen, G., Briscoe, T.: Corpus annotation for parser evaluation. arXiv preprint cs/9907013 (1999)

Author Index